Lecture Notes in Computational Vision and Biomechanics

Volume 35

Series Editors

João Manuel R. S. Tavares ⓘD, Departamento de Engenharia Mecânica, Faculdade de Engenharia, Universidade do Porto, Porto, Portugal
Renato Natal Jorge, Faculdade de Engenharia, Universidade do Porto, Porto, Portugal

Research related to the analysis of living structures (Biomechanics) has been carried out extensively in several distinct areas of science, such as, for example, mathematics, mechanical, physics, informatics, medicine and sports. However, for its successful achievement, numerous research topics should be considered, such as image processing and analysis, geometric and numerical modelling, biomechanics, experimental analysis, mechanobiology and enhanced visualization, and their application on real cases must be developed and more investigation is needed. Additionally, enhanced hardware solutions and less invasive devices are demanded. On the other hand, Image Analysis (Computational Vision) aims to extract a high level of information from static images or dynamical image sequences. An example of applications involving Image Analysis can be found in the study of the motion of structures from image sequences, shape reconstruction from images and medical diagnosis. As a multidisciplinary area, Computational Vision considers techniques and methods from other disciplines, like from Artificial Intelligence, Signal Processing, mathematics, physics and informatics. Despite the work that has been done in this area, more robust and efficient methods of Computational Imaging are still demanded in many application domains, such as in medicine, and their validation in real scenarios needs to be examined urgently. Recently, these two branches of science have been increasingly seen as being strongly connected and related, but no book series or journal has contemplated this increasingly strong association. Hence, the main goal of this book series in Computational Vision and Biomechanics (LNCV&B) consists in the provision of a comprehensive forum for discussion on the current state-of-the-art in these fields by emphasizing their connection. The book series covers (but is not limited to):

- Applications of Computational Vision and Biomechanics
- Biometrics and Biomedical Pattern Analysis
- Cellular Imaging and Cellular Mechanics
- Clinical Biomechanics
- Computational Bioimaging and Visualization
- Computational Biology in Biomedical Imaging
- Development of Biomechanical Devices
- Device and Technique Development for Biomedical Imaging
- Experimental Biomechanics
- Gait & Posture Mechanics
- Grid and High Performance Computing on Computational Vision and Biomechanics
- Image Processing and Analysis
- Image processing and visualization in Biofluids
- Image Understanding
- Material Models
- Mechanobiology

- Medical Image Analysis
- Molecular Mechanics
- Multi-modal Image Systems
- Multiscale Biosensors in Biomedical Imaging
- Multiscale Devices and BioMEMS for Biomedical Imaging
- Musculoskeletal Biomechanics
- Multiscale Analysis in Biomechanics
- Neuromuscular Biomechanics
- Numerical Methods for Living Tissues
- Numerical Simulation
- Software Development on Computational Vision and Biomechanics
- Sport Biomechanics
- Virtual Reality in Biomechanics
- Vision Systems
- Image-based Geometric Modeling and Mesh Generation
- Digital Geometry Algorithms for Computational Vision and Visualization

In order to match the scope of the Book Series, each book has to include contents relating, or combining both Image Analysis and mechanics.

Indexed in SCOPUS, Google Scholar and SpringerLink.

More information about this series at http://www.springer.com/series/8910

Jorge Belinha ·
Maria-Cristina Manzanares-Céspedes ·
António M. G. Completo

Editors

The Computational
Mechanics of Bone Tissue

Biological Behaviour, Remodelling
Algorithms and Numerical Applications

 Springer

Editors
Jorge Belinha
Department of Mechanical Engineering
School of Engineering
Polytechnic of Porto (ISEP)
Porto, Portugal

Maria-Cristina Manzanares-Céspedes
Departament de Patologia i Terapèutica
Experimental
University of Barcelona
Barcelona, Spain

António M. G. Completo
Department of Mechanical Engineering
University of Aveiro
Aveiro, Portugal

ISSN 2212-9391   ISSN 2212-9413   (electronic)
Lecture Notes in Computational Vision and Biomechanics
ISBN 978-3-030-37540-9   ISBN 978-3-030-37541-6   (eBook)
https://doi.org/10.1007/978-3-030-37541-6

This Springer imprint is published by the registered company Springer Nature Switzerland AG
The registered company address is: Gewerbestrasse 11, 6330 Cham, Switzerland

Preface

Bone tissue remains a highly researched topic investigated by many fields of science. Research in this topic has been continuous in recent decades, producing clinically and analytically relevant results. Generally, bone remodelling plays a vital role in the regulation of human skeleton. Since bone acts as a mineral repository of calcium and phosphorus, bone is responsible for the maintenance of mineral homeostasis. Highly reactive to its mechanical environment, bone tissue structurally adapts itself and induces the repair of its own damaged tissue. This self-repair process is commonly known as bone remodelling. Since the first observations of the bone tissue remodelling phenomenon, researchers have attempted to describe it, document it and predict it. With the advent of computer technology, several mathematical models and algorithms have been developed to simulate and to predict such a complex phenomenon.

Thus, with this book, a group of experts in bone tissue computational mechanics was gathered, aiming to combine their knowledge and insight in one single document. This book consists in three parts, covering the macro- and micro-scale biological mechanical behaviour of the bone tissue, the most recent bone tissue remodelling algorithms considering mechanical and chemical stimuli and the numerical simulation of bone and its adjacent structures.

Part I "Biological Characterization of Bone Tissue" includes two chapters, addressing the function, regulation, morphology, physiology, structure and quality of bone tissue. In the first chapter, M. T. Oliveira and J. C. Reis describe the tissue complexity, along with its response to external and internal stimuli. The overview goes from embryogenesis to endocrine regulation and bone remodelling. With this chapter, the reader becomes aware of the contribution of bone's cellular structure and tissue organization to bone remodelling. In the second chapter, J. M. D. A. Rollo and colleagues focus on bone quality assessment, providing a description of several methods and techniques to evaluate the micro-architecture of dried trabecular bones in vertebrae. The authors show the importance of such microstructural characterization in the evaluation and detection of osteoporosis.

Part II "Bone Remodelling Algorithms" focuses on numerical techniques to predict the transient response of bone tissue. Thus, in the first chapter of this part, M. C. Marques and colleagues present an extensive review on bone remodelling and regeneration models. Three distinct classifications of models are proposed and presented: the mechanoregulated models, the bioregulated models and the mechanobioregulated models. Furthermore, the chapter addresses the benefits of combining these mathematical models with advanced discretization meshless techniques. Next, it follows a chapter fully dedicated to the mathematical description of the dynamic behaviour of bone remodelling processes, involving physiological phenomena. Furthermore, R. M. Coelho and colleagues review the most recent models, highlighting those that include the main cellular processes, along with the biochemical control, and the pharmacokinetics/pharmacodynamics (PK/PD) of the most common treatments for diseases, such as cancer. In the end, they present their numerical results. In a subsequent chapter, Madalena M. A. Peyroteo and colleagues present a novel model to simulate the biological events during bone remodelling. For the first time, the model combines advanced discretization meshless techniques with temporal-spatial biological bone remodelling models. Furthermore, the proposed model enhances previous models by adding an additional spatial variable. In the last chapter of this second part, X. Wang and J. Fernandez present an innovative mechanostatistical approach to predict bone remodelling across scales. In this chapter, the authors propose a multiscale framework, bridging the gap between relevant spatial scales by passing data from different sources across a multitude of spatial scales to solve both organ-level and Haversian-level biomechanical states. The solutions stored in a database are then used by a statistical method to rapidly estimate the load-adaptation response.

Part III "Numerical Simulation of Bone Tissue and Adjacent Structures" consists of computational simulation of bone structures using well-established numerical techniques. In the first chapter of this last part, F. M. P. Almeida and António M. G. Completo present a review on the application of the finite element method to bone tissue analysis. From a historical point of view, the authors provide a survey on the most relevant works documented up to date, focusing on the kind of used elements, constitutive models and experimental validation.

In the second chapter of this third part, C. Bandeiras and António M.G. Completo show the performance of a computational approach for tissue engineered constructs. Since growth and remodelling (G&R) computational models are invaluable to interpret and predict the effects of experimental designs, the authors propose a computational framework to simulate the mechanical stimulation response of biotissues. In the last chapter, A. P. G. Castro and colleagues show a finite element method application of the intervertebral disc, compressed by two vertebrae. The findings are in accordance with different sources of experimental and numerical literature data.

Dedicated to the computational analysis of bone tissue mechanics, this book aims to contribute to the enhancement of the state of the art in bone tissue analysis, combining the knowledge of researchers from the biomedical and the engineering science fields. Intended readers include researchers interested in computational mechanics applications dealing with bone tissue structural analysis and/or remodelling response.

With this book, we hope to contribute to the enlargement of the scientific repository of this important and interesting field of computational biomechanics. We would like to finally thank all contributors for their effort and patience in making this book possible.

Porto, Portugal   Jorge Belinha
Barcelona, Spain   Maria-Cristina Manzanares-Céspedes
Aveiro, Portugal   António M. G. Completo

Acknowledgements

The editors acknowledge the support towards the organization and the publishing of this book to the following organizations:

- Instituto Politécnico do Porto (P. Porto)
- Instituto Superior de Engenharia do Porto (ISEP)—Departamento de Engenharia Mecânica
- Universidade de Aveiro—Departamento de Engenharia Mecânica
- Universitat de Barcelona—Faculty of Medicine and Health Sciences
- Instituto de Ciência e Inovação em Engenharia Mecânica e Engenharia Industrial (INEGI)
- TEMA—Centre for Mechanical Technology and Automation—Universidade de Aveiro
- Fundação para a Ciência e a Tecnologia (FCT)
- Project NORTE-01-0145-FEDER-000022—SciTech—Science and Technology for Competitive and Sustainable Industries, cofinanced by Programa Operacional Regional do Norte (NORTE2020), through Fundo Europeu de Desenvolvimento Regional (FEDER).

Contents

Computational Modelling of Tissue-Engineered Cartilage
Cátia Bandeiras and António M. G. Completo

A. P. G. Castro, P. Flores, J. C. P. Claro, António M. G. Completo
and J. L. Alves

Contributors

Francisco M. P. Almeida Department of Mechanical Engineering, University of Aveiro, Aveiro, Portugal

J. L. Alves Department of Mechanical Engineering, University of Minho, Guimarães, Portugal

Cátia Bandeiras Department of Mechanical Engineering, University of Aveiro, Aveiro, Portugal;
Institute of Bioengineering and Biotechnology, Instituto Superior Técnico, University of Lisbon, Lisbon, Portugal

Jorge Belinha Department of Mechanical Engineering, School of Engineering, Polytechnic of Porto (ISEP), Porto, Portugal

R. S. Boffa Departamento de Engenharia de Materiais, Universidade de São Paulo (USP), Escola de Engenharia de São Carlos, São Carlos, SP, Brazil

A. P. G. Castro IDMEC, Instituto Superior Técnico, University of Lisbon, Lisbon, Portugal

R. Cesar Departamento de Engenharia Mecânica, Universidade de São Paulo (USP), Escola de Engenharia de São Carlos, São Carlos, SP, Brazil

J. C. P. Claro Department of Mechanical Engineering, University of Minho, Guimarães, Portugal

Rui M. Coelho IDMEC, Instituto Superior Técnico, Universidade de Lisboa, Lisbon, Portugal

António M. G. Completo Departamento de Engenharia Mecânica, Universidade de Aveiro, Aveiro, Portugal

Joana da Costa Reis Escola de Ciências e Tecnologia, Universidade de Évora, Évora, Portugal

Lúcia Dinis Faculty of Engineering, Mechanical Engineering Department, University of Porto (FEUP), Porto, Portugal

R. Erbereli Departamento de Engenharia Mecânica, Universidade de São Paulo (USP), Escola de Engenharia de São Carlos, São Carlos, SP, Brazil

J. Fernandez Department of Engineering Science, Auckland Bioengineering Institute, Auckland, New Zealand

P. Flores Department of Mechanical Engineering, University of Minho, Guimarães, Portugal

T. P. Leivas Instituto de Ortopedia E Traumatologia, HCFMUSP-OIT—Hospital de Clinicas da Faculdade de Medicina, Universidade de São Paulo (USP), São Paulo, Brazil

M. C. Marques Institute of Mechanical Engineering and Industrial Management (INEGI), Porto, Portugal

R. Natal Jorge Faculty of Engineering, Mechanical Engineering Department, University of Porto (FEUP), Porto, Portugal

Joana P. Neto Instituto Superior Técnico, Universidade de Lisboa, Lisbon, Portugal

A. F. Oliveira Medical Teaching Department—CHP/HSA, Instituto de Ciencias Biomédicas Abel Salazar (ICBAS), Porto, Portugal

Maria Teresa Oliveira Escola de Ciências e Tecnologia, Universidade de Évora, Évora, Portugal

Madalena M. A. Peyroteo Institute of Mechanical Engineering and Industrial Management (INEGI), Porto, Portugal

J. M. D. A. Rollo Departamento de Engenharia de Materiais, Universidade de São Paulo (USP), Escola de Engenharia de São Carlos, São Carlos, SP, Brazil

D. C. Schwab DCS - English Consultancy Services, São Carlos, Brazil

Duarte Valério IDMEC, Instituto Superior Técnico, Universidade de Lisboa, Lisbon, Portugal

Susana Vinga INESC-ID, Instituto Superior Técnico, Universidade de Lisboa, Lisbon, Portugal;
Institute of Mechanical Engineering (IDMEC), IST, Lisbon, Portugal

X. Wang Auckland Bioengineering Institute, Auckland, New Zealand

Biological Characterization of Bone Tissue

Bone: Functions, Structure and Physiology

Joana da Costa Reis and Maria Teresa Oliveira

Abstract In this chapter, bone functions, regulation, morphological structure and physiology are revisited. Bone is a highly complex tissue, very sensitive and responsive to external and internal stimuli, and intimately intertwined with other organs. From embryogenesis to endocrine regulation and bone remodelling, a global assessment is presented. Considering the scope of this book, special emphasis is given to how cell structure and tissue organization modulate the response to mechanical stimuli.

1 Introduction

The deeply dynamic nature of bone may be missed by a less attentive eye. Bones are resilient and apparently quite rigid structures. They vary in shape, size and number and are divided in the axial and appendicular skeleton. Through life, they are subjected to loads and strains that temper their shape, with old matrices being replaced by newly formed ones, maintaining bone volume and strength. When trauma and fractures occur, bones are capable of healing if enough stability and proper alignment are guaranteed.

Osteogenesis, bone repair and remodelling are directed by the exchanges involving the environment, cell-to-cell interactions and cell–extracellular matrix.

Mechanical forces are crucial in early embryonic development. Morphogenesis is controlled through fluid flow mechanisms and by cellular contractility. Early embryo shaping depends on morula contraction determined by cohesivity. Multiple layers result in the development of endoderm, mesoderm and ectoderm in the

---

Joana da Costa Reis and Maria Teresa Oliveira contributed equally to this work.

J. da Costa Reis (✉) · M. T. Oliveira (✉)
Escola de Ciências e Tecnologia, Universidade de Évora, Largo dos Colegiais, Évora, Portugal
e-mail: jmfcr@uevora.pt

M. T. Oliveira
e-mail: teresoliveira@uevora.pt

© Springer Nature Switzerland AG 2020
J. Belinha et al. (eds.), *The Computational Mechanics of Bone Tissue*,
Lecture Notes in Computational Vision and Biomechanics 35,
https://doi.org/10.1007/978-3-030-37541-6_1

blastula [1–4]. Mechanical forces, cell geometry, and oriented cell division together orchestrate normal airway tube morphogenesis [5] and may help determine the neocortical organization [6]. Cells generate tension through contraction of actin-myosin cytoskeleton filaments, which are transmitted through cadherin-mediated adhesion sites to surrounding structures, these being either cells or extracellular matrix. The cytoskeletal conformation and cell shape stress-dependent changes regulate cell phenotype; interactions with the extracellular matrix are of paramount importance for cell phenotype [2]. Organ lateralization and asymmetry depend on unidirectional fluid flow, generated by specialized motor protein complex dynein. The fluid flow induces differences in key molecules' expression (such as the TGF-family signalling molecules) [7–10]. Lateralization may also depend on fluid shear, in the embryo, by acting on a group of non-motile cilia, coupled to calcium channels; fluid flow generated shear may cause the intracellular calcium concentrations to increase and initiate the cascade of events responsible for lateralization [11].

The mechanical environment is also a determining factor for vasculogenesis, angiogenesis [12, 13] and neuronal development [14–17].

The embryo mesoderm is constituted by spindle- or star-shaped cells called mesenchymal stem cells (MSCs). MSCs are the utmost pluripotential cells in the organism, originating different tissues such as the connective tissue, muscle, cardiovascular tissue and the whole skeletal system. Bone, cartilage, tendons and ligaments develop through mechanisms of proliferation, migration and differentiation, but also programmed cell death/apoptosis [18].

We now begin to address how complex bone is in its functions, how its macroarchitecture, microarchitecture and arrangement at molecular level play together remarkably, ensuring its responsiveness to external and internal stimuli and close entwining with other organs.

2 The Complexity Behind Simplicity

2.1 Bone Functions

Osseous tissue is the most rigid and resilient tissue of the body. Bone is composed of dense connective tissue; it is the primary skeleton component, thus providing structure, support and protection to vital organs, like the brain (skull), the spinal cord (vertebrae) and the heart and lungs (ribs and sternum). Vertebrae also participate in the spine shock absorbance—providing adequate load cushioning for the fibrocartilaginous joints at the intervertebral discs. Long bones provide structure, stability and, along with the joints, enable body movement—providing levers for the muscles.

Moreover, bones act as the major source of blood, since haematopoiesis occurs in their medullary cavity. In infants, the bone marrow of all long bones is capable of blood synthesis. With ageing, part of the red marrow turns into yellow fatty marrow, no longer capable of haematopoiesis. Functional red marrow in adults is limited to the vertebrae and the extremities of femur and tibia.

Bones also partake a vital role as:

- Mineral storage: mostly calcium, phosphate and magnesium; bone plays an important metabolic role, mediated by several hormones, regulating mineral homeostasis [19–21].
- Acid-base balance, as bone can buffer blood against extreme pH changes by absorbing or releasing alkaline salts, through bone cells' activity [22–24].
- Osteoblasts have been shown to produce growth factors, with production regulated by systemic hormones and local mechanical stress [25]. The bone matrix holds several growth factors such as insulin-like growth factors I and II, transforming growth factor beta, acidic and basic fibroblast growth factor, platelet-derived growth factor and bone morphogenetic proteins, released when resorption occurs [26].
- Adipose tissue storage (yellow bone marrow functions as a fatty acid/energy reserve) [27–29].
- Heavy metals and other foreign elements, after detoxification from the blood, are stored in bone and later excreted [30, 31].
- Bone functions as an endocrine organ, as it produces two known circulating hormones:

    a. Fibroblast Growth Factor 23 (FGF23): FGF23 was first described by Yamashita et al., and it is produced mainly by osteocytes [32, 33], but also by osteoblasts [34]. FG23 acts on the kidneys, inhibiting 1α-hydroxylation of vitamin D and promoting phosphorus excretion in urine [35–37]. FGF23 also decreases phosphorus absorption in the intestine, regulating inorganic phosphate metabolism and thus mineralization [35]. Serum calcium concentration regulates FGF23 production [38], thus making FGF23 into a calcium–phosphorus regulatory hormone [39]. Hence, FGF23 excess or deficiency results in anomalies of phosphate metabolism. Excess FGF23 hinders renal phosphate reabsorption and 1,25 dihydroxy vitamin D_3 [1,25(OH)$_2$D] production, causing hypophosphatemia and suppression of circulating 1,25(OH)$_2$D levels and, eventually, rachitic changes in bone [40]. These changes occur in autosomal dominant hypophosphatemic rickets and osteomalacia [41] and, in tumour-induced osteomalacia (TIO), a paraneoplastic syndrome [42]. In contrast, reductions in FGF23 cause tumoral calcinosis syndrome, characterized by hyperphosphatemia, increased 1,25(OH)$_2$D and soft tissue calcifications [40, 43]. An obligate FGF23 co-receptor was identified—Klotho [44]. Klotho is essential to activate FGF receptors and their signalling molecules. Secreted Klotho suppresses either by direct interaction or interference with receptors, the activity of several growth factors: insulin, insulin-like growth factor-1 (IGF-1) [45], Wnt [46] and TGF-β1 [47]. The FGF23-Klotho axis represents a specialized system responsible for the external and internal calcium and phosphorus balance in the bone, intestine and kidney. FGF23-Klotho axis works under parathormone regulation, with parathormone increasing serum FGF23 levels and directly promoting FGF23 expression by osteocytes [48, 49]; FGF23 exerts negative feedback by inhibiting the parathyroid glands [50, 51]. FGF23 production in the osteocyte may be inhibited by osteopontin [52].

b. Osteocalcin is a protein produced by osteoblasts in bone, and it is a major regulator of insulin secretion by direct action over the pancreatic β-cell. Osteocalcin also increases insulin sensitivity of peripheral tissues, e.g. muscles and liver, up-regulating glucose uptake and energy expenditure, thus contributing to glycaemia regulation [53–56]; it also reduces fat deposition by inducing adiponectin secretion by adipocytes [57, 58]. Blood osteocalcin levels are significantly lower in diabetic patients when compared to non-diabetic controls, and osteocalcin levels are inversely related to fat mass and blood glucose [59, 60]. Lastly, osteocalcin influences male fertility, by enhancing testosterone production by Leydig cells in the testes [61].

2.2   Bone Structure and Mechanical Behaviour

Bone is a composite material; the inorganic portion of bone comprising 70% (of which 95% is hydroxyapatite and 5% are impurities impregnated in hydroxyapatite), whilst 22–25% are organic (of which 94–98% are mainly collagen type I and other non-collagen proteins and 2–5% are cells); 5–8% is water [62].

Bone mechanical properties depend on porosity, composition, mineralization degree and organization of solid matrix. Therefore, the mechanical behaviour of an entire bone is highly dependent on its properties at a microscale [63, 64].

Bone can be classified according to its structural features at a microscopic level in woven and lamellar bones.

Woven bone is immature or pathologic, primary bone, and it is present in growth, fracture healing and diseases such as Paget's disease. Cells and matrix are laid randomly. Woven bone is formed during intramembranous, endochondral or rapid appositional bone growth. In large animals (whether reptiles, birds or mammals), woven bone with large vascular canals is rapidly deposited in the subperiosteal region. Canals are lined with osteoblasts that gradually deposit lamellae until the canal has a reduced diameter; the resulting structure is a primary Haversian system or osteon. The random distribution of its components explains woven bone's isotropy.

Lamellar bone is organized, mature bone, and it is morphologically classified into two different types: cortical or compact bone and cancellous or trabecular bone. Cortical and cancellous bone types differ in both structure and functional properties, but both are highly anisotropic.

The typical structure of a long bone, such as the femur or the humerus, comprises the cylindric shaft, or diaphysis, and the extremities, or epiphyses (Fig. 1). The outer surface is covered by a layer of dense connective tissue called periosteum, except for the areas of mobile articulation, covered with hyaline cartilage. The periosteum is highly vascularized and responsible for appositional bone growth. The endosteum

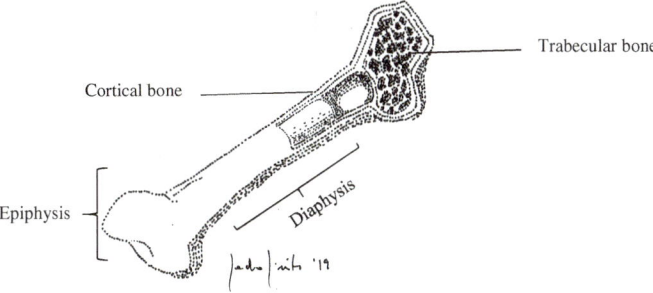

Fig. 1 Illustration of a long bone structure, showing the distribution of two different types of lamellar bone: cancellous and cortical compact bone

is a thin layer of connective tissue that lines the inner surface of the diaphysis, containing the medullary canal. The epiphysis consists of an outer layer of cortical bone surrounding the porous network formed by trabecular bone. Within the spaces between trabeculae lays red bone marrow [65]. Long bones, fundamental for load bearing and leverage, evolved as structures in which stiffness along the long axis was favoured.

Cortical bone (Fig. 2) accounts for approximately 80% of the skeletal mass. Cortical bone is vital to skeletal mechanical competence, both of long and flat bones. It is formed by tightly aligned collagen fibrils, making concentric lamellae. Each lamella is 2–3 μm thick and is arranged in distinct layers of parallel fibrils, each

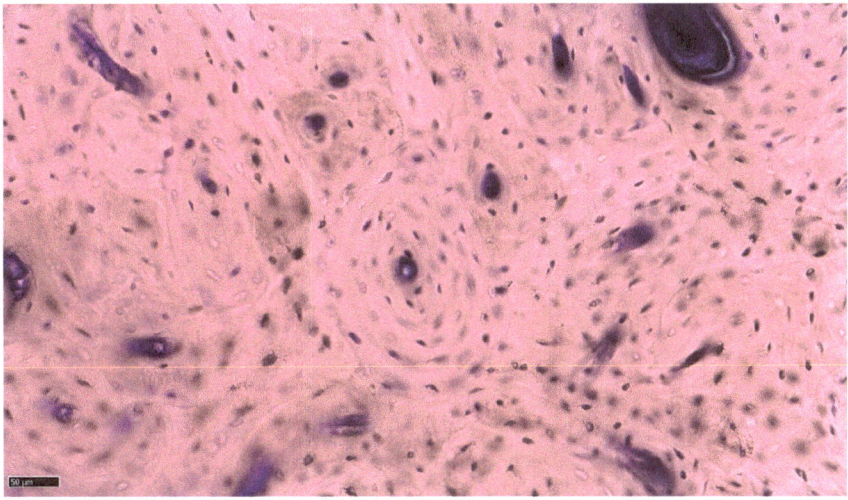

Fig. 2 Microphotograph of cortical bone in proximal tibia (undecalcified bone section of sheep tibia, Giemsa-Eosin, 40× magnification; slide digitalized using NanoZoomer SQ, Hamamatsu Photonics, Portugal). Haversian systems are evident, as are the concentric lamellae. Osteocytes are visible in their lacunae, in between lamellae

layer with a different fibril orientation [66]. Mineralization occurs by apatite crystals (mainly carbonated apatite) deposition within and around these fibrils. The lamellae form cylinders containing a hollow central canal where blood vessels and nerves run, composing the cortical bone microstructural unit, called Haversian system or osteon. From the centre of the osteon (Haversian canals), blood vessels form a three-dimensional network and penetrate the cortical bone layer perpendicularly (running within Volkman's channels) [67]. In between, the osteons are incomplete osteons, known as interstitial systems or interstitial bone.

Cancellous (or trabecular) bone is highly porous and adapted to compressive loads. The lamellae are organized in a parallel manner, forming trabeculae. These rod- and plate-shaped struts are organized into a flexible lattice with variable degrees of interconnectivity (Fig. 3).

The trabecular network is light and of utmost importance for load transfer in long bones, absorbing and distributing sudden stresses. In vertebrae, cancellous bone is the main load-bearing structure and essential for shock absorption. Trabeculae are approximately 200 μm thick and are orientated according to routine load-bearing direction [68]. This is evident in epiphyses and metaphyses of long bones, but also in the vertebrae and ribs. Trabeculae are covered by osteoblasts and bone-lining cells. Osteoblasts actively lay extracellular matrix (ECM), and bone-lining cells are in an inactive state. The metabolic rate of trabecular bone is higher than that of cortical bone and the remodelling phenomena more prominent. [18, 69].

Bone endures both compressive and tensile stresses. Bone is subjected to bending and torsion [62]. In humans, there is a large variation in strains, ranging from to 400 to 2000 μstrains or even as high as 4000 μstrains [62, 70, 71].

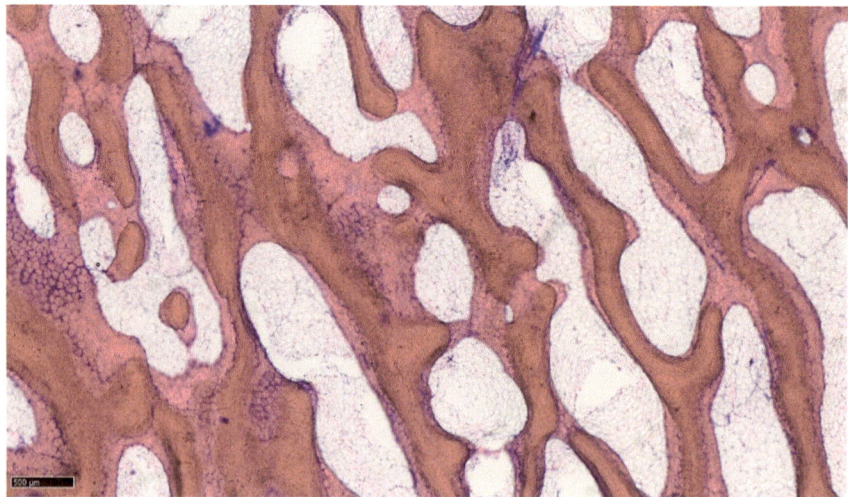

Fig. 3 Image of trabecular bone (undecalcified bone section of the proximal epiphysis of sheep's tibia, Giemsa-Eosin, 5× magnification; slide digitalized using NanoZoomer SQ, Hamamatsu Photonics, Portugal). The picture illustrates the sponge-like structure of cancellous bone

The bone exhibits a stress–strain response of sequential elastic and plastic responses. In its elastic region, no permanent damage is caused to the bone structure; if the stress increases, a gradual transition to a plastic response occurs. Post-yield deformations are permanent and cause trabecular fracture, cement lines and cracks. Crack formation and growth allow energy dissipation and are a powerful stimulus for bone remodelling in healthy bone.

The mineral component contributes to compression strength, while collagen fibrils are fundamental for tensile strength. The mineral phase is highly related to stiffness, whilst collagen is determinant for toughness [72]. A higher Young's modulus corresponds to less ductility and higher brittleness [73].

Bone material properties reflect, therefore, high functional specialization and depend on architecture, composition and component spatial distribution.

2.2.1   The Bone Matrix

Structure and material properties of the bone depend on collagen. The collagen I molecule is composed of three long peptide sequences, arranged helicoidally. Collagen is produced by osteoblasts and goes through several enzymatic modifications whilst still within the cell [74]. After leaving the cell, collagen molecule undergoes further cross-linking within itself and with other collagen molecules. Collagen chain mutations lead to diseases such as osteogenesis imperfecta [74, 75]. The triple tropocollagen units are aligned in fibrils and display a permanent dipole moment. Consequently, collagen acts as a piezoelectric and pyroelectric material and as an electromechanical transducer [76, 77]. The native polarity and the piezoelectric properties of collagen are associated with the mineralization process. Under compression, negative charges on the collagen surface become uncovered and attract calcium cations, which are then tailed by phosphate anions [77, 78]. Collagen can actively control mineralization, functioning in synergy with other non-collagenous proteins, inhibitors of hydroxyapatite nucleation. The positive net charge close to the C-terminal end of the collagen molecules promotes the infiltration of the fibrils with amorphous calcium phosphate; at the gap and overlap regions of the collagen molecule, the clusters of charged amino acids form nucleation sites and the amorphous calcium phosphate are changed into parallel oriented apatite crystals [79].

Non-collagenous proteins such as osteopontin, fibronectin, osteonectin and bone sialoprotein are present in much smaller quantities but are, nonetheless, essential for normal bone function and properties.

Osteopontin (OPN) is a non-collagenous glycoprotein, present in the bone matrix, binding to the cell surface and hydroxyapatite. It is mostly produced by proliferating pre-osteoblasts, osteoblasts and osteocytes, but also by fibroblasts, osteoclasts and macrophages [80, 81]. OPN intervenes in cell migration, adhesion and survival in diverse cell types. OPN is a key player in bone remodelling processes. Its production is modulated by mechanical loading, being up-regulated both by loading and by loading deprivation [81–83]. OPN has been proved to inhibit mineralization, but its deficiency significantly lessens bone fracture toughness and causes anomalous mineral distribution, leading to increased FGF23 production [52, 84–86].

Fibronectin mediates many cellular interactions with the ECM, playing an important part in cell adhesion, migration, growth and differentiation. It is determinant for vertebrate development and is mostly synthesized by osteoblast precursors and mature bone cells; it can also be produced at distant sites (such as the liver) and enters the systemic circulation. Some studies suggested that only circulating fibronectin exerts effects on the bone matrix [74, 87]. Fibronectin binds to collagen and may act as an extracellular scaffold, facilitating interactions of BMP1 with substrates [88]. Fibronectin may also be vital for the osteogenic differentiation of mesenchymal cells [89, 90].

Osteonectin or secreted protein acidic and rich in cysteine (SPARC) is secreted by osteoblasts during bone formation, and it is one of the most abundant non-collagenous proteins in the bone matrix. Osteonectin is a regulator of bone mineralization; its attachment to collagen can inhibit or promote mineral formation. It interacts also with apatite through its N-terminal domain, inhibiting crystal growth [91]. Osteonectin knockout mice suffer from osteopenia due to osteoblasts and osteoclasts defective function and low bone turnover. Changes in the osteonectin encoding gene have also been linked to idiopathic osteoporosis and osteogenesis imperfecta [92].

Thrombospondin-2 (TSP-2) is another matricellular protein that also exerts its effects on osteoblast proliferation and function, being involved in MSCs adhesion and migration; it has also influence on angiogenesis and tumour growth and metastization [93–96]. TSP-2 likely participates in bone remodelling, since it promotes osteoclastogenesis through the RANKL-dependent pathway [95, 96].

Bone sialoprotein (BSP) is a highly glycosylated and sulphated phosphoprotein that is found almost exclusively in mineralized connective tissues [97]. BSP knockout mice have higher trabecular bone mass and reduced amounts of cortical bone; they also present a very low turnover. BSP defective mice maintain unloading bone response, as opposite to OPN knockout mice [98]. The absence of BSP also leads to changes in the growth plates, decreased bone length and delayed ossification [99]. BSP and OPN are part of the small integrin-binding ligand N-linked glycoproteins (SIBLING) family, and the recent studies suggest the interplay in between these proteins is determinant in bone biology [100].

Proteoglycan (PG) encoding genes are expressed in skeletal and non-skeletal tissues but with stronger expression in bone, joints and liver. PGs are a large family of molecules and perform many biological functions. PGs help to structure bone by mediating collagen secretion and fibril organization; they also act as mineralization inhibitors. PGs also modulate cytokines and growth factors' biological activity in bone [101]. In bone, PrG4 gene expression is under the control of PTH [102].

From the reviewed above, it becomes clear that non-collagenous and collagen matrix proteins are fundamental for bone morphology and material properties, interacting with each other and with cells and responding to stimuli generated locally or systemically. It is also evident that matrix components have a multiplicity of functions. The role of a molecule is modulated by changes in its structure and by interactions with other substances.

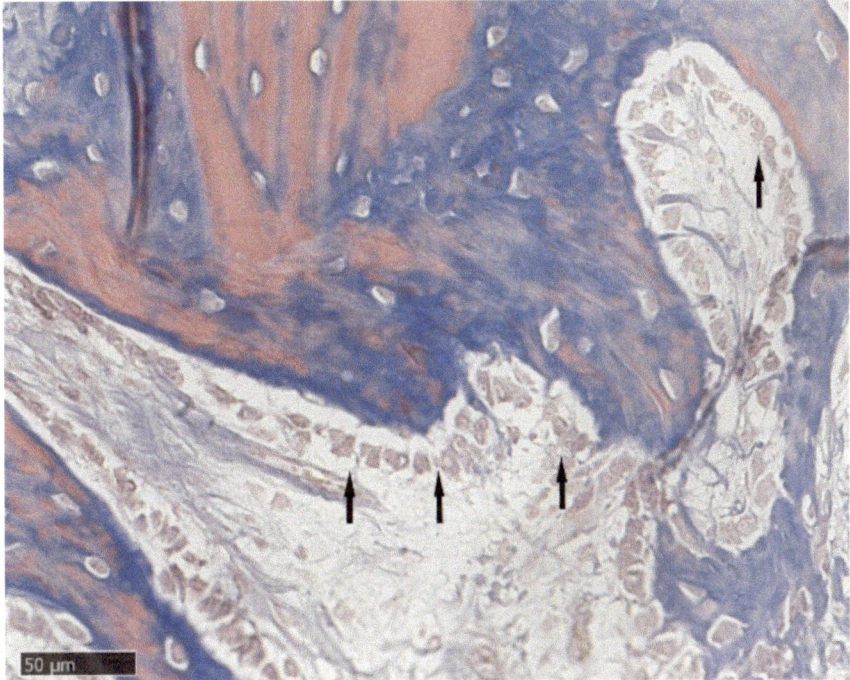

Fig. 4 Osteoblasts are cuboidal cells that when actively deposing matrix on bone surfaces (microphotograph, decalcified sheep bone section, Mason trichrome, magnification 40×; black arrows point line of active osteoblasts). When quiescent, osteoblasts appear as flat bone-lining cells

2.2.2 Bone Cell Population

Mature bone contains three core cell populations: osteoblasts, osteocytes and osteoclasts.

Osteoblasts

Osteoblasts arise from MSCs, sharing a common background with chondrocytes, myoblasts and fibroblasts. Osteoblasts differentiate under the influence of a variety of hormones, cytokines and the local mechanical environment [103]. These cells, when active, are cuboidal/round (Fig. 4), with specific features consistent with their secretory functions, such as prominent Golgi complexes and endoplasmic reticulum (with multiple vesicles and vacuoles); these are even more evident during matrix secretion and early stages of mineralization [104].

Osteoblasts can also remain on bone surfaces as flat bone-lining cells, in a quiescent state, with few apparent cell organelles. During the osteoblast maturation process, there are increased levels of expression of pro-collagen, osteopontin and osteocalcin; bone sialoprotein seems to be more strongly expressed at intermediate phases of differentiation [105, 106]. Osteoblast differentiation is impaired when gap junctions are inhibited, suggesting communication to neighbouring cells is essential for differentiation [107]. Osteoblasts produce non-mineralized matrix—osteoid— that becomes gradually mineralized, wherein they become trapped and some differentiate into osteocytes. Runx2 induces the expression of major bone matrix protein genes in vitro. Runx2 expression is up-regulated in pre-osteoblasts, being maximal in immature osteoblasts and down-regulated in mature osteoblasts. Although Runx2 is weakly expressed in undifferentiated mesenchymal cells, it induces their osteogenic commitment [108]. Once Runx2 is activated, cells undergo the three stages of differentiation, with the synthetization of different molecules: in stage 1, the cells proliferate and express fibronectin, collagen, TGFβ receptor 1, and osteopontin; during stage, osteoblast will differentiate and act on the extracellular matrix through alkaline phosphatase and collagen; at stage 3, the osteoblast will assume its characteristic cuboidal shape and secrete significant amounts of osteocalcin. Osteocalcin will promote matrix mineralization [109]. Osteoblast differentiation is influenced by $1,25(OH)_2D_3$ and mechanical stimuli, amongst other factors [110].

Osteocytes

Osteocytes are the most abundant cells of bone, comprising more than 90% of the osteoblast lineage and contributing to bone formation and resorption [111, 112]. They are fully differentiated osteoblasts embedded in the mineralized matrix, inside the osteocytic lacunae. Lacunae are located between the lamellae and connected with surrounding lacunae by a canalicular system (Fig. 5). Osteocytes have long dendritic cell processes (50–60 by cell) that lay within the canaliculi. The extremities of the cell processes connect osteocytes amongst themselves and allow contact with osteoblasts and bone-lining cells [18, 113, 114]. The resulting functional syncytium shares a common environment [115].

Osteocytes have no matrix secretion functions; however, they are responsible for sensing changes in the bone structure and commanding bone remodelling.

Pre-osteoblasts and osteoblasts are less responsive to fluid shear stress than osteocytes. Mechanosensitivity seems to increase during differentiation. However, osteoblasts are able to modulate their response according to the mechanical stimuli intensity [62]. Osteocyte functions include mechanosensing and maintaining bone matrix [116, 117]. The sensation of electrical signals may be one of the functions of osteocytes, and electrical signals mediated by osteocytes may regulate the cell behaviour in bone tissue [118]. Flexoelectric fields are generated by fractures in the bone mineral and may be large enough to induce osteocyte apoptosis and initiate bone remodelling [119].

The same mechanical stimulus may cause a different response in osteocytes according to their cell body shape [120]. Recently, a study reports that osteocyte plasma membrane disruptions, caused by mechanical loading, act as triggering mechanosensing events, both in vitro and in vivo [121].

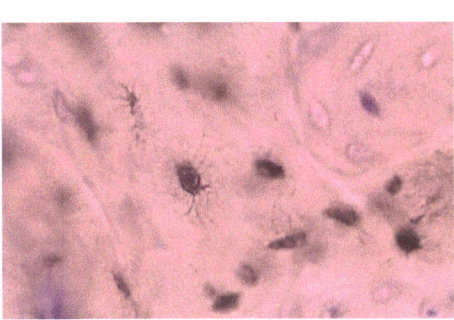

Fig. 5 Detail of microphotograph of an undecalcified bone section of sheep tibia, Giemsa-Eosin, on the left, showing osteocytes (Giemsa-Eosin, 40× magnification and 200% zoom; slide digitalized using NanoZoomer SQ, Hamamatsu Photonics, Portugal). The canaliculi where cell processes run are observable. The image on the right illustrates a simplified version of the resulting three-dimensional syncytium

Osteocytes' early response to mechanical loading results in vesicular ATP release by exocytosis, tuned according to the magnitude of the stimulus [122]. Mechanical stimulation of osteocytes also causes fluctuations in intracellular calcium levels; these are responsible for calcium-dependent actin contraction and release of extracellular vesicles containing bone regulatory proteins [123]. In fact, osteocytes respond to mechanical stimuli by producing various messenger molecules, such as nitric oxide and prostaglandins, namely prostaglandin E2 (PGE2) [117, 124, 125]. This response is dependent on the function of stretch-activated calcium channels [126], although reserves of intracellular calcium also contribute [123]. PGE2 has anabolic effects, stimulating osteoblast activity and new bone formation [127]. Nitric oxide inhibits bone resorption, by suppressing osteoclast formation and increasing the expression of osteoprotegerin [128, 129].

The lifespan of osteocytes is highly variable and likely associated with the rate of bone remodelling, depending on mechanical and environmental factors such as hormones; osteocytes apoptosis may be inhibited or induced by a variety of physiological and pathological conditions. Osteocyte apoptosis may be induced by biological effectors such as hormones, without being accompanied by increased osteoclastogenesis [130–134].

Young osteocytes are polarized towards the mineralization front, just like osteoblasts are, with the nucleus remaining close to vessels [104]. As lamellar bone matures, the osteocytes tend to spread their processes perpendicular to the longitudinal axis of trabeculae and long bones and appear as flattened cells. In immature bone,

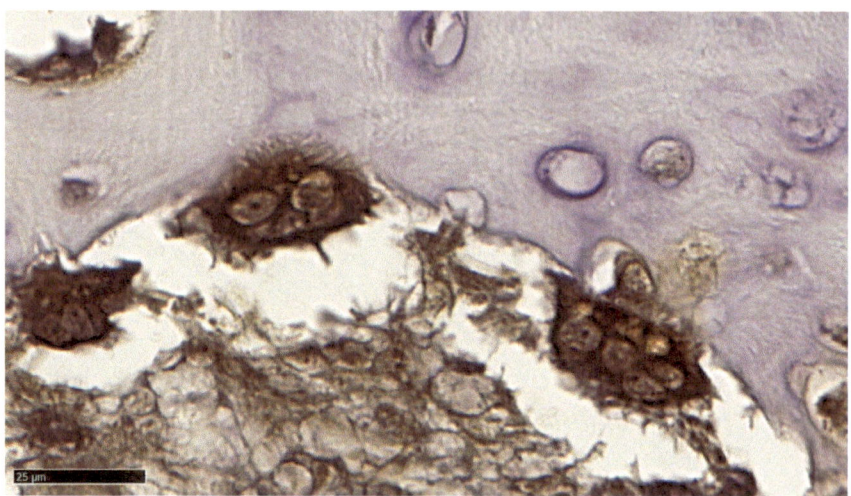

Fig. 6 A microphotograph of TRAP positive osteoclasts firmly attached to the bone surface. The ruffled border membrane is visible in direct contact with bone. This is the resorbing organelle; along its enlarged ruffled contact surface, proton pumps lower the local pH, dissolving hydroxyapatite

plump osteocytes with randomly distributed processes predominate [130]. Osteocyte density is closely related to bone architecture and thus to its mechanical behaviour [135].

Ageing has been correlated with smaller canaliculi, in lower numbers per lacuna, leading ultimately to reduced mechanosensitivity in the aged individual [136, 137].

Osteoclasts

Osteoclasts are multinucleated cells and belong to the same lineage as macrophages and monocytes (Fig. 6). Like macrophages, osteoclasts are able to merge and form multinucleated cells and to phagocytize [138]. The cell precursor may differentiate into either an osteoclast or a macrophage. The differentiation path depends on the progenitor cell being exposed to a receptor activator of several ligands (receptor activator of nuclear factor-κB ligand—RANKL, osteoprotegerin and osteoclast differentiation factor—ODF) or to colony-stimulating factors related to immune system [139–141].

The osteoclast presents distinctive functional features:

- osteoclasts can attach firmly to the bone surface, isolating the area under the cell membrane from its surroundings; the membrane domain responsible for the isolation of the resorption site is called sealing zone [142, 143];
- osteoclasts acidify the mineral matrix by the action of protons pumps at the ruffled border membrane, a resorbing organelle; the lowering of the pH causes the dissolution of the hydroxyapatite crystals [144–146];

- osteoclasts are capable of synthesizing and secreting enzymes such as tartrate-resistant acid phosphatase (TRAP) and cathepsins in a directional manner; the proteases secreted by osteoclasts cleave the organic matrix; through the combined action of lysosome enzymes, matrix metalloproteinases and the pH reduction, bone is resorbed [147, 148];
- osteoclasts can phagocytize the resultant organic debris and minerals, removing them from the resorption lacunae, through a transcytosis process [149, 150].

The bone resorption process begins with differentiation and recruitment of osteo-clast precursors, which merge and originate matured multinucleated bone-resorbing osteoclasts. Bone resorption begins when the osteoclast attaches to the mineralized bone matrix through the interaction of integrins with matrix proteins, like osteopontin and bone sialoprotein, previously laid down by osteoblasts [143].

2.3 Regulation of Bone Metabolism (Modelling/Remodelling)

The bone cell populations are responsible for bone remodelling and repair. These processes are regulated systemically by hormones, neuropeptides and other mediators and locally by cytokines and growth factors [151, 152].

2.3.1 Parathormone (PTH), Vitamin D and Calcitonin

The bone mineral metabolism (calcium and phosphorus) is regulated by parathor-mone (PTH), calcitonin, FGF23 and vitamin D.

PTH is a peptide hormone produced by the parathyroid glands in response to low levels of extracellular ionized calcium, detected by specific cell-surface calcium-sensing receptors located in the parathyroid glandular tissue. High levels of PTH increase the number of osteoclasts and trigger resorption of bone matrix, with con-sequent release of calcium phosphate and increasing calcemia. This mechanism has developed as a protection against acute hypocalcemia. Inversely, low levels of PTH cause the elevation of osteoblast numbers. PTH also acts on osteoblasts' receptors, stimulating proliferation and differentiation and inhibiting apoptosis [153]. PTH also regulates kidney function by impairing phosphate reabsorption and promoting its excretion, by stimulating calcium reabsorption and up-regulating a hydroxylase enzyme (CYP27B1), thus promoting 1,25(OH)2 vitamin D3 synthesis [154].

Circulating hormonal metabolite, $1\alpha,25$-dihydroxy vitamin D3 (1,25(OH)2D3), enhances several physiological functions, including intestinal calcium and phosphate absorption, bone phosphate and calcium resorption, and renal calcium and phosphate reabsorption, which results in a rise in the blood calcium and phosphate, required for bone passive mineralization of unmineralized bone matrix to occur [155, 156]. Additionally, 1,25(OH)2D3 stimulates differentiation of osteoblasts and the expres-sion of several bone proteins, like bone-specific alkaline phosphatase, osteocalcin, osteonectin, osteoprotegerin and other cytokines, and influences the proliferation and apoptosis of other bone cells, including hypertrophic chondrocytes [157]. This may help explain why endogenous PTH levels can have anabolic and catabolic effects and are associated with differential skeletal effects on cortical and trabecular bones [158].

Calcitonin is produced by parafollicular cells of the thyroid, in direct response to extracellular calcium, through the same sensor that regulates the production of PTH. It inhibits matrix resorption, promotes calcium and phosphate excretion, thus reducing calcium and phosphate serum levels; calcitonin inhibits osteoclast mobility and the secretion of proteolytic enzymes [159, 160].

2.3.2 Growth Hormone (GH)

Growth hormone or somatotropin is secreted in pulses by the anterior pituitary gland, inducing bone longitudinal growth [161]. It also induces organs such as the liver and the skeleton to synthesize somatomedins that influence growth, such as insulin-like growth factor 1 (IGF-1) and 2 (IGF-2) [162]. The chondrocytes in the epiphyseal plate are stimulated not only by IGF1 and IGF2 but also directly by GH; proliferative and hypertrophic chondrocytes also secrete IGFs; IGF-1 acts inhibiting further GH secretion [163, 164].

According to Ohlsson et al. [162], GH action in bone remodelling follows a "biphasic model": initially, it increases bone resorption, causing bone loss, followed by a phase of increased bone formation. When bone formation is more stimulated than bone resorption (transition point), the bone mass increases. A net increase of bone mass will be seen after 12–18 months of GH treatment in GH-deficient adults [165]. GH increases bone growth, by increasing both periosteal and endocortical bone formation, bone mineral content (BMC) and bone mineral density (BMD). GH acts synergistically with PTH to increase bone growth and bone formation, bone density and mass and to decrease bone resorption [166].

2.3.3 Insulin and Insulin-like Growth Factors (IGF-1 and IGF-2)

IGF-1 stimulates chondrocyte proliferation in the growth plate, thus playing a crucial role in longitudinal bone growth [167]. It is also involved in the formation of trabecular bone [168]. Insulin and IGF-1 have anabolic effects over the osteoblast and promote bone development, mainly through the activation of Akt and ERK signalling pathways; also, IGF-1 is capable of inducing osteoblasts in vivo proliferation whilst inhibiting the gene expression of osteocalcin, a marker for differentiating osteoblasts; insulin enhances osteocalcin expression but has no effect on osteoblast proliferation [169]. Additionally, insulin indirectly enhances Runx2 expression, a regulator of osteoblast differentiation [55, 169]. A study with insulin-deficient type I diabetic mice showed that these mice presented a decreased expression of Runx2 and the Runx2-regulated genes, like osteocalcin and collagen type I, and a secondary decrease in bone formation. Bone loss was restored after insulin treatment, which increased Runx2 expression and the expression of related genes [170].

Likewise, IGF-2 potentiates BMP-9-induced osteogenic differentiation and bone formation [171] through PI3K/AKT signalling. Moreover, a recent study in mice aortas showed that IGF-2 induces the expression of miR-30e, in a feedback loop. miR-30e is a major down-regulator of osteogenic differentiation of MSCs and smooth muscle cells [172].

2.3.4   Sex Steroids (Oestrogen and Testosterone)

Bone metabolism is strongly influenced by sex steroids. Oestrogen is an important regulator of skeletal development and homeostasis, both in men and women, exerting direct and indirect effects on the skeleton [173–175]. Indirectly, it influences, for example, the calcium intestinal absorption [176, 177] and secretion [178], and the calcium renal excretion; oestrogen also influences the secretion of PTH [179, 180]. Oestrogen maintains bone homeostasis by inhibiting osteoblast and osteocyte apoptosis [134, 181, 182], and it inhibits osteoclast formation and activity, inducing osteoclast apoptosis [183–187]. Oestrogen deficiency causes bone loss and osteoporosis [188].

Androgens are also important to bone homeostasis. However, their role is likely more important during growth and contributes, via the GH/IGF system, to bone formation at the periosteum [189]. Androgens contribute to the maintenance of cancellous bone mass and integrity, regardless of age or gender [190, 191]. Androgen-deprivation therapy has negative effects on bone mineral density; these effects can be partially delayed by exercise, in the lumbar vertebrae but not in the hip [192].

2.3.5   Thyroid Hormones

The skeleton is a target tissue for thyroid hormones, namely for thyroid hormone 3,5,3′-L-triiodothyronine (T3). Thyroid hormones influence bone growth during early development and adult bone turnover and maintenance. They act both directly, by stimulating bone resorption and formation, and indirectly, by enhancing the effects of growth hormone over tissues. Hypothyroidism causes impaired bone formation and growth delay; thyrotoxicosis is a recognized cause of secondary osteoporosis, and abnormal thyroid hormone signalling has been recognized as an osteoarthritis' risk factor [193]. T3 stimulates osteoblast proliferation and differentiation, with bone matrix secretion, modification and mineralization. Thus, bone turnover is increased by thyroid hormones, which is confirmed by increased biochemical markers of bone turnover, such as osteocalcin and bone-specific alkaline phosphatase [194–196], and therefore, bone loss can occur [160, 197]. Thyroid-stimulating hormone (TSH), produced by the hypophysis, has direct effects on bone turnover [198], and TSH receptors have been found on osteoblasts and osteoclasts, although available data does not allow conclusions on whether TSH inhibits, increases or does not affect osteoblast differentiation and function [193]. Still, recombinant TSH showed antiresorptive effects in ovariectomized rats [198, 199] and lower TSH levels—with no apparent association with free T4 levels—have been related to hip fracture risk, supporting the idea that TSH effect on the skeleton may be independent of free T4, though its action on dedicated membrane receptors can be up-regulated by modulators [196, 200].

2.3.6   Leptin ("Satiety" Hormone)

Leptin is produced mainly in adipose tissue, and it is a regulator of food intake and energy expenditure through its effects on the central nervous system (CNS). Its influence in bone metabolism probably follows two pathways: a central pathway, activating the sympathetic nervous system that inhibits bone formation, and a peripheral pathway, promoting bone formation through leptin receptors on osteoblastic cells [201, 202]. Leptin inhibits osteoclast generation [203], promotes the decrease in cancellous bone and increases in cortical bone, thus enhancing bone enlargement [204, 205]; it also increases osteoblast number and activity, acting primarily through the peripheral pathways [206]. Another study showed that leptin increases bone mineral content and density, especially at the lumbar spine [207]. However, in the ovine foetus, leptin infusion caused increased femur porosity and connectivity density, and vertebral trabecular thickness whilst leptin receptor antagonist infusion decreased trabecular spacing and increased trabecular number, degree of anisotrophy and connectivity density in the lumbar vertebrae; effects differed in females and males [208]. Leptin also increases the expression of IGF-1 receptor and IGF-1 receptor messenger [209]. During infancy and childhood, leptin and IGF-1 were associated with body composition in preterm-born children. The same study also describes leptin association with bone parameters in early infancy, but not in childhood [210]. These results suggest leptin role on bone metabolism and architecture may vary with gender, age and interaction with other hormones and factors. Leptin is also a key up-regulator of FGF23 secretion [211], and it has been described as a direct enhancer of parathormone secretion [212].

2.3.7   Bone Morphogenetic Proteins (BMPs)

BMPs are a group of 15 growth factors also known by cytokines, which belong to the transforming growth factor β (TGF-β) superfamily, with the ability to induce the formation of bone [213] and cartilage [214]. BMPs play a major role in the regulation of osteoblast lineage-specific differentiation and later bone formation [215]. Alterations in BMPs activity are often associated with a great variety of clinical pathologies, like skeletal and extra-skeletal anomalies, autoimmune, cancer and cardiovascular diseases [216]. BMPs crosstalk with several other major signalling pathways, e.g. Wnt, Akt/mTOR, miRNA, among others, having Runx2 as a key integrator [216, 217]. Among all BMPs, BMP9 has stronger osteogenic inductive activity over MSCs [215,

218, 219]; BMP9 also acts synergistically with TGF-β and GH to enhance bone formation [216, 220, 221]. In addition to BMP9, other BMPs also have shown the ability to induce osteogenesis in vivo, such as BMP2, BMP6 and BMP7 [222–224], with recombinant human BMP2 and BMP7 already being commercialized with the purpose of enhancing bone healing [225]. However, recent studies indicate the existence of age-related differences in BMP2-mediated bone regeneration, including relative dose sensitivity [226]. Contrariwise, BMP3 is known to be a negative regulator of bone formation and BMP4 has been shown to decrease trabecular bone formation in a murine model [219, 227].

2.4   Bone Remodelling and Cell Interchange

Healthy bone, both cortical and trabecular, is continuously remodelling, a dynamic process with bone resorption and formation. Bone remodelling is modulated by mechanical loading, blood calcium levels and a wide range of paracrine and endocrine factors.

The bone remodelling process depends on the coordinate actions of osteoblasts, osteoclasts, osteocytes and osteoblast-derived bone-lining cells, along with other cells, such as macrophages and immune cells. The ensemble constitutes the "basic multicellular unit" (BMU) or "bone remodelling unit" (BRU). In the BMU, the amount of bone lysis achieved by osteoclasts is equal to the amount of bone produced by osteoblasts. The balance between osteoblastic and osteoclastic activity is known as coupling. Frost proposed that bone longitudinal growth, modelling and BMU-based remodelling activities were modulated by a "mechanostat", a mechanism modulating bone mass in the function of mechanical use, in which BMUs would play a central role, along with bone longitudinal growth and modelling (bone formation). Bone modelling was thus considered as an adaptative response to overloading and remodelling as a response to underloading, with given strain set points for each process [228].

Osteoclasts and osteoblasts within the BMU may function under the control of other cell types, since osteoblasts and osteoclasts may perform their functions in the absence of each other [229, 230]. Cells from the osteoblast lineage express receptors for cytokines and other local secreted factors that stimulate osteoclast formation [231]. The BMU can be inhibited by old age, drugs, endocrine, metabolic or inflammatory diseases.

Regardless of the triggering stimulus, osteoclast formation depends on RANKL. Osteoblasts express membrane-bond RANKL, and this regulatory molecule interacts with a receptor (receptor activator of nuclear factor-κB—RANK), expressed on the surface of osteoclast precursors. The RANK activation by RANKL is essential for the fusion of the osteoclast precursor cells and osteoclast formation [232].

Both down-regulation and up-regulation of RANKL expression by osteoblasts under similar mechanical stimulation have been described [233, 234]. Osteoblasts subjected to different mechanical stimuli respond by an increase in RANKL-bound and a decrease in soluble RANKL secretion [235].

The RANKL/RANK coordinated effects can only be understood by adding osteoprotegerin to the axis. Cells from osteoblastic lineage produce osteoprotegerin (OPG). OPG is soluble and blocks the interaction between RANKL and RANK, acting as a decoy receptor for RANKL. OPG thus inhibits osteoclast formation and induces osteoclast apoptosis [236]. Osteoblasts, in addition, secrete macrophage colony-stimulating factor-1 (M-CSF-1); M-CSF-1 promotes osteoclast precursor proliferation and RANK expression [237, 238]. Osteoblast-like cells' cultures mechanically stimulated may respond by a decrease in the production of OPG, without a change in the RANKL production, with a consequent increase in the ratio of RANKL/OPG. This could translate into increased bone remodelling. However, subjecting osteoclast-like cells to the same mechanical stimuli regimen, decreased TRAP and a period of stimulation of one minute at 0.3 Hz frequency, a decrease in cell fusion and resorption activity was observed [239].

RANKL expression by osteoblast lineage cells is enhanced when microdamage within the bone matrix occurs. Microdamage may occur under physiological bone loading and in pathological conditions. The presence of microcracks is sensed by osteocytes and may induce osteocyte apoptosis; osteocyte apoptosis may also be induced by disuse and is closely correlated with higher bone remodelling levels [131, 240–245].

Pulsating fluid flow (PFF)-treated osteocyte cultures conditioned the culture medium, inhibiting osteoclast formation and decreasing in vitro bone resorption. These effects have not been detected in the medium from PFF-treated fibroblast cultures [246]. In osteocytes subjected to PFF, nitric oxide is involved in the up- and down-regulation of at least two apoptosis-related genes (Bcl-2 and caspase-3, with antiapoptotic protective and pro-apoptotic functions, respectively) [247]. Nitric oxide (NO) is a second messenger molecule produced in response to mechanical stimulation of osteoblasts and osteocytes, and other cell types such as endothelial cells, with a large variety of biological functions [248–251].

Osteocytes, thus, regulate osteoclastogenesis and osteoclast activity through soluble factors and messenger molecules.

Other pathways are relevant for osteoblasts, osteocytes and osteoclasts interweaved regulation, such as the Notch signalling pathway. In osteocytes, the Notch receptor's activation induces OPG and Wnt signalling, decreasing cancellous bone remodelling and inducing cortical bone formation [252]. Wnt/Lrp5 signalling in osteocytes has been considered as a key pathway for bone response to loading [253].

2.5 Bone Mechanotransduction

Bone mechanotransduction, essential in health and disease states, is not yet fully understood. The elements involved in transduction include the ECM, cell–cell adhesions, cell–ECM adhesions, cell membrane components, specialized surface processes, nuclear structures and cytoskeleton.

2.5.1   The Cell Membrane Elements, Cell–Cell and ECM–Cell Adhesions

Cell membrane-associated mechanotransduction mechanisms depend on the integrity of the phospholipid bilayer. Mechanotransduction pathways are disrupted if membrane cholesterol is depleted, inhibiting the response to hydrostatic and fluid shear stress [254, 255]. Cytoskeleton actin polymerization and assembly are influenced by membrane cholesterol levels [256, 257]. However, it has been proposed that actin polymerization during synaptic vesicle recycling is influenced by vesicular cholesterol, but not plasma membrane cholesterol, as suggested by a study wherein the inhibition of actin polymerization by the extraction of vesicular cholesterol resulted in the dispersal of synaptic vesicle proteins [258]. But even with a functional cell membrane, if integrin binding is impaired, actin cytoskeleton will not re-organize in response to shear stress [259]. However, nanometre- to micron-sized tears, reparable defects in the cell plasma membrane promote particle flux across the cell membrane, namely Ca^{2+} influx [121].

Integrins are cell adhesion receptors, heterodimers of non-covalently associated 18α and 8β subunits, in mammals, that can combine to generate 24 different receptors with different binding properties and different tissue distributions [260, 261]. These subunits possess an extracellular portion with several domains, able to bind to large multi-adhesive ECM molecules, which in turn bind to other ECM molecules, growth factors, cytokines and matrix-degrading proteases [260]. Integrins were first acknowledged as bridging the ECM and the cell cytoskeleton, including the actin cytoskeleton but also the intermediate filament network, essentially vimentin and laminin [262]. Cells use multiple mechanisms to sense and respond to mechanical stress applied to integrins [263]. Recruitment of vimentin has been shown to depend on integrin $\beta3$ subunits, underpinning the relationship between the various cytoskeletal elements and integrins [264]. The cytoplasmatic portions of integrin β subunit bind to talin, which can also directly bind to vinculin and actin filaments [265]. On the other hand, integrin $\alpha4$ subunit binds to paxillin [266], a protein that integrates sites of cell adhesion to the ECM.

Integrins allow communication between structures in the interior and outside of the cell, in a bidirectional way. The inside-out signalling turns the integrin extracellular domains into the active conformation. In the outside-in pathway, when an integrin binds to the extracellular ligand, it clusters with other bound integrins, forming focal adhesions, highly organized intracellular complexes; these are connected to the cytoskeleton. The focal adhesions integrate a range of different molecules, including the cytoplasmic portions of the clustered integrins, proteins of the cytoskeleton and signalling molecules [265, 267]. Initial adhesions to substrates are characterized by punctuating areas at the limits of lamellipodia, usually known as focal complexes. Focal adhesions are the mature form of cell–matrix adhesion, with an elongated shape, and are associated with bundles of actin and myosin (stress fibres). There is a specialized form of focal contact, in which integrin binds to fibronectin fibrils and tensin but with low levels of tyrosine kinases [268, 269]. Most focal adhesions also contain several types of signalling molecules like tyrosine phosphatases and tyrosine kinases and adaptor proteins [267, 270–272].

Matrix proteins may also modulate cell adhesion; connective tissue growth factor (CTGF), which is a matrix protein, enhances osteoblast adhesion (via $\alpha V \beta_1$ integrin) and cell proliferation, by inducing cytoskeletal reorganization and Rac1 activation [273]. Another matrix protein—osteoactivin—also modulates osteoblast adhesion, differentiation and function, stimulating alkaline phosphatase (ALP) activity, osteocalcin production, nodule formation, and matrix mineralization [274]. $\alpha 5 \beta 1$ integrin interacts with its high-affinity ligand CRRETAWAC, enhancing the Wnt/β-catenin signalling mechanism to promote osteoblast differentiation independently of cell adhesion [275].

Cell adhesion and mechanical stimulation depend on integrin mediation [276]. Forces applied to integrin receptors cause local adhesion proteins to be recruited, and the cell adapts by making the integrin–cytoskeleton linkages more rigid; myosin II contraction makes the cell apply tension to the substrate [277]. Different signalling pathways are triggered by sensed stress through integrin receptors. Sequential expression of integrin ligands (osteopontin, fibronectin and bone sialoprotein) in response to mechanical stimulation of osteoblasts has been described [278]. Bonds between integrin and ligands become stronger in the presence of cell tension [279].

Osteocytes are highly specialized in their interaction with ECM; osteocyte cell bodies express $\beta 1$ integrins while cell processes express $\beta 3$ integrins, the latter in a punctuate distribution [280–283]. Thi et al. [284] identified the cell processes as the mechanosensory organs in osteocytes. It has been demonstrated that integrin $\alpha V \beta 3$ is essential for the maintenance of osteocyte cell processes and also for mechanosensation and mechanotransduction by osteocytes, by ATP release that triggers calcium signalling [285, 286]. $\beta 1$ integrins have shown to regulate specific aspects of mechanotransduction, namely the cortical osteocyte response to disuse [283]. In osteoblasts, a mechanical load applied to $\beta 1$ integrin subunit results in calcium influx [287], independently from gap junctions [288]. Another study showed that ERK1/2 activation by strain prevented osteocyte apoptosis but required the integrin/cytoskeleton/Src/ERK signalling pathway activation [133].

Apart from integrin, other membrane proteins are responsible for the conduction of mechanical stimuli. Cadherins, which connect to the cytoskeleton, also mediate force-induced calcium influx [289, 290] and participate in the Wnt/β-catenin pathway [291]. In osteoblasts, it has been suggested that GPI-anchored proteins may play an important role in mechanosensing, by demonstrating that the overexpression of GPI-PLD, an enzyme that can specifically cleave GPI-anchored proteins from cell membranes, inhibits flow-induced intracellular calcium mobilization and ERK1/2 activation in MC3T3-E1 cells [255]. Ephrins (ligands) and Ephs (receptors) contribute to cell–cell interactions between osteoclasts and osteoblasts, helping to regulate bone resorption and formation, and appear to be necessary for hMSC differentiation [292, 293]. Lastly, another family of proteins—galectins—is also involved in regulating osteogenesis; for example, Gal-3, which is expressed both by osteo-cytes and osteoblasts, plays a significant role as a modulator of major signalling pathways, such as Wnt signalling, MAPK and PI3K/AKT pathways [294]; Gal-8 induces RANKL expression by osteoblasts and osteocytes, osteoclastogenesis and bone mass reduction in mice [295]; and Gal-9 induces osteoblast differentiation through the CD44/Smad signalling pathway in the absence of bone morphogenetic proteins (BMPs) [296].

Gap junctions are transmembrane channels that connect the cytoplasm of adja-cent cells. Only small metabolites, ions and signalling molecules like calcium and cAMP pass through these channels since the molecular weight must be lower than 1 kDa [297, 298]. Gap junctions are essential for bone mechanosensation since in osteoblastic cells, the PGE2 production induced by fluid flow is dependent on intact gap junctions; if these are disturbed, PGE2 production does not occur [288, 299]. Mice lacking Cx43 gap junctions in osteoblasts and/or osteocytes exhibit increased osteocyte apoptosis, endocortical resorption and periosteal bone formation [300].

2.5.2 Primary Cilia

In different cell types, different structures ensure recognition of mechanical stimuli; kidney epithelial cells possess a single microvillar projection on their apical surface (primary cilia). A similar structure was described in osteoblasts and osteoblast-like cells [301, 302]. Primary cilia originate in the centrosome and project from the sur-face of bone cells; its deflection during flow indicates that they have the potential to sense fluid flow. These cilia deflect upon application of 0.03 Pa steady fluid flow and recoil after cessation of flow [303, 304]. In bone, primary cilia translate fluid flow into cellular responses, independently of Ca^{2+} flux and stretch-activated ion channels [303]. It has been demonstrated in vitro that, apart from mediating the up-regulation of specific osteogenic genes, primary cilia are also chief mediators of oscillatory fluid flow-induced extracellular calcium deposition, thereby playing an essential role in load-induced mineral matrix deposition [301]. A study using knock-out mice of Kif3a, which results in defective primary cilia, showed that primary cilia are essential for the ability of pre-osteoblasts to sense strain-related mechanical stimuli at a healing bone–implant interface, inducing osteoblast further differentia-tion [305]; using the same animal model, another study shown primary cilia were paramount for MSCs to sense mechanical signals and enhance osteogenic lineage commitment in vivo [306]. Primary cilia must also be present in osteocytes for pulsed electromagnetic fields to inhibit osteocyte-mediated osteoclastogenesis and inhibit osteocyte apoptosis, modulate cytoskeletal distribution and decrease RANKL/OPG expression [95, 96].

Concerning osteocytes, there is still conflicting information regarding in vivo expression of cilia. Their role as mechanosensors depends on the type and number of cells with cilia and on the local mechanical environment. The incidence of primary cilia in osteocytes has been described as of 4%; this may indicate that cilia function as mechanosensors on a selected number of cells or that cilia function in concert with other mechanosensing mechanisms [307].

2.5.3   The Cytoskeleton

The cell cytoskeleton network is coupled to the ECM through specific transmembrane receptors. Integrins connect to the cytoskeleton through focal adhesions that gather actin-associated proteins such as talin, vinculin, paxillin and zyxin. Both paxillin and zyxin belong to a group of LIM domain structural proteins, which have been suggested as mechanoresponders responsible for regulating stress fibres assembly, repair and remodelling in response to changing forces [308]. Focused stresses applied to the surface of the cellular membrane are transferred across the network of cell adhesions, microfilaments and microtubules and affect distant cellular sites such as the mitochondria and nucleus or the cell membrane on the opposite side. The transmission of strain towards the ECM stimulates structural changes at a higher organization level, making it stronger [309, 310].

The cell deformation in consequence of applied stress does not correspond to the predicted behaviour of an isotropic viscoelastic material; the interior of the cell, the cytoskeleton, is anisotropic. The intricate network of microtubules and microfilaments, how it spreads and is connected to the point of applied force, may result in structures away from the load application point to be further displaced than closer ones; displacements towards the origin of the compressive stimulus are also possible. Behaving in an anisotropic way, cells can respond to an external force according to its magnitude and direction [311–313]. An intact cytoskeleton is necessary for the rendering of applied forces into mitochondria movements. Since mitochondria are semi-autonomous organelles, highly dynamic, the distress caused by mechanical stimuli exerts biological effects on their function [313], both in health and disease [314].

It is, therefore, logical that mechanical properties of the ECM affect the behaviour of cells from osteoblastic lineage, with mature focal adhesions and a more organized actin cytoskeleton associated with more rigid substrates, suggesting that controlling substrate compliance enables control over differentiation [315] and that this influence on differentiation is independent of protein tethering and substrate porosity [316].

Other factors are determinant for cell fate. A recent in vitro study showed similar patterns in cell growth, differentiation and gene expression in human osteoblasts and endothelial cells when implanted in two different ceramic scaffolds—β-tricalciumphosphate and calcium-deficient hydroxyapatite. These scaffolds had different chemical and physical characteristics, with results suggesting that the interaction between different cell types and scaffold materials is crucial for growth, differentiation and long-term outcomes of tissue-engineered constructs [317]. It has also been highlighted the importance of surface roughness of the biomaterials in osteogenic differentiation and the contribution of specific integrin subunits in mediating cell response to different materials [318]; additionally, the application of synthetic integrin-binding peptidomimetic ligands (αVβ3- or α5β1-selective) to a titanium graft enhanced cell adhesion, proliferation, differentiation and ALP expression in vitro osteoblast-like cells, resulting in a higher mineralization on the surfaces coated with the ligands [319].

The biochemical nature of the substrate, its rigidity and spatial organization is recognized by cells through signalling from molecular complexes that are integrin-based.

In most anchorage-dependent cells, cell spreading on ECM is required for cell progression and growth; increasing cytoskeletal tension results in cell flattening, a rise in actin bundling and bucking of microtubules. Spread cells can transfer most of the load to the ECM.

The cell shape is influenced by how the cytoskeleton organizes its elements and it is determinant for cell function. For example, osteocyte morphology and alignment differ in two types of bone, fibula and calvaria, probably due to different mechanical loading patterns, which influence the cytoskeletal structure and thus cell shape [320]. Also, osteocyte and lacunae morphology may vary in pathological bone conditions, and these morphological variations may be an adaptation to the differences in matrix properties and, thus, different bone strain levels under similar stimuli [321]. Osteocyte morphology is characterized by long dendritic-like processes, cell shape also assumed by osteoblast MC3T3 cells cultured in 3D; however, differences in cytoskeleton elements in the processes of these two cell types may indicate differences in function; microtubules are predominant on osteoblasts' processes while actin ensures integrity of osteocytes' cytoplasmatic projections [322]. Osteocyte sensitivity to mechanical load applied to the microparticles varies between those attached to the cell bodies and the ones attached to the cell processes: a much smaller displacement of the second ones is needed to cause an intracellular calcium influx that rapidly propagates to the cell body; if local stimulus is applied to the cell body, the reaction is slower and a higher displacement is needed to cause the calcium transient [323].

Osteoblasts, osteoid–osteocytes and mature osteocytes have different mechanical properties. The elastic modulus is higher in the cell periphery than in the perinuclear region; the elastic modulus in both regions decreases as bone cells mature. These differences in elastic modulus probably depend on the number of actin filaments, as it has been shown in other cell types. Furthermore, focal adhesion area is smaller in mature osteocytes, when comparing to osteoblasts. If peptides containing RGD sequence are added to culture medium, both the focal adhesion area and the elastic modulus of osteoblasts decreases whilst remaining unaffected in osteocytes [324].

2.6   Mechanotransduction Mechanisms

The multitude of cellular structures, messenger substances, environmental factors and levels of organization of the organs involved in the mechanotransduction mechanisms in distinct cells and tissues makes it extremely complex to understand, predict and replicate how responses are composed at cellular, organ and living organism levels.

2.6.1   Strain, Frequency and Loading Duration

Bone remodelling is influenced by strain magnitude, frequency and loading duration. Wolff developed mathematical equations for trabeculae orientation and thickness prediction according to load [325]. Later, Turner enunciated three essential rules critical for bone remodelling [326]:

1. Remodelling is determined by dynamic loading, not by static loading;
2. Short periods of loading quickly trigger a response; prolonging loading times any further diminishes the magnitude of bone cell response;
3. Bone cells have memory and accommodate to routine loading, diminishing the amplitude of the response triggered by a same repeated stimulus.

Increasing loading frequency increased strain-related bone deposition in vivo, whilst decreasing the threshold for osteogenesis and bone formation [327]. Human osteoblasts subjected to strains varying from 0.8 to 3.2% respond to higher strain with increased expression of osteocalcin, type I collagen and Cbfa1/Runx2, and to lower strain magnitudes with an increase of alkaline phosphatase activity [328].

Bone formation depends on strain magnitude [329], along with the number of loading cycles at low frequencies [330]. Frost theorized that a minimum effective strain level was necessary to trigger bone formation, above 3000 micro-strain [228]. Strain distribution is also paramount for skeletal adaptation. Unusual strain distribution will rapidly trigger an osteogenic response, as suggested by the extensive periosteal and endosteal bone proliferation described by Rubin and Lanyon [331] in a study conducted in poultry. Rest periods between loading cycles also intensify osteogenic response [332] and maximize cell response [333]. During active exercise, peak strains in long bones may be high, but strains as low as 0.15% are enough to ensure osteoblast recruitment in vivo [331]. Human bone marrow stem cells show variable early osteogenic differentiation and gene expression accordingly with load and frequency regimen of cyclic hydrostatic pressure; osteogenic differentiation on the long term occurred under mechanical stimulation, independently of load magnitude and frequency, within the tested physiological ranges [334].

The adaptation of cortical bone is correlated with frequency, although not linearly; the changes in geometry are more significant with higher frequency, with a plateau for frequencies past 10 Hz [335].

Other mechanisms apart from direct deformation of cells are involved in bone mechanical stimulation. Bone's canalicular system is filled with fluid. Simulation

of osteogenic load levels has produced higher shear stresses due to fluid displacement in the canaliculi. The fluid flows within the canalicular system, wherein the
osteocytes extend their cell processes, reinforcing osteocyte processes as the main
mechanosensing organ in mature bone cells [336]. Multiple canaliculi intersect at
points (canalicular joints); these occur with a density similar to that of lacunae and
represent areas of enlarged space, with consequences on fluid flow variables such as
fluid mass and velocity [337]. Microstructural changes associated with osteoporosis
reduce interstitial fluid flow around osteocytes in the lacunar–canalicular system of
cortical bone, impairing mechanosensation [338].

As reviewed previously, shear stresses resulting from fluid flow cause calcium
influx through mechanosensitive channels [339, 340]. Calcium influx occurs is
osteoblasts in response to oscillatory fluid flow [288].

Fluid also carries electrically charged particles. The resulting fluid flow phenomena are common to other biological tissues but not limited to living structures. The
fact that fluid flow changes interfacial chemistry has been recognized; the flow of
fresh water along the surfaces disturbs the equilibrium of dissolved ions, changing
the surface charge and the molecular orientation of the water at the interface [341].
Likewise, when bone is deformed, a thin sheet of fluid with particles with charge
opposite to that of the matrix and bone cells is formed [342]; when a non-uniform
mechanical load is applied to the bone structure, the ions in the fluid move away
from the matrix. Therefore, the displacement of the electrically charged fluid creates
an electrical field aligned with the fluid flow. This causes an electrical potential,
and the phenomenon is known as strain-generated bone streaming potential and has
been described in bone [342–345]. The density of matrix fixed charges influences
the magnitude of the generated streaming potential [346], so the mechanosensory
ability along bone may vary and, ultimately, influence dynamic stiffness.

2.6.2 Bone Piezoelectricity and Flexoeletricity

Fukada and Yasuda first described bone piezoelectrical properties in 1957. In dry
bone samples submitted to a compressive load, an electrical potential was generated,
an occurrence explained by the direct piezoelectric effect [347]. In connective tissues,
such as bone, skin, tendon and dentine, the dipole moments are related to the collagen
fibres, composed by strongly polar protein molecules aligned [76, 348, 349].

Recently, it has been suggested that hydroxyapatite flexoelectricity is the main
source of bending-induced polarization in cortical bone [119].

The architecture of bone itself, with its aligned lamellae, contributes to the
existence of potentials through the bone structure [348].

Bone piezoelectric constants, i.e. the polarization generated per unit of mechanical stress, change with moisture content, maturation state (immature bone has lower
piezoelectric constants when comparing to mature bone) and architectural organization (altered areas, such as bone neoplasia osteosarcoma, show lower values) [350].
In dentin, piezoelectric constants are higher when moisture contents increase, also
behaving anisotropically; tubule orientation determined piezoelectricity, stronger

parallel to the tubules [351]. Wet bone also behaves as a piezoelectric material [347, 350, 352].

Bone piezoelectrical properties have risen interest, in the context of bone physiology and electromechanics. It has been related to bone remodelling mechanisms and to streaming potential mechanisms [353, 354]. Using a piezoelectric substrate and the piezoelectric converse effect were tested in vitro and in vivo with promising results, mechanically stimulating osteoblastic cells and bone, suggesting the potential for clinical application [355, 356]. The development of new synthetic scaffolds is a new emergent field for the bone tissue engineering industry. Hydroxyapatite/barium titanate [357] or polycaprolactone/barium titanate composites [358] with piezoelectric coefficients dependent on distribution and density of barium titanate particles aim to improve cell adhesion and differentiation. A wide range of biomaterials with piezoelectric properties, with potential application for bone regeneration, is available [359]. Mesenchymal cell differentiation in 3Dpiezoelectric scaffolds can be modulated by the voltage output (or streaming potential); lower voltage output scaffolds promoted chondrogenic differentiation while high voltage output promoted osteogenic differentiation [360]. Electromechanical stimulation also promoted improved differentiation when compared to mechanical load alone [360].

Due to the potential impact on therapeutic approaches to bone remodelling and healing, more and more research is being conducted on bioinspired approaches that consider piezoelectric bone properties.

Acknowledgements This work has been partially supported by the European Commission under the 7th Framework Programme through the project Restoration, grant agreement CP-TP 280575-2 and through Portugal 2020/Alentejo 2020, grant POCI-01-0145-FEDER-032486. The support from Hamamatsu Photonics in providing the NanoZoomer SQ is also gratefully acknowledged. The authors would also like to thank Mr. Pedro Félix Pinto for the artwork included in this chapter that he so kindly prepared and made available.

References

1. de Vries WN, Evsikov AV, Haak BE et al (2004) Maternal β-catenin and E-cadherin in mouse development. Development 131:4435–4445
2. Ingber DE (2006) Mechanical control of tissue morphogenesis during embryological development. Dev Biol 50:255–266
3. Oster GF, Murray JD, Harris AK (1983) Mechanical aspects of mesenchymal morphogenesis. J Embryol Exp Morphol 78:83–125
4. Takeichi M (1988) The cadherins: cell-cell adhesion molecules controlling animal morphogenesis. Development 102:639–655
5. Tang Z, Hu Y, Wang Z et al (2018) Mechanical forces program the orientation of cell division during airway tube morphogenesis. Dev Cell 44:313–325
6. Foubet O, Trejo M, Toro R (2018) Mechanical morphogenesis and the development of neocortical organisation. Cortex
7. Cartwright JHE, Piro O, Tuval I (2004) Fluid-dynamical basis of the embryonic development of left-right asymmetry in vertebrates. Proc Natl Acad Sci USA 101:7234–7239

8. Collignon J, Varlet I, Robertson EJ (1996) Relationship between asymmetric nodal expression and the direction of embryonic turning. Nature 381:155–158

9. Nakamura T, Mine N, Nakaguchi E et al (2006) Generation of robust left-right asymmetry in the mouse embryo requires a self-enhancement and lateral-inhibition system. Dev Cell 11:495–504

10. Okada Y, Nonaka S, Tanaka Y et al (1999) Abnormal nodal flow precedes situs inversus in iv and inv mice. Mol Cell 4:459–468

11. McGrath J, Somlo S, Makova S et al (2003) Two populations of node monocilia initiate left-right asymmetry in the mouse. Cell 114:61–73

12. Patwari P, Lee RT (2008) Mechanical control of tissue morphogenesis. Circ Res 103:234–243

13. Schmidt A, Brixius K, Bloch W (2007) Endothelial precursor cell migration during vasculogenesis. Circ Res 101:125–136

14. Anava S, Greenbaum A, Ben Jacob E et al (2009) The regulative role of neurite mechanical tension in network development. Biophys J 96:1661–1670

15. Bray D (1979) Mechanical tension produced by nerve cells in tissue culture. J Cell Sci 37:391–410

16. Dennerll TJ, Lamoureux P, Buxbaum RE, Heidemann SR (1989) The cytomechanics of axonal elongation and retraction. J Cell Biol 109:3073–3083

17. le Noble F, Klein C, Tintu A et al (2008) Neural guidance molecules, tip cells, and mechanical factors in vascular development. Cardiovasc Res 78:232–241

18. Carter DR, Beaupré GS (2001) Skeletal tissue histomorphology and mechanics. Skeletal function and form. Cambridge University Press, Cambridge, pp 31–52

19. Belanger LF (1969) Osteocytic osteolysis. Calcif Tissue Res 4:1–12

20. Teti A, Zallone A (2009) Do osteocytes contribute to bone mineral homeostasis? Osteocytic osteolysis revisited. Bone 44:11–16

21. Zallone A, Teti A, Primavera MV, Pace G (1983) Mature osteocytes behaviour in a repletion period: the occurrence of osteoplastic activity. Basic Appl Histochem 27:191–204

22. Green J, Kleeman CR (1991) The role of bone in the regulation of systemic acid-base balance. Kidney Int 39:9–26

23. Arnett T (2003) Regulation of bone cell function by acid-base balance. Proc Nutr Soc 62:511–520

24. Bushinsky DA, Krieger NS (2015) Acid-base balance and bone health. In: Holick MF, JNieves NW (eds) Nutrition and bone health. Humana Press Springer, New York, pp 335–357

25. Baylink DJ, Finkelman RD, Mohan S (1993) Growth factors to stimulate bone formation. J Bone Miner Res 8:S565–S572

26. Linkhart TA, Mohan S, Baylink DJ (1996) Growth factors for bone growth and repair: IGF, TGFβ and BMP. Bone 19:S1–S12

27. Krings A, Rahman S, Huang S et al (2012) Bone marrow fat has brown adipose tissue characteristics, which are attenuated with aging and diabetes. Bone 50:546–552

28. Rosen CJ, Ackert-Bicknell C, Rodriguez JP, Pino AM (2009) Marrow fat and the bone microenvironment: developmental, functional, and pathological implications. Crit Rev Eukaryot Gene Expr 19:109–124

29. Suchacki KJ, Cawthorn WP, Rosen CJ (2016) Bone marrow adipose tissue: formation, function and regulation. Curr Opin Pharmacol 28:50–56

30. Roelofs-Iverson RA, Mulder DW, Elveback LR et al (1984) ALS and heavy metals: a pilot case-control study. Neurology 34:393

31. Sharma B, Singh S, Siddiqi NJ (2014) Biomedical implications of heavy metals induced imbalances in redox systems. BioMed Research Int 2014:640754

32. Rhee Y, Bivi N, Farrow E et al (2011) Parathyroid hormone receptor signaling in osteocytes increases the expression of fibroblast growth factor-23 in vitro and in vivo. Bone 49:636–643

33. Yamashita T, Yoshioka M, Itoh N (2000) Identification of a novel fibroblast growth factor, FGF-23, preferentially expressed in the ventrolateral thalamic nucleus of the brain. Biochem Biophys Res Commun 277:494–498

34. Masuyama R, Stockmans I, Torrekens S et al (2006) Vitamin D receptor in chondrocytes promotes osteoclastogenesis and regulates FGF23 production in osteoblasts. J Clin Invest 116:3150–3159
35. Fukumoto S, Martin TJ (2009) Bone as an endocrine organ. Trends Endocrinol Metab 20:230–236
36. Haussler MR, Whitfield GK, Kaneko I et al (2012) The role of vitamin D in the FGF23, klotho, and phosphate bone-kidney endocrine axis. Rev Endocr Metab Disord 13:57–69
37. Shimada T, Hasegawa H, Yamazaki Y et al (2004) FGF-23 is a potent regulator of vitamin D metabolism and phosphate homeostasis. J Bone Miner Res 19:429–435
38. David V, Dai B, Martin A et al (2013) Calcium regulates FGF-23 expression in bone. Endocrinology 154:4469–4482
39. Rodriguez-Ortiz ME, Lopez I, Muñoz-Castañeda JR et al (2012) Calcium deficiency reduces circulating levels of FGF23. J Am Soc Nephrol 23:1190–1197
40. Fukumoto S, Yamashita T (2007) FGF23 is a hormone-regulating phosphate metabolism-unique biological characteristics of FGF23. Bone 40:1190–1195
41. ADHR Consortium (2000) Autosomal dominant hypophosphataemic rickets is associated with mutations in FGF23. Nat Genet 26:345–348
42. Shimada T, Mizutani S, Muto T et al (2001) Cloning and characterization of FGF23 as a causative factor of tumor-induced osteomalacia. Proc Natl Acad Sci USA 98:6500–6505
43. Lyles KW, Halsey DL, Friedman NE, Lobaugh B (1988) Correlations of serum concentrations of 1,25-dihydroxyvitamin D, phosphorus, and parathyroid hormone in tumoral calcinosis. J Clin Endocrinol Metab 67:88–92
44. Urakawa I, Yamazaki Y, Shimada T et al (2006) Klotho converts canonical FGF receptor into a specific receptor for FGF23. Nature 444:770–774
45. Kurosu H, Yamamoto M, Clark JD et al (2005) Suppression of aging in mice by the hormone Klotho. Science 309:1829–1833
46. Liu H, Fergusson MM, Castilho RM et al (2007) Augmented Wnt signaling in a mammalian model of accelerated aging. Science 317:803–806
47. Doi S, Zou Y, Togao O et al (2011) Klotho inhibits transforming growth factor-β1 (TGF-β1) signaling and suppresses renal fibrosis and cancer metastasis in mice. J Biol Chem 286:8655–8665
48. López I, Rodríguez-Ortiz ME, Almadén Y et al (2011) Direct and indirect effects of parathyroid hormone on circulating levels of fibroblast growth factor 23 in vivo. Kidney Int 80:475–482
49. Quarles LD (2012) Role of FGF23 in vitamin D and phosphate metabolism: implications in chronic kidney disease. Exp Cell Res 318:1040–1048
50. Ben-Dov IZ, Galitzer H, Lavi-Moshayoff V et al (2007) The parathyroid is a target organ for FGF23 in rats. J Clin Invest 117:4003–4008
51. Krajisnik T, Bjorklund P, Marsell R et al (2007) Fibroblast growth factor-23 regulates parathyroid hormone and 1alpha-hydroxylase expression in cultured bovine parathyroid cells. J Endocrinol 195:125–131
52. Paloian NJ, Leaf EM, Giachelli CM (2016) Osteopontin protects against high phosphate-induced nephrocalcinosis and vascular calcification. Kidney Int 89:1027–1036
53. Ferron M, Hinoi E, Karsenty G, Ducy P (2008) Osteocalcin differentially regulates β cell and adipocyte gene expression and affects the development of metabolic diseases in wild-type mice. Proc Natl Acad Sci USA 105:5266–5270
54. Ferron M, Wei J, Yoshizawa T et al (2010) Insulin signaling in osteoblasts integrates bone remodelling and energy metabolism. Cell 142:296–308
55. Fulzele K, Riddle RC, DiGirolamo DJ et al (2010) Insulin receptor signaling in osteoblasts regulates postnatal bone acquisition and body composition. Cell 142:309–319
56. Lee NK, Karsenty G (2008) Reciprocal regulation of bone and energy metabolism. Trends Endocrinol Metab 19:161–166
57. Reid IR, Ames R, Evans MC et al (1992) Determinants of total body and regional bone mineral density in normal postmenopausal women—a key role for fat mass. J Clin Endocrinol Metab 75:45–51

58. Ribot C, Tremollieres F, Pouilles JM et al (1987) Obesity and postmenopausal bone loss: the influence of obesity on vertebral density and bone turnover in postmenopausal women. Bone 8:327–331

59. Kindblom JM, Ohlsson C, Ljunggren Ö et al (2009) Plasma osteocalcin is inversely related to fat mass and plasma glucose in elderly Swedish men. J Bone Miner Res 24:785–791

60. Pittas AG, Harris SS, Eliades M et al (2009) Association between serum osteocalcin and markers of metabolic phenotype. J Clin Endocrinol Metab 94:827–832

61. Oury F, Sumara G, Sumara O et al (2011) Endocrine regulation of male fertility by the skeleton. Cell 144:796–809

62. Sommerfeldt D, Rubin C (2001) Biology of bone and how it orchestrates the form and function of the skeleton. Eur Spine J 10:S86–S95

63. Augat P, Schorlemmer S (2006) The role of cortical bone and its microstructure in bone strength. Age Ageing 35(suppl_2):ii27–ii31

64. Rho J-Y, Kuhn-Spearing L, Zioupos P (1998) Mechanical properties and the hierarchical structure of bone. Med Eng Phys 20:92–102

65. Van De Graaff K (2001) Skeletal system: introduction and the axial skeleton. In: Lange M, Tibbetts K, Queck K (eds) Human Anatomy, 6th edn. McGraw-Hill College, Boston, pp 131–171

66. Weiner S, Traub W, Wagner HD (1999) Lamellar bone: structure-function relations. J Struct Biol 126:241–255

67. Meyer U, Wiesmann HP (2006) Bone and cartilage. In: Schröder G (ed) Bone and cartilage engineering, 1st edn. Springer, Berlin, pp. 7–46

68. Oftadeh R, Perez-Viloria M, Villa-Camacho JC et al (2015) Biomechanics and mechanobiology of trabecular bone: a review. J Biomech Eng 137:010802

69. Currey JD (2003) The many adaptations of bone. J Biomech 36:1487–1495

70. Burr DB, Milgrom C, Fyhrie D et al (1996) In vivo measurement of human tibial strains during vigorous activity. Bone 18:405–410

71. Duncan RL, Turner CH (1995) Mechanotransduction and the functional response of bone to mechanical strain. Calcif Tissue Int 57:344–358

72. Zioupos P, Currey JD, Hamer AJ (1999) The role of collagen in the declining mechanical properties of aging human cortical bone. J Biomed Mater Res A 45:108–116

73. Turner CH (2006) Bone strength: current concepts. Ann N Y Acad Sci 1068:429–446

74. Young MF (2003) Bone matrix proteins: their function, regulation, and relationship to osteoporosis. Osteoporos Int 14:35–42

75. Bodian DL, Chan T-F, Poon A et al (2009) Mutation and polymorphism spectrum in osteogenesis imperfecta type II: implications for genotype-phenotype relationships. Hum Mol Gen 18:463–471

76. Fukada E, Yasuda I (1964) Piezoelectric effects in collagen. Jpn J Appl Phys 3:117–121

77. Noris-Suárez K, Lira-Olivares J, Ferreira AM et al (2007) In vitro deposition of hydroxyapatite on cortical bone collagen stimulated by deformation-induced piezoelectricity. Biomacromol 8:941–948

78. Ferreira AM, González G, González-Paz RJ et al (2009) Bone collagen role in piezoelectric mediated remineralization. Acta Microsc 18:278–286

79. Nudelman F, Pieterse K, George A et al (2010) The role of collagen in bone apatite formation in the presence of hydroxyapatite nucleation inhibitors. Nat Mater 9:1004–1009

80. Ashizawa N, Graf K, Do YS et al (1996) Osteopontin is produced by rat cardiac fibroblasts and mediates A (II)-induced DNA synthesis and collagen gel contraction. J Clin Invest 98:2218–2227

81. Perrien DS, Brown EC, Aronson J et al (2002) Immunohistochemical study of osteopontin expression during distraction osteogenesis in the rat. J Histochem Cytochem 50:567–574

82. Gross TS, King KA, Rabaia NA et al (2005) Upregulation of osteopontin by osteocytes deprived of mechanical loading or oxygen. J Bone Miner Res 20:250–256

83. Harter LV, Hruska KA, Duncan RL (1995) Human osteoblast-like cells respond to mechanical strain with increased bone matrix protein production independent of hormonal regulation. Endocrinology 136:528–535

84. Fisher LW, Torchia DA, Fohr B et al (2001) Flexible structures of SIBLING proteins, bone sialoprotein, and osteopontin. Biochem Biophys Res Commun 280:460–465

85. Jahnen-Dechent W, Schäfer C, Ketteler M et al (2008) Mineral chaperones: a role for fetuin-A and osteopontin in the inhibition and regression of pathologic calcification. J Mol Med 86:379–389

86. Thurner PJ, Chen CG, Ionova-Martin S et al (2010) Osteopontin deficiency increases bone fragility but preserves bone mass. Bone 46:1564–1573

87. Bentmann A, Kawelke N, Moss D et al (2010) Circulating fibronectin affects bone matrix, whereas osteoblast fibronectin modulates osteoblast function. J Bone Miner Res 25:706–715

88. Huang G, Zhang Y, Kim B et al (2009) Fibronectin binds and enhances the activity of bone morphogenetic protein 1. J Biol Chem 284:25879–25888

89. Kang Y, Georgiou AI, MacFarlane RJ et al (2017) Fibronectin stimulates the osteogenic differentiation of murine embryonic stem cells. J Tissue Eng Regen Med 11:1929–1940

90. Linsley C, Wu B, Tawil B (2013) The effect of fibrinogen, collagen type I, and fibronectin on mesenchymal stem cell growth and differentiation into osteoblasts. Tissue Eng Part A 19:1416–1423

91. Matlahov I, Iline-Vul T, Abayev M et al (2015) Interfacial mineral–peptide properties of a mineral binding peptide from osteonectin and bone-like apatite. Chem Mater 27:5562–5569

92. Rosset EM, Bradshaw AD (2016) SPARC/osteonectin in mineralized tissue. Matrix Biol 52:78–87

93. Delany AM, Hankenson KD (2009) Thrombospondin-2 and SPARC/osteonectin are critical regulators of bone remodelling. J Cell Commun Signal 3:227–238

94. Delany AM, Amling M, Priemel M et al (2000) Osteopenia and decreased bone formation in osteonectin-deficient mice. J Clin Invest 105:915–923

95. Wang M, Chao CC, Chen PC et al (2019) Thrombospondin enhances RANKL-dependent osteoclastogenesis and facilitates lung cancer bone metastasis. Biochem Pharmacol 166:23–32

96. Wang P, Tang C, Wu J et al (2019) Pulsed electromagnetic fields regulate osteocyte apoptosis, RANKL/OPG expression, and its control of osteoclastogenesis depending on the presence of primary cilia. J Cell Physiol 234:10588–10601

97. Ganss B, Kim RH, Sodek J (1999) Bone sialoprotein. Crit Rev Oral Biol Med 10:79–98

98. Malaval L, Wade-Guéye NM, Boudiffa M et al (2008) Bone sialoprotein plays a functional role in bone formation and osteoclastogenesis. J Exp Med 205:1145–1153

99. Holm E, Aubin JE, Hunter GK et al (2015) Loss of bone sialoprotein leads to impaired endochondral bone development and mineralization. Bone 71:145–154

100. Bouleftour W, Juignet L, Verdière L et al (2019) Deletion of OPN in BSP knockout mice does not correct bone hypomineralization but results in high bone turnover. Bone 120:411–422

101. Lamoureux F, Baud'huin M, Duplomb L et al (2007) Proteoglycans: key partners in bone cell biology. BioEssays 29:758–771

102. Novince CM, Michalski MN, Koh AJ et al (2012) Proteoglycan 4: a dynamic regulator of skeletogenesis and parathyroid hormone skeletal anabolism. J Bone Miner Res 27:11–25

103. Nakamura H (2007) Morphology, function, and differentiation of bone cells. J Hard Tissue Biol 16:15–22

104. Palumbo C (1986) A three-dimensional ultrastructural study of osteoid-osteocytes in the tibia of chick embryos. Cell Tissue Res 246:125–131

105. Bellows CG, Reimers SM, Heersche JNM (1999) Expression of mRNAs for type-I collagen, bone sialoprotein, osteocalcin, and osteopontin at different stages of osteoblastic differentiation and their regulation by 1,25 dihydroxyvitamin D_3. Cell Tissue Res 297:249–259

106. Bellows CG, Heersche JNM (2001) The frequency of common progenitors for adipocytes and osteoblasts and of committed and restricted adipocyte and osteoblast progenitors in fetal rat calvaria cell populations. J Bone Miner Res 16:1983–1993

107. Schiller PC, D'Ippolito G, Balkan W et al (2001) Gap-junctional communication is required for the maturation process of osteoblastic cells in culture. Bone 28:362–369

108. Komori T (2019) Regulation of Proliferation, Differentiation and Functions of Osteoblasts by Runx2. Int J Mol Sci 20:1694–1705
109. Rutkovskiy A, Stensløkken KO, Vaage IJ (2016) Osteoblast differentiation at a glance. Med Sci Mon Basic Res 22:95–106
110. van der Meijden K, Bakker AD, van Essen HW et al (2016) Mechanical loading and the synthesis of 1, 25 (OH) 2 D in primary human osteoblasts. J Steroid Biochem Mol Biol 156:32–39
111. van Bezooijen RL, Roelen BA, Visser A et al (2004) Sclerostin is an osteocyte-expressed negative regulator of bone formation, but not a classical BMP antagonist. J Exp Med 199:805–814
112. Bonewald LF (2007) Osteocyte messages from a bony tomb. Cell Metab 5:410–411
113. Jiang JX, Siller-Jackson AJ, Burra S (2007) Roles of gap junctions and hemichannels in bone cell functions and in signal transmission of mechanical stress. Front Biosci 12:1450–1462
114. Tate ML Knothe, Adamson JR, Tami AE, Bauer TW (2004) The osteocyte. Int J Biochem Cell Biol 36:1–8
115. Tate ML Knothe (2003) Whither flows the fluid in bone? An osteocyte's perspective. J Biomech 36:1409–1424
116. Burger EH, Klein-Nulend J (1999) Mechanotransduction in bone—role of the lacuno-canalicular network. FASEB J 13:101–112
117. Mullender M, El Haj AJ, Yang Y et al (2004) Mechanotransduction of bone cells in vitro: mechanobiology of bone tissue. Med Biol Eng Comput 42:14–21
118. Huang CP, Chen XM, Chen ZQ (2008) Osteocyte: the impresario in the electrical stimulation for bone fracture healing. Med Hypotheses 70:287–290
119. Vasquez-Sancho F, Abdollahi A, Damjanovic D et al (2018) Flexoelectricity in bones. Adv Mater 30:1705316
120. van Oers RF, Wang H, Bacabac RG (2015) Osteocyte shape and mechanical loading. Curr Osteoporos Rep 13:61–66
121. Yu K, Sellman DP, Bahraini A et al (2018) Mechanical loading disrupts osteocyte plasma membranes which initiates mechanosensation events in bone. J Orthop Res 36:653–662
122. Kringelbach TM, Aslan D, Novak I et al (2015) Fine-tuned ATP signals are acute mediators in osteocyte mechanotransduction. Cell Signal 27:2401–2409
123. Morrell AE, Brown GN, Robinson ST et al (2018) Mechanically induced Ca 2+ oscillations in osteocytes release extracellular vesicles and enhance bone formation. Bone Res 6:6
124. Cherian PP, Cheng B, Gu S et al (2003) Effects of mechanical strain on the function of gap junctions in osteocytes are mediated through the prostaglandin EP2 receptor. J Biol Chem 278:43146–43156
125. Klein-Nulend J, Helfrich MH, Sterck JGH et al (1998) Nitric oxide response to shear stress by human bone cell cultures is endothelial nitric oxide synthase dependent. Biochem Biophys Res Commun 250:108–114
126. Rawlinson SCF, Pitsillides AA, Lanyon LE (1996) Involvement of different ion channels in osteoblasts' and osteocytes' early responses to mechanical strain. Bone 19:609–614
127. Jee WSS, Mori S, Li XJ, Chan S (1990) Prostaglandin E2 enhances cortical bone mass and activates intracortical bone remodelling in intact and ovariectomized female rats. Bone 11:253–266
128. Fan X, Roy E, Zhu L et al (2004) Nitric oxide regulates receptor activator of nuclear factor κB ligand and osteoprotegerin expression in bone marrow stromal cells. Endocrinology 145:751–759
129. Kasten TP, Collin-Osdoby P, Patel N et al (1994) Potentiation of osteoclast bone-resorption activity by inhibition of nitric oxide synthase. Proc Natl Acad Sci USA 91:3569–3573
130. Hirose S, Li M, Kojima T et al (2007) A histological assessment on the distribution of the osteocytic lacunar canalicular system using silver staining. J Bone Miner Metab 25:374–382
131. Jilka RL, Noble B, Weinstein RS (2013) Osteocyte apoptosis. Bone 54:264–271
132. Lee KC, Jessop H, Suswillo R et al (2004) The adaptive response of bone to mechanical loading in female transgenic mice is deficient in the absence of oestrogen receptor-alpha and -beta. Endocrinology 182:193–201

133. Plotkin LI, Mathov I, Aguirre JI et al (2005) Mechanical stimulation prevents osteocyte apoptosis: requirement of integrins, Src kinases, and ERKs. Am J Physiol 289:633–643

134. Tomkinson A, Reeve J, Shaw RW, Noble BS (1997) The death of osteocytes via apoptosis accompanies estrogen withdrawal in human bone. J Clin Endocrinol Metab 82:3128–3135

135. Metz LN, Martin RB, Turner AS (2003) Histomorphometric analysis of the effects of osteocyte density on osteonal morphology and remodelling. Bone 33:753–759

136. Milovanovic P, Zimmermann EA, Hahn M et al (2013) Osteocytic canalicular networks: morphological implications for altered mechanosensitivity. ACS Nano 7:7542–7551

137. Okada S, Yoshida S, Ashrafi S, Schraufnagel D (2002) The Canalicular Structure of Compact Bone in the Rat at Different Ages. Microsc Microanal 8:104–115

138. Rubin J, Greenfield EM (2005) Osteoclast: origin and differentiation. In: Farach-Carson MC, Bronner F, Rubin J (eds) Bone resorption. Springer, London, pp 1–23

139. Asagiri M, Takayanagi H (2007) The molecular understanding of osteoclast differentiation. Bone 40:251–264

140. Nakagawa N, Kinosaki M, Yamaguchi K et al (1998) RANK is the essential signaling receptor for osteoclast differentiation factor in osteoclastogenesis. Biochem Biophys Res Commun 253:395–400

141. Takayanagi H (2008) Regulation of osteoclastogenesis and osteoimmunology. Bone 42:S40

142. Marchisio PC, Cirillo D, Naldini L et al (1984) Cell-substratum interaction of cultured avian osteoclasts is mediated by specific adhesion structures. J Cell Biol 99:1696–1705

143. Väänänen HK, Horton M (1995) The osteoclast clear zone is a specialized cell-extracellular matrix adhesion structure. J Cell Sci 108:2729–2732

144. Baron R, Neff L, Louvard D, Courtoy PJ (1985) Cell-mediated extracellular acidification and bone resorption: evidence for a low pH in resorbing lacunae and localization of a 100-kD lysosomal membrane protein at the osteoclast ruffled border. J Cell Biol 101:2210–2222

145. Blair HC, Teitelbaum SL, Ghiselli R, Gluck S (1989) Osteoclastic bone resorption by a polarized vacuolar proton pump. Science 245:855–857

146. Rousselle AV, Heymann D (2002) Osteoclastic acidification pathways during bone resorption. Bone 30:533–540

147. Littlewood-Evans A, Kokubo T, Ishibashi O et al (1997) Localization of cathepsin K in human osteoclasts by in situ hybridization and immunohistochemistry. Bone 20:81–86

148. Vääräniemi J, Halleen JM, Kaarlonen K et al (2004) Intracellular machinery for matrix degradation in bone-resorbing osteoclasts. J Bone Miner Res 19:1432–1440

149. Salo J, Lehenkari P, Mulari M et al (1997) Removal of osteoclast bone resorption products by transcytosis. Science 276:270–273

150. Yamaki M, Nakamura H, Takahashi N et al (2005) Transcytosis of calcium from bone by osteoclast-like cells evidenced by direct visualization of calcium in cells. Arch Biochem Biophys 440:10–17

151. Harada SI, Rodan GA (2003) Control of osteoblast function and regulation of bone mass. Nature 423:349–355

152. Karsenty G, Kronenberg HM, Settembre C (2009) Genetic control of bone formation. Annu Rev Cell Dev Biol 25:629–648

153. Siddiqui JA, Partridge NC (2016) Physiological bone remodeling: systemic regulation and growth factor involvement. Physiology (Bethesda) 31:233–245

154. Murayama A, Takeyama K, Kitanaka S et al (1998) The promoter of the human 25-hydroxyvitamin D3 1 alpha-hydroxylase gene confers positive and negative responsiveness to PTH, calcitonin, and $1\alpha,25(OH)_2D_3$. Biochem Biophys Res Commun 249:11–16

155. Haussler MR, Whitfield GK, Haussler CA et al (1998) The nuclear vitamin D receptor: biological and molecular regulatory properties revealed. J Bone Miner Res 13:325–349

156. Saini RK, Kaneko I, Jurutka PW et al (2013) 1, 25-dihydroxyvitamin d3 regulation of fibroblast growth factor-23 expression in bone cells: evidence for primary and secondary mechanisms modulated by leptin and interleukin-6. Calcif Tissue Int 92:339–353

157. Clarke B (2008) Normal bone anatomy and physiology. Clin J Am Soc Nephrol 3:S131–S139

158. Hong AR, Lee JH, Kim JH et al (2019) Effect of endogenous parathyroid hormone on bone geometry and skeletal microarchitecture. Calcif Tissue Int 104:382–389
159. Boissy P, Saltel F, Bouniol C et al (2002) Transcriptional activity of nuclei in multinucleated osteoclasts and its modulation by calcitonin. Endocrinology 143:1913–1921
160. Hadjidakis DJ, Androulakis II (2006) Bone remodelling. Ann NY Acad Sci 1092:385–396
161. Isaksson OG, Jansson JO, Gause IA (1982) Growth hormone stimulates longitudinal bone growth directly. Science 216:1237–1239
162. Ohlsson C, Bengtsson BA, Isaksson OG et al (1998) Growth hormone and bone. Endocrine Rev 19:55–79
163. Ranke MB, Wit JM (2018) Growth hormone—past, present and future. Nat Rev Endocrinol 14:285–300
164. Wu S, Yang W, De Luca F (2015) Insulin-like growth factor-independent effects of growth hormone on growth plate chondrogenesis and longitudinal bone growth. Endocrinology 156:2541–2551
165. Kuzma M, Kuzmova Z, Zelinkova Z et al (2014) Impact of the growth hormone replacement on bone status in growth hormone deficient adults. Growth Horm IGF Res 24:22–28
166. Guevarra MS, Yeh JK, Castro Magana M, Aloia JF (2010) Synergistic effect of parathyroid hormone and growth hormone on trabecular and cortical bone formation in hypophysectomized rats. Hormone Res Paediatr 73:248–257
167. Lupu F, Terwilliger JD, Lee K et al (2001) Roles of growth hormone and insulin-like growth factor 1 in mouse postnatal growth. Dev Biol 229:141–162
168. Zhang M, Xuan S, Bouxsein ML et al (2002) Osteoblast-specific knockout of the insulin-like growth factor (IGF) receptor gene reveals an essential role of IGF signaling in bone matrix mineralization. J Biol Chem 277:44005–44012
169. Zhang W, Shen X, Wan C et al (2012) Effects of insulin and insulin-like growth factor 1 on osteoblast proliferation and differentiation: differential signalling via Akt and ERK. Cell Biochem Funct 30:297–302
170. Fowlkes JL, Bunn RC, Liu L et al (2008) Runt-related transcription factor 2 (RUNX2) and RUNX2-related osteogenic genes are down-regulated throughout osteogenesis in type 1 diabetes mellitus. Endocrinology 149:1697–1704
171. Chen JH, Liu C, You L, Simmons CA (2010) Boning up on Wolff's Law: mechanical regulation of the cells that make and maintain bone. J Biomech 43:108–118
172. Ding W, Li J, Singh J et al (2015) miR-30e targets IGF2-regulated osteogenesis in bone marrow-derived mesenchymal stem cells, aortic smooth muscle cells, and ApoE −/− mice. Cardiovasc Res 106:131–142
173. Turner RT, Riggs BL, Spelsberg TC (1994) Skeletal effects of estrogen. Endocr Rev 15:275–300
174. Khosla S, Monroe DG (2018) Regulation of bone metabolism by sex steroids. Cold Spring Harb Perspect Med 8:a031211
175. Prince RL (1994) Counterpoint: estrogen effects on calcitropic hormones and calcium homeostasis. Endocr Rev 15:301–309
176. Liel Y, Shany S, Smirnoff P, Schwartz B (1999) Estrogen increases 1, 25-dihydroxyvitamin D receptors expression and bioresponse in the rat duodenal mucosa 1. Endocrinology 140:280–285
177. ten Bolscher M, Netelenbos JC, Barto R, van Buuren LM (1999) Estrogen regulation of intestinal calcium absorption in the intact and ovariectomized adult rat. J Bone Miner Res 14:1197–1202
178. Draper CR, Edel MJ, Dick IM et al (1997) Phytoegens reduce bone loss and bone resorption in oophorectomized rats. J Nut 127:1795–1799
179. Robinson LJ, Yaroslavskiy BB, Griswold RD et al (2009) Estrogen inhibits RANKL-stimulated osteoclastic differentiation of human monocytes through estrogen and RANKL-regulated interaction of estrogen receptor-α with BCAR1 and Traf6. Exp Cell Res 315:1287–1301

180. Väänänen HK (2005) Mechanism of osteoclast mediated bone resorption-rationale for the design of new therapeutics. Adv Drug Deliv Rev 57:959–971
181. Emerton KB, Hu B, Woo AA et al (2010) Osteocyte apoptosis and control of bone resorption following ovariectomy in mice. Bone 46:577–583
182. Kousteni S, Chen JR, Bellido T et al (2002) Reversal of bone loss in mice by nongenotropic signaling of sex steroids. Science 298:843–846
183. Faloni APDS, Sasso-Cerri E, Rocha FRG et al (2012) Structural and functional changes in the alveolar bone osteoclasts of estrogen-treated rats. J Anat 220:77–85
184. Faloni APS, Sasso-Cerri E, Katchburian E, Cerri PS (2007) Decrease in the number and apoptosis of alveolar bone osteoclasts in estrogen-treated rats. J Periodont Res 42:193–201
185. Hughes DE, Dai A, Tiffee JC et al (1996) Estrogen promotes apoptosis of murine osteoclasts mediated by TGF-β. Nature Med 2:1132–1136
186. Khosla S, Oursler MJ, Monroe DG (2012) Estrogen and the skeleton. Trends Endocrinol Metab 23:576–581
187. Rodan GA, Martin TJ (2000) Therapeutic approaches to bone diseases. Science 289:1508–1514
188. Riggs BL, Khosla S, Melton LJ (1998) A unitary model for involutional osteoporosis: estrogen deficiency causes both type I and type II osteoporosis in postmenopausal women and contributes to bone loss in aging men. J Bone Miner Res 13:763–773
189. Almeida M, Laurent MR, Dubois V et al (2016) Estrogens and androgens in skeletal physiology and pathophysiology. Physiol Rev 97:135–187
190. Compston JE (2001) Sex steroids and bone. Physiol Rev 81:419–447
191. Vanderschueren D, Vandenput L, Boonen S et al (2004) Androgens and bone. Endocr Rev 25:389–425
192. Taaffe DR, Galvão DA, Spry N et al (2019) Immediate versus delayed exercise in men initiating androgen deprivation: effects on bone density and soft tissue composition. BJU Int 123:261–269
193. Bassett JD, Williams GR (2016) Role of thyroid hormones in skeletal development and bone maintenance. Endocr Rev 37:135–187
194. El Hadidy M, Ghonaim M, El Gawad S, El Atta MA (2011) Impact of severity, duration, and etiology of hyperthyroidism on bone turnover markers and bone mineral density in men. BMC Endocr Disord 11:15
195. Harvey RD, McHardy KC, Reid IW et al (1991) Measurement of bone collagen degradation in hyperthyroidism and during thyroxine replacement therapy using pyridinium cross-links as specific urinary markers. J Clin Endocrinol Metab 72:1189–1194
196. Waring AC, Harrison S, Fink HA et al (2013) A prospective study of thyroid function, bone loss, and fractures in older men: The MrOS study. J Bone Miner Res 28:472–479
197. Britto JM, Fenton AJ, Holloway WR et al (1994) Osteoblasts mediate thyroid hormone stimulation of osteoclastic bone resorption. Endocrinology 134:169–176
198. Abe E, Marians RC, Yu W et al (2003) TSH is a negative regulator of skeletal remodelling. Cell 115:151–162
199. Sun L, Vukicevic S, Baliram R et al (2008) Intermittent recombinant TSH injections prevent ovariectomy-induced bone loss. Proc Natl Acad Sci USA 105:4289–4294
200. Neumann S, Eliseeva E, Boutin A et al (2018) Discovery of a positive allosteric modulator of the thyrotropin receptor: potentiation of thyrotropin-mediated preosteoblast differentiation in vitro. J Pharmacol Exp Ther 364:38–45
201. Chen XX, Yang T (2015) Roles of leptin in bone metabolism and bone diseases. J Bone Miner Metab 33:474–485
202. Shi Y, Yadav VK, Suda N et al (2008) Dissociation of the neuronal regulation of bone mass and energy metabolism by leptin in vivo. Proc Natl Acad Sci USA 105:20529–20533
203. Holloway WR, Collier FM, Aitken CJ et al (2002) Leptin inhibits osteoclast generation. J Bone Miner Res 17:200–209
204. Elefteriou F, Takeda S, Ebihara K et al (2004) Serum leptin level is a regulator of bone mass. Proc Natl Acad Sci USA 101:3258–3263

205. Hamrick MW, Ferrari SL (2008) Leptin and the sympathetic connection of fat to bone. Osteoporos Int J Establ Result Coop Eur Found Osteoporos Natl Osteoporos Found USA 19:905–912
206. Turner RT, Kalra SP, Wong CP et al (2013) Peripheral leptin regulates bone formation. J Bone Miner Res 28:22–34
207. Mantzoros CS, Magkos F, Brinkoetter M et al (2011) Leptin in human physiology and pathophysiology. Am J Physiol Endocrinol Metab 301:E567–E584
208. DeBlasio MJ, Lanham SA, Blache D et al (2018) Sex-and bone-specific responses in bone structure to exogenous leptin and leptin receptor antagonism in the ovine fetus. Am J Physiol Regul Integr Comp Physiol 314:R781–R790
209. Maor G, Rochwerger M, Segev Y, Phillip M (2002) Leptin acts as a growth factor on the chondrocytes of skeletal growth centres. J Bone Miner Res 17:1034–1043
210. Ruys CA, van de Lagemaat M, Lafeber HN et al (2018) Leptin and IGF-1 in relation to body composition and bone mineralization of preterm-born children from infancy to 8 years. Clin Endocrinol 89:76–84
211. Tsuji K, Maeda T, Kawane T et al (2010) Leptin stimulates fibroblast growth factor 23 expression in bone and suppresses renal $1\alpha 25$-dihydroxyvitamin D_3 synthesis in leptin-deficient ob/ob mice. J Bone Miner Res 25:1711–1723
212. López I, Pineda C, Raya AI et al (2016) Leptin directly stimulates parathyroid hormone secretion. Endocrine Abstracts 41:GP144
213. Urist MR (1965) Bone: formation by autoinduction. Science 150:893–899
214. Kobayashi T, Lyons KM, McMahon AP, Kronenberg HM (2005) BMP signaling stimulates cellular differentiation at multiple steps during cartilage development. Proc Natl Acad Sci USA 102:18023–18027
215. Beederman M, Lamplot JD, Nan G et al (2013) BMP signaling in mesenchymal stem cell differentiation and bone formation. J Biomed Sci Eng 6:32–52
216. Rahman MS, Akhtar N, Jamil HM et al (2015) TGF-/BMP signaling and other molecular events: regulation of osteoblastogenesis and bone formation. Bone Res 3:15005
217. Lin GL, Hankenson KD (2011) Integration of BMP, Wnt, and notch signaling pathways in osteoblast differentiation. J Cell Biochem 112:3491–3501
218. Kang Q, Song WX, Luo Q et al (2008) A comprehensive analysis of the dual roles of BMPs in regulating adipogenic and osteogenic differentiation of mesenchymal progenitor cells. Stem Cells Dev 18:545–559
219. Kang Q, Sun MH, Cheng H et al (2004) Characterization of the distinct orthotopic bone-forming activity of 14 BMPs using recombinant adenovirus-mediated gene delivery. Gene Ther 11:1312–1320
220. Huang E, Zhu G, Jiang W et al (2012) Growth hormone synergizes with BMP9 in osteogenic differentiation by activating the JAK/STAT/IGF1 pathway in murine multilineage cells. J Bone Miner Res 27:1566–1575
221. Li RD, Deng ZL, Hu N et al (2012) Biphasic effects of TGFβ1 on BMP9-induced osteogenic differentiation of mesenchymal stem cells. BMB Rep 45:509–514
222. Cheng H, Jiang W, Phillips FM et al (2003) Osteogenic activity of the fourteen types of human bone morphogenetic proteins (BMPs). J Bone Joint Surg Am 85:1544–1552
223. Franceschi RT, Wang D, Krebsbach PH, Rutherford RB (2000) Gene therapy for bone formation: in vitro and in vivo osteogenic activity of adenovirus expressing BMP-7. Ann Arbor 1001:48109–1078
224. Jane JA Jr, Dunford BA, Kron A et al (2002) Ectopic osteogenesis using adenoviral bone morphogenetic protein (BMP)-4 and BMP-6 gene transfer. Mol Ther 6:464–470
225. Carreira AC, Lojudice FH, Halcsik E et al (2014) Bone morphogenetic proteins facts, challenges, and future perspectives. J Dent Res 93:335–345
226. Cheng A, Krishnan L, Tran L et al (2019) The effects of age and dose on gene expression and segmental bone defect repair after BMP-2 Delivery. JBMR Plus 3:e100681-11
227. Holien T, Westhrin M, Moen SH et al (2018) BMP4 gene therapy inhibits myeloma tumor growth, but has a negative impact on bone. Blood 132:1928

228. Frost HM (1987) Bone "mass" and the "mechanostat": a proposal. Anat Rec 219:1–9
229. Corral DA, Amling M, Priemel M et al (1998) Dissociation between bone resorption and bone formation in osteopenic transgenic mice. Proc Natl Acad Sci USA 95:13835–13840
230. Kong Y-Y, Feige U, Sarosi I et al (1999) Activated T cells regulate bone loss and joint destruction in adjuvant arthritis through osteoprotegerin ligand. Nature 402:304–309
231. Suda T, Takahashi N, Udagawa N et al (1999) Modulation of osteoclast differentiation and function by the new members of the tumor necrosis factor receptor and ligand families. Endocr Rev 20:345–357
232. Miyamoto T, Suda T (2003) Differentiation and function of osteoclasts. Keio J Med 52:1–7
233. Fan X, Rahnert JA, Murphy TC et al (2006) Response to mechanical strain in an immortalized pre-osteoblast cell is dependent on ERK1/2. J Cell Physiol 207:454–460
234. Kreja L, Liedert A, Hasni S, Claes L, Ignatius A (2008) Intermittent mechanical strain increases RANKL expression in human osteoblasts. J Biomech 41:S462
235. Kim DW, Lee HJ, Karmin JA et al (2006) Mechanical loading differentially regulates membrane-bound and soluble RANKL availability in MC3T3-E1 cells. Ann NY Acad Sci 1068:568–572
236. Liu W, Xu C, Zhao H et al (2015) Osteoprotegerin induces apoptosis of osteoclasts and osteoclast precursor cells via the fas/fas ligand pathway. PLoS ONE 10:e0142519
237. Arai F, Miyamoto T, Ohneda O et al (1999) Commitment and differentiation of osteoclast precursor cells by the sequential expression of c-Fms and receptor activator of nuclear factor κb (RANK) receptors. J Exp Med 190:1741–1754
238. Romas E, Sims NA, Hards DK et al (2002) Osteoprotegerin reduces osteoclast numbers and prevents bone erosion in collagen-induced arthritis. Am J Pathol 161:1419–1427
239. Kadow-Romacker A, Hoffmann JE, Duda G et al (2009) Effect of mechanical stimulation on osteoblast- and osteoclast-like cells in vitro. Cells Tissues Organs 190:61–68
240. Bentolila V, Boyce TM, Fyhrie DP et al (1998) Intracortical remodelling in adult rat long bones after fatigue loading. Bone 23:275–281
241. Mann V, Huber C, Kogianni G et al (2006) The influence of mechanical stimulation on osteocyte apoptosis and bone viability in human trabecular bone. J Musculoskelet Neuronal Interact 6:408–417
242. Martin RB (2007) Targeted bone remodelling involves BMU steering as well as activation. Bone 40:1574–1580
243. Mori S, Burr DB (1993) Increased intracortical remodelling following fatigue damage. Bone 14:103–109
244. Noble BS, Peet N, Stevens HY et al (2003) Mechanical loading: biphasic osteocyte survival and targeting of osteoclasts for bone destruction in rat cortical bone. Am J Physiol 284:934–943
245. Verborgt O, Tatton NA, Majeska RJ, Schaffler MB (2002) Spatial distribution of Bax and Bcl-2 in osteocytes after bone fatigue: complementary roles in bone remodelling regulation? J Bone Miner Res 17:907–914
246. Tan SD, de Vries TJ, Kuijpers-Jagtman AM et al (2007) Osteocytes subjected to fluid flow inhibit osteoclast formation and bone resorption. Bone 41:745–751
247. Tan SD, Bakker AD, Semeins CM et al (2008) Inhibition of osteocyte apoptosis by fluid flow is mediated by nitric oxide. Biochem Biophys Res Commun 369:1150–1154
248. Rössig L, Haendeler J, Hermann C et al (2000) Nitric oxide down-regulates MKP-3 mRNA levels. J Biol Chem 275:25502–25507
249. Smalt R, Mitchell FT, Howard RL, Chambers TJ (1997) Induction of NO and prostaglandin E2 in osteoblasts by wall-shear stress but not mechanical strain. Am J Physiol 273:751–758
250. Zaman G, Pitsillides AA, Rawlinson SCF et al (1999) Mechanical strain stimulates nitric oxide production by rapid activation of endothelial nitric oxide synthase in osteocytes. J Bone Miner Res 14:1123–1131
251. van'T Hof RJ, Ralston SH (2001) Nitric oxide and bone. Immunology 103:255–261
252. Canalis E, Adams DJ, Boskey A et al (2013) Notch signaling in osteocytes differentially regulates cancellous and cortical bone remodelling. J Biol Chem 288:25614–25625

253. Bullock WA, Pavalko FM, Robling AG (2019) Osteocytes and mechanical loading: the Wnt connection. Orthod Craniofac Res 22:175–179

254. Ferraro JT, Daneshmand M, Bizios R, Rizzo V (2004) Depletion of plasma membrane cholesterol dampens hydrostatic pressure and shear stress-induced mechanotransduction pathways in osteoblast cultures. Am J Physiol Cell Physiol 286:831–839

255. Xing Y, Gu Y, Xu LC et al (2011) Effects of membrane cholesterol depletion and GPI-anchored protein reduction on osteoblastic mechanotransduction. J Cell Physiol 226:2350–2359

256. Klausen TK, Hougaard C, Hoffmann EK, Pedersen SF (2006) Cholesterol modulates the volume-regulated anion current in Ehrlich-Lettre ascites cells via effects on Rho and F-actin. Am J Physiol Cell Physiol 291:757–771

257. Qi M, Liu Y, Freeman MR, Solomon KR (2009) Cholesterol-regulated stress fibre formation. J Cell Biochem 106:1031–1040

258. Dason JS, Smith AJ, Marin L, Charlton MP (2014) Cholesterol and F-actin are required for clustering of recycling synaptic vesicle proteins in the presynaptic plasma membrane. J Physiol 592:621–633

259. Radel C, Rizzo V (2005) Integrin mechanotransduction stimulates caveolin-1 phosphorylation and recruitment of Csk to mediate actin reorganization. Am J Physiol 288:936–945

260. Barczyk M, Carracedo S, Gullberg D (2010) Integrins. Cell Tissue Res 339:269–280

261. Hynes RO (2002) Integrins: bidirectional, allosteric signaling machines. Cell 110:673–687

262. Nievers MG, Schaapveld RQJ, Sonnenberg A (1999) Biology and function of hemidesmosomes. Matrix Biol 18:5–17

263. Matthews BD, Overby DR, Mannix R, Ingber DE (2006) Cellular adaptation to mechanical stress: role of integrins, Rho, cytoskeletal tension and mechanosensitive ion channels. Journal Cell Sci 119:508–518

264. Bhattacharya R, Gonzalez AM, DeBiase PJ et al (2009) Recruitment of vimentin to the cell surface by β3 integrin and plectin mediates adhesion strength. J Cell Sci 122:1390–1400

265. Cram EJ, Schwarzbauer JE (2004) The talin wags the dog: new insights into integrin activation. Trends Cell Biol 14:55–57

266. Brown MC, Perrotta JA, Turner CE (1996) Identification of LIM3 as the principal determinant of paxillin focal adhesion localization and characterization of a novel motif on paxillin directing vinculin and focal adhesion kinase binding. J Cell Biol 135:1109–1123

267. Geiger B, Spatz JP, Bershadsky AD (2009) Environmental sensing through focal adhesions. Nature Rev Mol Cell Biol 10:21–33

268. El-Hoss J, Arabian A, Dedhar S, St-Arnaud R (2014) Inactivation of the integrin-linked kinase (ILK) in osteoblasts increases mineralization. Gene 533:246–252

269. Katz B-Z, Zamir E, Bershadsky A et al (2000) Physical state of the extracellular matrix regulates the structure and molecular composition of cell-matrix adhesions. Mol Biol Cell 11:1047–1060

270. Parsons JT (1996) Integrin-mediated signalling: regulation by protein tyrosine kinases and small GTP-binding proteins. Curr Opin Cell Biol 8:146–152

271. Teo BKK, Wong ST, Lim CK et al (2013) Nanotopography modulates mechanotransduction of stem cells and induces differentiation through focal adhesion kinase. ACS Nano 7:4785–4798

272. Yamada KM, Geiger B (1997) Molecular interactions in cell adhesion complexes. Curr Opin Cell Biol 9:76–85

273. Hendesi H, Barbe MF, Safadi FF et al (2015) Integrin mediated adhesion of osteoblasts to connective tissue growth factor (CTGF/CCN2) induces cytoskeleton reorganization and cell differentiation. PLoS ONE 10(2):e0115325

274. Moussa FM, Hisijara IA, Sondag GR et al (2014) Osteoactivin promotes osteoblast adhesion through HSPG and αvβ1 integrin. J Cell Biochem 115:1243–1253

275. Saidak Z, Le Henaff C, Azzi S et al (2015) Wnt/β-catenin signaling mediates osteoblast differentiation triggered by peptide-induced α5β1 integrin priming in mesenchymal skeletal cells. J Biol Chem 290:6903–6912

276. Carvalho RS, Schaffer JL, Gerstenfeld LC (1998) Osteoblasts induce osteopontin expression in response to attachment on fibronectin: demonstration of a common role for integrin

receptors in the signal transduction processes of cell attachment and mechanical stimulation. J Cell Biochem 70:376–390

277. Riveline D, Zamir E, Balaban NQ et al (2001) Focal contacts as mechanosensors: externally applied local mechanical force induces growth of focal contacts by an mDia1-dependent and ROCK-independent mechanism. J Cell Biol 153:1175–1185

278. Carvalho RS, Bumann A, Schaffer JL, Gerstenfeld LC (2002) Predominant integrin ligands expressed by osteoblasts show preferential regulation in response to both cell adhesion and mechanical perturbation. J Cell Biochem 84:497–508

279. Friedland JC, Lee MH, Boettiger D (2009) Mechanically activated integrin switch controls α5β1 function. Science 323:642–644

280. Litzenberger JB, Kim JB, Tummala P, Jacobs CR (2010) β1 integrins mediate mechanosensitive signaling pathways in osteocytes. Calcif Tissue Int 86:325–332

281. Litzenberger JB, Tang WJ, Castillo AB, Jacobs CR (2009) Deletion of β1 integrins from cortical osteocytes reduces load-induced bone formation. Cell Mol Bioeng 2:416–424

282. McNamara LM, Majeska RJ, Weinbaum S et al (2009) Attachment of osteocyte cell processes to the bone matrix. Anat Rec 292:355–363

283. Phillips JA, Almeida EA, Hill EL et al (2008) Role for β1 integrins in cortical osteocytes during acute musculoskeletal disuse. Matrix Biol 27:609–618

284. Thi MM, Suadicani SO, Schaffler MB et al (2013) Mechanosensory responses of osteocytes to physiological forces occur along processes and not cell body and require αVβ3 integrin. Proc Natl Acad Sci USA 110:21012–21017

285. Haugh MG, Vaughan TJ, McNamara LM (2015) The role of integrin α V β 3 in osteocyte mechanotransduction. J Mech Behav Biomed Mater 42:67–75

286. Cabahug-Zuckerman P, Stout RF Jr et al (2018) Potential role for a specialized β3 integrin-based structure on osteocyte processes in bone mechanosensation. J Orthop Res 36:642–652

287. Pommerenke H, Schmidt C, Durr F et al (2002) The mode of mechanical integrin stressing controls intracellular signaling in osteoblasts. J Bone Miner Res 17:603–611

288. Saunders MM, You J, Trosko JE et al (2001) Gap junctions and fluid flow response in MC3T3-E1 cells. Am J Physiol—Cell Physiol 281:1917–1925

289. Gillespie PG, Walker RG (2001) Molecular basis of mechanosensory transduction. Nature 413:194–202

290. Kazmierczak P, Sakaguchi H, Tokita J et al (2007) Cadherin 23 and protocadherin 15 interact to form tip-link filaments in sensory hair cells. Nature 449:87–91

291. Marie PJ, Hay E (2013) Cadherins and Wnt signalling: a functional link controlling bone formation. BoneKEy Rep 2:4

292. Matsuo K, Otaki N (2012) Bone cell interactions through Eph/ephrin: bone modeling, remodelling and associated diseases. Cell Adh Migr 6:148–156

293. Tamma R, Zallone A (2012) Osteoblast and osteoclast crosstalks: from OAF to Ephrin. Inflamm Allergy-Drug Targets 11:196–200

294. Nakajima K, Kho DH, Yanagawa T et al (2016) Galectin-3 in bone tumor microenvironment: a beacon for individual skeletal metastasis management. Cancer Metastasis Rev 35:333–346

295. Vinik Y, Shatz-Azoulay H, Vivanti A et al (2015) The mammalian lectin galectin-8 induces RANKL expression, osteoclastogenesis, and bone mass reduction in mice. Elife 4:e05914

296. Tanikawa R, Tanikawa T, Hirashima M et al (2010) Galectin-9 induces osteoblast differentiation through the CD44/Smad signaling pathway. Biochem Biophys Res Commun 394:317–322

297. Flagg-Newton J, Simpson I, Loewenstein WR (1979) Permeability of the cell-to-cell membrane channels in mammalian cell junctions. Science 205:404–407

298. Steinberg TH, Civitelli R, Geist ST et al (1994) Connexin43 and connexin45 form gap junctions with different molecular permeabilities in osteoblastic cells. EMBO J 13:744–750

299. Saunders MM, You J, Zhou Z et al (2003) Fluid flow-induced prostaglandin E2 response of osteoblastic ROS 17/2.8 cells is gap junction-mediated and independent of cytosolic calcium. Bone 32:350–356

300. Bivi N, Condon KW, Allen MR (2012) Cell autonomous requirement of connexin 43 for osteocyte survival: consequences for endocortical resorption and periosteal bone formation. J Bone Miner Res 27:374–389

301. Delaine-Smith RM, Sittichokechaiwut A, Reilly GC (2014) Primary cilia respond to fluid shear stress and mediate flow-induced calcium deposition in osteoblasts. FASEB J 28:430–439

302. Myers KA, Rattner JB, Shrive NG, Hart DA (2007) Osteoblast-like cells and fluid flow: cytoskeleton-dependent shear sensitivity. Biochem Biophys Res Commun 364:214–219

303. Malone AMD, Anderson CT, Tummala P et al (2007) Primary cilia mediate mechanosensing in bone cells by a calcium-independent mechanism. Proc Natl Acad Sci USA. 104:13325–13330

304. Xiao Z, Zhang S, Mahlios J et al (2006) Cilia-like structures and polycystin-1 in osteoblasts/osteocytes and associated abnormalities in skeletogenesis and Runx2 expression. J Biol Chem 281:30884–30895

305. Leucht P, Monica SD, Temiyasathit S et al (2013) Primary cilia act as mechanosensors during bone healing around an implant. Med Eng Phys 35:392–402

306. Chen JC, Hoey DA, Chua M et al (2016) Mechanical signals promote osteogenic fate through a primary cilia-mediated mechanism. FASEB J 30:1504–1511

307. Coughlin TR, Voisin M, Schaffler MB et al (2015) Primary cilia exist in a small fraction of cells in trabecular bone and marrow. Calcif Tissue Int 96:65–72

308. Smith MA, Hoffman LM, Beckerle MC (2014) LIM proteins in actin cytoskeleton mechanoresponse. Trends Cell Biol 24:575–583

309. Wang N, Ingber DE (1994) Control of cytoskeletal mechanics by extracellular matrix, cell shape, and mechanical tension. Biophys J 66:2181–2189

310. Wang N, Butler JP, Ingber DE (1993) Mechanotransduction across the cell surface and through the cytoskeleton. Science 260:1124–1127

311. del Álamo JC, Norwich GN, Yshuan JL et al (2008) Anisotropic rheology and directional mechanotransduction in vascular endothelial cells. Proc Natl Acad Sci USA 105:15411–15416

312. Hu S, Chen J, Fabry B et al (2003) Intracellular stress tomography reveals stress focusing and structural anisotropy in cytoskeleton of living cells. Am J Physiol 285:1082–1090

313. Silberberg YR, Pelling AE, Yakubov GE et al (2008) Mitochondrial displacements in response to nanomechanical forces. J Mol Recognit 21:30–36

314. Koike M, Nojiri H, Ozawa Y et al (2015) Mechanical overloading causes mitochondrial superoxide and SOD2 imbalance in chondrocytes resulting in cartilage degeneration. Sci Rep 5:11722

315. Khatiwala CB, Peyton SR, Putnam AJ (2006) Intrinsic mechanical properties of the extracellular matrix affect the behavior of pre-osteoblastic MC3T3-E1 cells. Am J Physiol Cell Physiol 290:1640–1650

316. Wen JH, Vincent LG, Fuhrmann A et al (2014) Interplay of matrix stiffness and protein tethering in stem cell differentiation. Nat Mater 13:979–987

317. Ritz U, Götz H, Baranowski A et al (2016) Influence of different calcium phosphate ceramics on growth and differentiation of cells in osteoblast–endothelial co-cultures. J Biomed Mater Res B Appl Biomater 105:1950–1962

318. Olivares-Navarrete R, Rodil SE, Hyzy SL et al (2015) Role of integrin subunits in mesenchymal stem cell differentiation and osteoblast maturation on graphitic carbon-coated microstructured surfaces. Biomaterials 51:69–79

319. Fraioli R, Rechenmacher F, Neubauer S et al (2015) Mimicking bone extracellular matrix: Integrin-binding peptidomimetics enhance osteoblast-like cells adhesion, proliferation, and differentiation on titanium. Colloids Surf B 128:191–200

320. Vatsa A, Breuls RG, Semeins CM et al (2008) Osteocyte morphology in fibula and calvaria—is there a role for mechanosensing? Bone 43:452–458

321. Van Hove RP, Nolte PA, Vatsa A et al (2009) Osteocyte morphology in human tibiae of different bone pathologies with different bone mineral density—is there a role for mechanosensing? Bone 45:321–329

322. Murshid SA, Kamioka H, Ishihara Y et al (2007) Actin and microtubule cytoskeletons of the processes of 3D-cultured MC3T3-E1 cells and osteocytes. J Bone Miner Metab 25:259

323. Adachi T, Aonuma Y, Tanaka M et al (2009) Calcium response in single osteocytes to locally applied mechanical stimulus: Differences in cell process and cell body. J Biomech 42:1989–1995

324. Sugawara Y, Ando R, Kamioka H et al (2008) The alteration of a mechanical property of bone cells during the process of changing from osteoblasts to osteocytes. Bone 43:19–24

325. Prendergast PJ, Huiskes R (1995) The biomechanics of Wolff's law: recent advances. Ir J Med Sci 164:152–154

326. Turner CH (1998) Three rules for bone adaptation to mechanical stimuli. Bone 23:399–407

327. Hsieh Y-F, Turner CH (2001) Effects of loading frequency on mechanically induced bone formation. J Bone Miner Res 16:918–924

328. Zhu J, Zhang X, Wang C et al (2008) Different magnitudes of tensile strain induce human osteoblasts differentiation associated with the activation of ERK1/2 phosphorylation. Int J Mol Sci 9:2322–2332

329. Mosley JR, March BM, Lynch J, Lanyon LE (1997) Strain magnitude related changes in whole bone architecture in growing rats. Bone 20:191–198

330. Cullen DM, Smith RT, Akhter MP (2001) Bone-loading response varies with strain magnitude and cycle number. J Appl Physiol 91:1971–1976

331. Rubin CT, Lanyon LE (1984) Regulation of bone formation by applied dynamic loads. J Bone Joint Surg 66:397–402

332. Srinivasan S, Ausk BJ, Poliachik SL et al (2007) Rest-inserted loading rapidly amplifies the response of bone to small increases in strain and load cycles. J Appl Physiol 102:1945–1952

333. Pereira AF, Shefelbine SJ (2014) The influence of load repetition in bone mechanotransduction using poroelastic finite-element models: the impact of permeability. Biomechan Model Mechanobiol 13:215–225

334. Stavenschi E, Corrigan MA, Johnson GP et al (2018) Physiological cyclic hydrostatic pressure induces osteogenic lineage commitment of human bone marrow stem cells: a systematic study. Stem Cell Res Ther 9:276

335. Warden SJ, Turner CH (2004) Mechanotransduction in the cortical bone is most efficient at loading frequencies of 5–10 Hz. Bone 34:261–270

336. Verbruggen SW, Vaughan TJ, McNamara LM (2014) Fluid flow in the osteocyte mechanical environment: a fluid–structure interaction approach. Biomechan model mechanobiol 13:85–97

337. Wittig NK, Laugesen M, Birkbak ME et al (2019) Canalicular junctions in the osteocyte lacuno-canalicular network of cortical bone. ACS Nano 13:6421–6430

338. Gatti V, Azoulay EM, Fritton SP (2018) Microstructural changes associated with osteoporosis negatively affect loading-induced fluid flow around osteocytes in cortical bone. J Biomech 66:127–136

339. Nauli SM, Alenghat FJ, Luo Y et al (2003) Polycystins 1 and 2 mediate mechanosensation in the primary cilium of kidney cells. Nature Genet 33:129–137

340. Praetorius HA, Frokiaer J, Nielsen S, Spring KR (2003) Bending the primary cilium opens Ca^{2+}-sensitive intermediate-conductance K^+ channels in MDCK Cells. J Membr Biol 191:193–200

341. Waychunas GA (2014) Disrupting dissolving ions at surfaces with fluid flow. Science 344:1094–1095

342. Gross D, Williams WS (1982) Streaming potential and the electromechanical response of physiologically-moist bone. J Biomech 15:277–295

343. Frijns A, Huyghe J, Wijlaars M (2005) Measurements of deformations and electrical potentials in a charged porous medium. In: Gladwell GML, Huyghe J, Raats PA, Cowin SC (eds) IUTAM symposium on physicochemical and electromechanical interactions in porous media, vol 125, pp 133–139. Springer, Dordrecht

344. Hong J, Ko S, Khang G, Mun M (2008) Intraosseous pressure and strain generated potential of cylindrical bone samples in the drained uniaxial condition for various loading rates. J Mater Sci Mater Med 19:2589–2594

345. Pienkowski D, Pollack SR (1983) The origin of stress-generated potentials in fluid-saturated bone. J Orthop Res 1:30–41
346. Iatridis J, Laible J, Krag M (2003) Influence of fixed charge density magnitude and distribution on the intervertebral disc: applications of a poroelastic and chemical electric (PEACE) model. J Biomech Eng 125:12–24
347. Fukada E, Yasuda I (1957) On the piezoelectric effect of bone. J Phys Soc Jpn 12:1158–1162
348. Elmessiery MA (1981) Physical basis for piezoelectricity of bone matrix. IEE Proc A 128:336–346
349. Halperin C, Mutchnik S, Agronin A et al (2004) Piezoelectric effect in human bones studied in nanometer scale. Nano Lett 4:1253–1256
350. Marino AA, Becker RO (1974) Piezoelectricity in bone as a function of age. Calcif Tissue Int 14:327–331
351. Wang T, Feng Z, Song Y, Chen X (2007) Piezoelectric properties of human dentin and some influencing factors. Dent Mater 23:450–453
352. Reinish GB, Nowick AS (1975) Piezoelectric properties of bone as functions of moisture content. Nature 253:626–627
353. Ahn AC, Grodzinsky AJ (2009) Relevance of collagen piezoelectricity to "Wolff's Law": a critical review. Med Eng Phys 31:733–741
354. Ramtani S (2008) Electro-mechanics of bone remodelling. Int J Eng Sci 46:1173–1182
355. Frias C, Reis J, e Silva FC et al (2010) Polymeric piezoelectric actuator substrate for osteoblast mechanical stimulation. J Biomech 43:1061–1066
356. Reis J, Frias C, Canto e Castro C, Botelho ML, Marques AT, Simões JA, Capela e Silva F, Potes J (2012) A new piezoelectric actuator induces bone formation in vivo: a preliminary study. BioMed Res Int 613403
357. Zhang Y, Chen L, Zeng J et al (2014) Aligned porous barium titanate/hydroxyapatite composites with high piezoelectric coefficients for bone tissue engineering. Mater Sci Eng C 39:143–149
358. Liu J, Gu H, Liu Q, Ren L, Li G (2019) An intelligent material for tissue reconstruction: The piezoelectric property of polycaprolactone/barium titanate composites. Mater Lett 236:686–689
359. Jacob J, More N, Kalia K, Kapusetti G (2018) Piezoelectric smart biomaterials for bone and cartilage tissue engineering. Inflamm Regen 38:2
360. Damaraju SM, Shen Y, Elele E, Khusid B, Eshghinejad A, Li J, Jaffe M, Arinzeh TL (2017) Three-dimensional piezoelectric fibrous scaffolds selectively promote mesenchymal stem cell differentiation. Biomaterials 149:51–62

Bone Quality Assessment at the Atomic Scale

J. M. D. A. Rollo, R. S. Boffa, R. Cesar, R. Erbereli, D. C. Schwab and T. P. Leivas

Abstract The assessment of osteoporosis regarding bone mass and microarchitecture "quality" contributes in determining fracture risk. Therefore, the crystalline structure of hydroxyapatite may indicate the quality of trabecular bones through the identification of crystallite sizes, microhardness and microdeformation values and calcium and phosphorous proportions in the three types of bones: normal, osteopenic, and osteoporotic. Nine L1 vertebrae-dried trabecular bones from human cadavers were used. The characterization of the three types of bones was made through scanning electron microscopy, EDS, microhardness, and X-ray diffractometry with the

J. M. D. A. Rollo (✉) · R. S. Boffa
Departamento de Engenharia de Materiais, Universidade de São Paulo (USP), Escola de Engenharia de São Carlos, Av. Trabalhador Saocarlense, 400—Parque Arnold Schimidt, 13565-590 São Carlos, SP, Brazil
e-mail: tfase@sc.usp.br

R. S. Boffa
e-mail: boffa3266@hotmail.com

R. Cesar · R. Erbereli
Departamento de Engenharia Mecânica, Universidade de São Paulo (USP), Escola de Engenharia de São Carlos, Av. Trabalhador Saocarlense, 400 – Parque Arnold Schimidt, 13565-590 São Carlos, SP, Brazil
e-mail: reinaldofisica@gmail.com

R. Erbereli
e-mail: rogerio.erbereli@usp.br

D. C. Schwab
DCS - English Consultancy Services, Rua São Sebastião, 1795 – Sala 7, 13560-230 São Carlos, Brazil
e-mail: dcschwab@gmail.com

T. P. Leivas
Instituto de Ortopedia E Traumatologia, HCFMUSP-OIT—Hospital de Clinicas da Faculdade de Medicina, Universidade de São Paulo (USP), Rua Dr. Ovidio Pires de Campos, 333 – Cerqueira Cesar, 05403-010 São Paulo, Brazil
e-mail: tomazpuga@gmail.com

© Springer Nature Switzerland AG 2020
J. Belinha et al. (eds.), *The Computational Mechanics of Bone Tissue*,
Lecture Notes in Computational Vision and Biomechanics 35,
https://doi.org/10.1007/978-3-030-37541-6_2

Rietveld refinement method. The results show that the microstructural characterization possibilities the identification of the three types of bones: normal, osteopenic, and osteoporotic, allowing the detection of osteoporosis based on bone quality.

1 Introduction

Osteoporosis is defined by the National Institutes of Health as a skeleton disorder characterized by compromised bone resistance and high fracture risk [1].

Professionals have assumed that all patients with very low T-scores (bone mass measurement) have osteoporosis. Values higher than -1.0 are considered normal, between -1.0 and -2.5 osteopenic and below -2.5 osteoporotic. However, the T-score derives from a specific population; therefore, in other populations, the T-score has its problems. Since the bone mineral density (BMD) is a limited fracture risk indicator, the clinical and scientific interest has increased in the complementary analysis that could improve the fracture risk prediction [2–6].

A normal BMD does not guarantee that a fracture will not happen and, reciprocally, for a BMD in the osteoporotic level, fractures will be more probable, but not impossible to prevent. Due to these paradoxes in treatment, the term became popular in the early nineties and, since then, the concept of bone resistance amplified to more than just density, also aggregating characteristics related to bone quality. There are many properties representing bone quality, among them there is the crystalline structure of the inorganic part of the bone (hydroxyapatite crystals) [1, 7–17].

Analyzing the crystalline structure in the atomic scale (crystallite size, calcium and phosphorous parts, microdeformation), visually (scanning electron microscopy), and mechanically (microhardness), it is expected to identify relations, for dried trabecular bones indicating their condition regarding bone quality.

Even though it is not satisfactory, the evaluation for osteoporosis considers only bone quantity (BMD). It is important to evaluate bone quality with the analysis of the inorganic part of the bone through the microstructural characterization of hydroxyapatite crystals.

2 Hydroxyapatite

Biological hydroxyapatite (HA) is considered the structural model for the mineral phase of the bone, and it presents imperfections, different from the HA found in rocks. The ions on the crystal surface are hydrated, generating a layer of ions and water called hydration cover, which facilitates the exchange of ions between the crystal and the interstitial fluid. It may have multiple substitutions and deficiencies in all ionic sites. Among the impurities of the apatite crystals, the most noticeable is the replacement of B type carbonates from the HA (CHA) in the phosphates groups, also presenting replacements of potassium, magnesium, strontium and sodium for calcium ions, chlorides and fluorides for the hydroxyl groups. These impurities may alter the crystalline structure, reducing the crystallinity, affecting the elasticity and the bone resistance of the apatite. The size of the crystal and the bone mineral crystallinity may also be altered due to certain diseases and therapies [18–28].

Therefore, the HA may have a varied composition. Calcium-deficient hydroxyapatite (CDHA), or non-stoichiometric, can be obtained in low temperatures, with a composition expressed as $Ca_{10-x}(HPO_4)_x(PO_4)_{6-x}(OH)_{2-x}$, where x varies from 0 to 1: 0 for non-stoichiometric HA and 1 for complete CDHA [29]

Pure HA presents a molar reason of 1.67, as shown in Table 1.

Table 1 Main calcium phosphates

Name	Formula	Ca/P	Mineral	Symbol
Monocalcium phosphate monohydrate	$Ca(H_2PO_4)_2 . H_2O$	0.50	–	MCPM
Dicalcium phosphate	$CaHPO_4$	1.00	Monetite	DCP
Dicalcium phosphate dihydrate	$CaHPO_4 . 2H_2O$	1.00	Brushite	DCPD
Octacalcium phospate	$Ca_8H_2(PO_4)_6 . 5H_2O$	1.33	–	OCP
Precipitated hydroxyapatite	$Ca_{10-x}(HPO_4)_x(PO_4)_{6-x}(OH)_{2-x}$	1.50–1.67	–	PHA
Tricalcium phosphate	$Ca_9(HPO_4)(PO_4)_5(OH)$	1.5	–	TCP
Amorphous calcium phosphate	$Ca_3(PO_4)_2 . nH_2O^a$	1.5	–	ACP
Monocalcium phosphate	$Ca(H_2PO_4)_2$	0.50	–	MCP
α—Tricalcium phosphate	$\alpha\text{-}Ca_3(PO_4)_2$	1.5	–	α-TCP
β—Tricalcium phosphate	$\beta\text{-}Ca_3(PO_4)_2$	1.50	–	β–TCP
Sintered hydroxyapatite	$Ca_5(PO_4)_3OH$	1.67	Hydroxyapatite	HA
Oxyapatite	$Ca_{10}(PO_4)_6O$	1.67	–	OXA
Tetracalcium phosphate	$Ca_4(PO_4)_2O$	2.00	Hilgenstockite	TetCP
Carbonated apatite	$Ca_{8.8}(HPO_4)_{0.7}(PO_4)_{4.5}(CO_3)_{0.7}(OH)_{1.3}$		Dahlite	CAP

The higher the molar reason Ca/P, the lower will be the solubility of the material, but this rate is also influenced regarding its chemical composition, local pH, temperature, particle sizes, and crystallinity [30–33].

3 Rietveld Refinement Method

The X-ray diffraction characterization methods are used for the indexation of the crystalline phases, unit cell refinement, crystallite size determination, net microdeformation, quantitative analysis of the phases, etc. [34, 35].

The structure of a typical diffraction standard in the powder can be described by the positions, intensities, and forms of the multiple Bragg reflections. Each of these components holds information about the crystalline structure of the material, sample properties, and the instrumental standards, as seen in Table 2 [36].

The size of the crystallite and the residual tension (microdeformations) may then be analyzed by the format of the peak, more specifically by its width, also taking into consideration the instrumental nature and the specific conditions for each experiment (diffractometer slot width, band wavelength generated by the source, angular divergence of the beams, etc.) [36].

Once a routine is established for the calculus of the profile, it is necessary to choose the refinement method to be adopted. The most commonly accepted strategy is the minimum squares technique, which aims to minimize the sum of the squares of the differences between the theoretical model and the data obtained in the measurements, adjusting values of the parameters present in the theory in order to find the ideal values for these parameters.

In 1967–69, Rietveld presented a refinement method of crystalline structures. The Rietveld method is a powerful tool for the structural analysis of most crystalline materials in powder form, which is used nowadays to solve all problems mentioned before (unit cell refinement, crystallite size determination, net microdeformation, quantitative analysis of the phases) using the minimum squares technique. For its application, the diffraction data is used as it lefts the diffractometer, without any sort

Table 2 Possible information to be analyzed by the X-ray generated diffractogram [36]

Standard component	Crystalline structure	Specimen property	Instrumental parameter
Peak position	Unitary cell parameters	Absorption; porosity	Radiation (wavelength); sample alignment
Peak intensity	Atomic parameters	Preferred orientation; Absorption; porosity	Geometry and configuration
Peak format	Crystallinity; disarray; defects	Crystallite size; tension	X-ray conditioning

of alteration, which follows the scientific criteria that no modifications must be done to the observations for them to be analyzed [37, 38].

At first, the method was only applied in materials analyzed by neutrons diffraction. Later on, after some adaptations, its application was made possible for measures obtained by X-ray diffraction as well. For the X-ray, there are no simple functions, different from the neutrons, the radiation with which the method was originally developed has peaks modeled by Gaussian. The peaks are closer to Gaussian for small spreading angles and closer to Lorentzian for large angles. Therefore, the pseudo-Voigt function is used, a normalized linear combination of a Gaussian with a Lorentzian as shown in Eq. (1): [39–42].

$$pV(x) = \eta L(X) + (1 - \eta)G(X) \tag{1}$$

where η is a refined parameter that determines the percentage of contribution of each function; L is the Lorentzian; G the Gaussian function; $pV(x)$ is the pseudo-Voigt function; x and X are variables. Define also: $pV(x)$, x and X.

The structure refinement by the Rietveld method consists of applying the minimum squares method in order to find the structural parameter values to make the calculated intensity and the measured intensity agree in the best possible way. The minimum squares paradigm considers that the best adjustment between a group of N values obtained experimentally $y_i^E(x_i)$ and a model function $f(a, x)$, which depends on M refinable parameters a_i, is obtained when the residue χ^2 is minimized, this residue is defined by Eq. (2):

$$\chi^2 = \sum_{i=1}^{N} \frac{\left| y_i^E - f(a, x_i) \right|^2}{\sigma_i^2} \tag{2}$$

where the vector a represents the parameters to be refined, which are its components. This function has an important statistical significance. It is possible to show that finding the minimum value for χ^2 is equivalent to finding the refined values of the parameters for which there is a higher probability of the model function $f(a, x_i)$ to coincide with the experimental data y_i^E [43].

The parameters refined by the minimum squares method are divided into two categories: structural parameters are those that measure the characteristics in the sample; the net parameters and the instrumental parameters are those which do not depend on the characteristics of the sample but on the experimental conditions [44].

It is important to highlight that the refinement must follow a sequence, where the first parameters to be refined using a standard sample are usually the instrumental ones: GW, GV, etc. Right after the structural parameters are adjusted, which are related to the sample: net parameters (a, b, c) and, when necessary, the angles of the unit cell (α, β and γ); atomic positions (x, y, z), thermal parameters (isotropic or anisotropic) and last, GU and GP which together with GW and GV are parameters for adjusting width to the half-height expressed by the Cagliotti formula shown in Eq. (3): [45].

Table 3 Terms used by the software GSAS

Acronym	Description
GU, GV, GW, GP	Coefficients in the Cagliotti formula (modified by Scherrer enlargement) from the Gaussian component for the peak width. Where GV and GW are related to the enlargement related to the equipment and GP and GU are related to the enlargement of the crystallite size itself
LX	Coefficient from the Lorentzian component for peak width by the isotropic size of the crystallite
Sfec and ptec	Coefficients of the Lorentzian width, but are used only for anisotropic enlargement. Stec is associated with the component Y and ptec with the component X
Uisos	Atomic dislocations (Uisos)—thermal vibration parameters
S/L and H/L	Geometric terms that describe peak enlargement at low angles, in general $2\theta < 15°$, with CuKα radiation (and also for the enlargement of peaks at high angles $2\theta > 165°$). These terms may be calculated by dividing the width of the diverging and receptive slots by the diffractometer radius. These terms are generally the same and must be fixed for a specific (known) group of slots and diffractometer radius, that is, not for instrumental parameters
Trns	Correction for sample transparency in the Bragg-Brentano geometry. It is the inverse of the absorption effect and for the Bragg-Brentano geometry it is zero for high absorbing samples and different from zero for the low or intermediate absorbing samples
Shft	Correction in the peak position for the vertical displacement of the sample in the Bragg-Brentano geometry. It depends highly on the coupling of the sample to the diffractometer, so special care must be taken to prevent this sort of mistake
Eta	mixed factor Gaussian-Lorentzian
SXXX, SYYY, and SZZZ	Lorentzian anisotropic enlargement factors of microtension

$$H_k^2 = U \tan^2 \theta_k + V \tan \theta_k + W \qquad (3)$$

Define V and W, θ_k and U. The definitions of the terms used by the GSAS software used in the refinement according to its manual are described in Table 3.

3.1 Crystallite Size and Microdeformations: Scherrer Equation and Williamson-Hall Graph

In the diffraction peaks enlargement analysis method, it is understood that the total enlargement has three components: crystallites size, residual microdeformations, and instrumental parameters, and each one can be identified separately, as shown by Dehlinger and Kochendörfer [46]. The particle size for each sample was calculated after the parameters V and W, taken from a standard sample, were refined in order to find the instrumental parameters, which would be used later to nullify the instrumental influence for the peak width calculation. The enlargement of the resulting peak from the residual microdeformation β_s can be expressed by Eq. (4):

$$\beta_s = 4\,\varepsilon\,\tan\theta \tag{4}$$

where ε is the microdeformation and θ is the Bragg angle. The enlargement component originated by the crystallites, β_c, can be expressed by Scherrer's equation (Eq. 5): [47].

$$\beta_c = \frac{k'\lambda}{D\,\cos\theta} \tag{5}$$

where k' is a constant, which depends on the reflection and form of the crystal (which usually has value 1), λ is the radiation wavelength, D is the crystallite size and θ is the Bragg angle. Width at total half-height β already corrected will be the sum of the two equations, represented in Eq. (6):

$$\beta = \beta_s + \beta_c = \frac{k'\lambda}{D\,\cos\theta} + 4\,\varepsilon\,\tan\theta \tag{6}$$

where corrected β is represented by Eq. (7):

$$\beta = \sqrt{\beta_{exp}^2 - \beta_{inst}^2} \tag{7}$$

And it will be the same as H_k from the Cagliotti formula. Equation (6) can be rewritten as Eq. (8):

$$\frac{\beta\cos\theta}{\lambda} = \frac{k'}{D} + \frac{4\varepsilon}{\lambda}\sin\theta \tag{8}$$

which will be the equation used to create the Williamson–Hall graph of a first-degree equation, as represented in Fig. 1, where $Y = \frac{\beta\cos\theta}{\lambda}$ and $X = \sin\theta$ and Williamson and Hall [48].

The linear coefficient of the line will provide the value for the size of the crystallite, Eq. (9):

$$D = \frac{1}{b} \tag{9}$$

And the angular coefficient will provide the microdeformations, Eq. (10):

$$\varepsilon = \frac{a\lambda}{4} \tag{10}$$

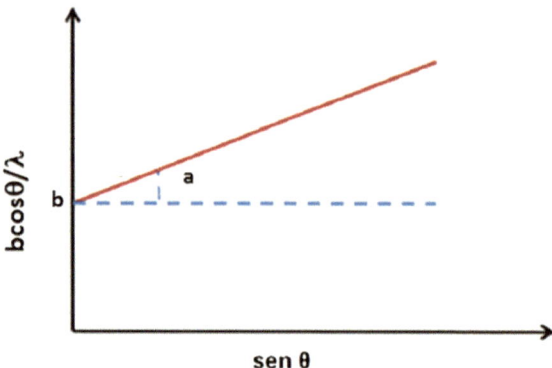

Fig. 1 Representation of Williamson–Hall graph

Properly define/recall parameters Fig. 1a, b. Tension in a material may have two distinct effects over the diffractogram. Due to a uniform effort, compressive or distensive (macrotension), the distances between the atomic plans may become higher or lower, causing the occasional displacement in the position of the peaks.

Uniform efforts are related to the distension and compression simultaneous forces, which result in an enlargement of the diffracted peaks in its original position. This phenomenon, called microtension in crystallites or microdeformation, may be related to different causes: displacements, vacancies, defects, shear planes, thermal expansions, and contractions. Microdeformation may be understood as a relative variation of the net parameter or the interplanar distance caused by the effects mentioned above.

4   Ultrasonometry of the Calcaneus

A great number of comparative studies evaluate the relation between the quantitative ultrasonometry (QU) and the BMD measured by the gold standard dual-energy X-ray absorption (DEXA). QU values may reflect bone density or its architecture or even other bone properties besides density. Previous studies have shown a good relationship between the QU measured in the calcaneus and the BMD also measured in the calcaneus by DEXA, but a poorer correlation between the QU measured in the calcaneus and the BMD from other parts. However, despite the low correlation with other parts of the skeleton, QU proved to be a valid tool for predicting osteoporotic fractures, independent from associations with bone density. The calcaneus is the most popular part for many reasons: It is formed by 90% trabecular bone, having higher bone remodeling than the cortical bone due to the surface/volume reason; it is accessible, the lateral surfaces are relatively plain and parallel, reducing repositioning errors [49–57].

The bone ultrasonometry is based in the determination of the parameters related to two properties that are modified after passing through the material: the bone ultrasound attenuation (BUA) and the apparent speed of wave propagation or speed of sound (SOS), which may be evaluated in different regions such as the tibia, metacarpus, calcaneus, and phalanges [58].

The BUA is based on the loss of energy to the environment, and it includes many variables such as absorption, spreading, diffraction, refraction, and conversion [59, 60].

In commercial apparatus, speed measurements are taken by transmission methods, in which a transducer acts as a transmitter and a second one as a receptor, applied to quantitative measures of bone density. In this method, the speed of the ultrasound in the medium can be calculated by dividing the distance by the corresponding time. In our case, quantitative measures were made in the calcaneus by the use of the transmission method and, for calculating speed, the speed in time of flight (TOF) method, with a fixed separation of the transducers (axial method).

$$\text{speed TOF} = \frac{x \, V_w}{x - (\Delta t \, V_w)} \tag{11}$$

In this equation, x is the thickness of the calcaneus including soft tissue, V_w is the speed of the ultrasound in water, and Δt is the difference in time of transit with and without the sample.

5 Materials and Methods

5.1 Ethical Aspects

Sample collection followed the procedures established and approved by the research protocol number 4408/11 of the Research Ethics Committee of the Medicine School in the University of São Paulo, as well as the protocol number 336231 of the Research Ethics Committee of the Federal University of São Carlos.

5.2 Materials

Lumbar vertebrae (L1) were surgically removed from human cadavers, three normal, three osteopenic, and three osteoporotic, 12 h post-mortem, in the Clinical Hospital Morgue, in the city of São Paulo.

5.3 Samples Pre-selection

Before the material was collected, the samples were pre-divided into three groups: normal, osteopenic, and osteoporotic, through QU, using the equipment Achilles InSight (GE Medical Systems Lunar). This pre-selection was conducted as the cadavers arrived at the morgue, before the necropsy technicians started the autopsy. Both feet of each cadaver were placed in the equipment, and triple measurements were taken, calculating the average for each foot and the total average.

5.4 Sample Preparation and Characterization

At the Institute of Orthopedics and Traumatology (IOT) in the Clinical Hospital (CH) in the city of São Paulo, more specifically in the biomechanics laboratory, the samples were extracted axially from vertebral frozen corpses using a trephine drill after the surgical procedure of dissection, and they were standardized in cylindrical format of (10×20) mm. Bone marrow was extracted by a washing process, and the sample was kept humid with serum, frozen at $-20\ °C$.

The samples analyzed by MEV (Zeiss Leo 40—Cambridge, England) at the São Carlos Chemistry Institute had no bone marrow and were dried in a cylindrical shape of (5×10) mm, and covered with carbon and gold (approximately 20 nm).

5.5 DRX and Rietveld Method Parameters

Powder DRX has a potency of 40 kV and 40 mA, copper radiation $k - \alpha$, and wavelength $\lambda = 1.54056$ Å. The samples were analyzed at an angle of $\theta/2\theta$ of 20 at 70°, pace of 0.02° of 10 s each, followed by the use of the software EVA for the search match.

The diffractograms obtained by the DRX were refined by the Rietveld method through the software GSAS using the interface EXPGUI. The program ConvX was used to convert the data into a format accepted by the GSAS and the ICSD in order to obtain the file.cif (database standard diffractogram) of hydroxyapatite. A quartz standard was used to calibrate the software to the equipment, refining unit cells, atomic displacements (Uisos), scale factor, background radiation, shft, GW, GV, GU, LX, S/L, and H/L. In order to refine the diffractograms, the shifted Chebyshev function was used with ten terms to consider the anisotropy of the bone (Fig. 2) and the unit cell (Fig. 3), the atomic displacements (Uisos) (Fig. 3), the scale factor (Fig. 4), the background radiation (Fig. 2), the shft (Fig. 5), the GU (Fig. 5), the GP (Fig. 5), the ptec (Fig. 5), the sfec (Fig. 5), and the LX (Fig. 5) were refined in cycles of 10 (Fig. 6) with Marquardt damping (Fig. 6) equals to 1 and the effects caused by

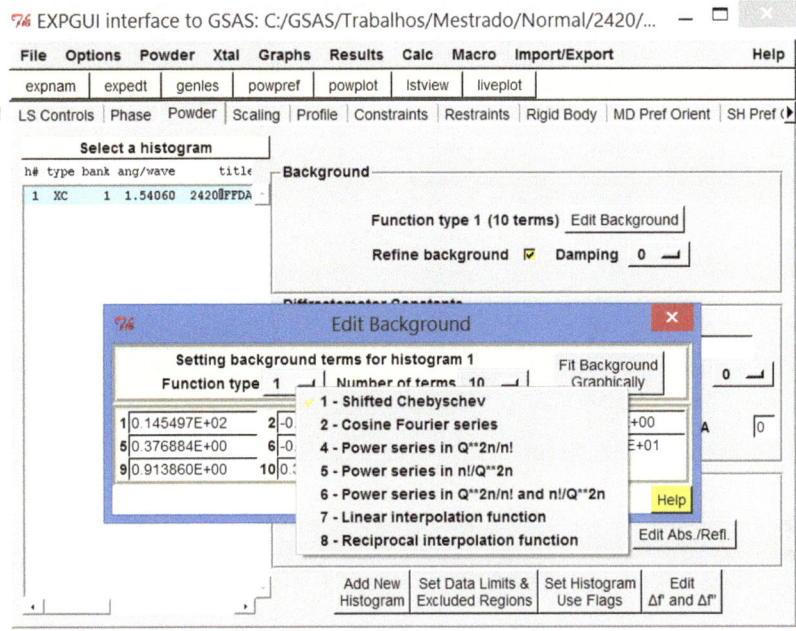

Fig. 2 Use of the function shifted Chebyshev and refinement of background radiation

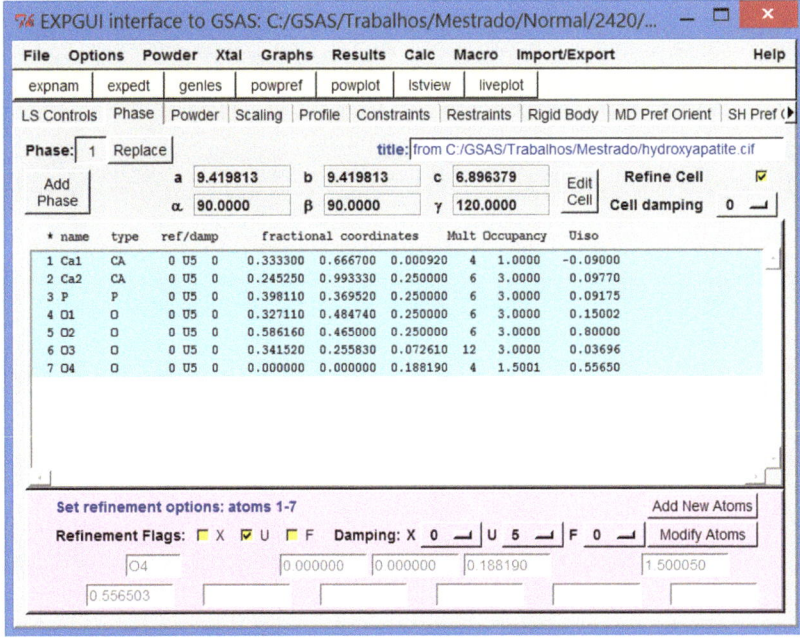

Fig. 3 Refinement of unit cells of the atomic displacements (Uisos)

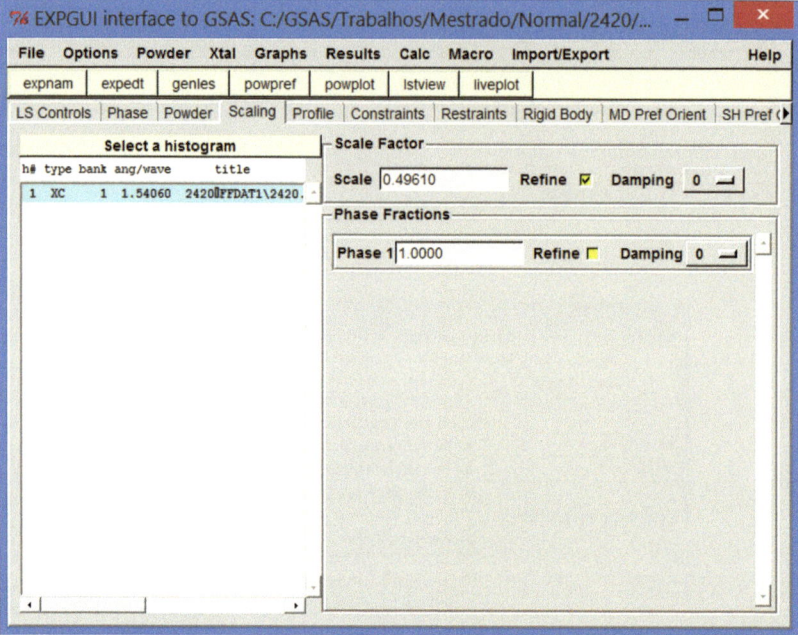

Fig. 4 Scale factor refinement

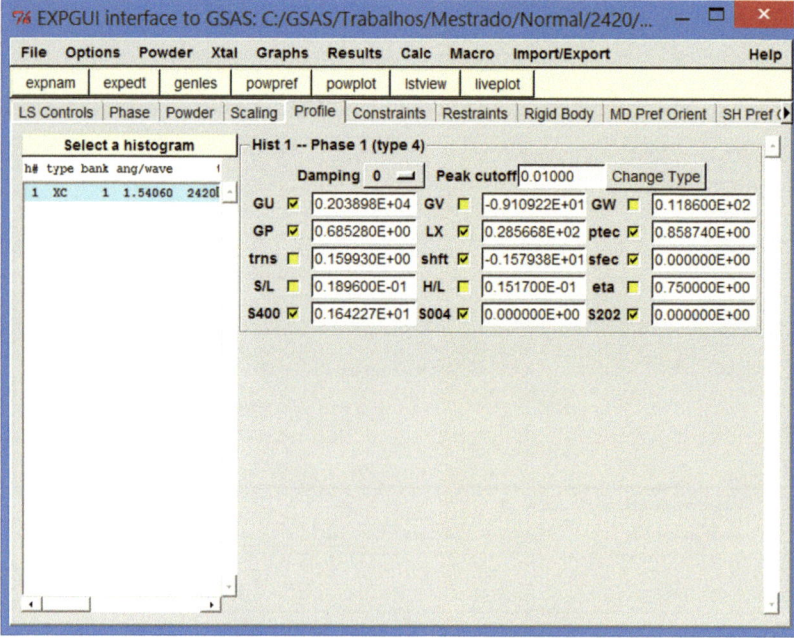

Fig. 5 GU, GP, LX, ptec, shft, sfec, SXXX, SYYY, and SZZZ refinement

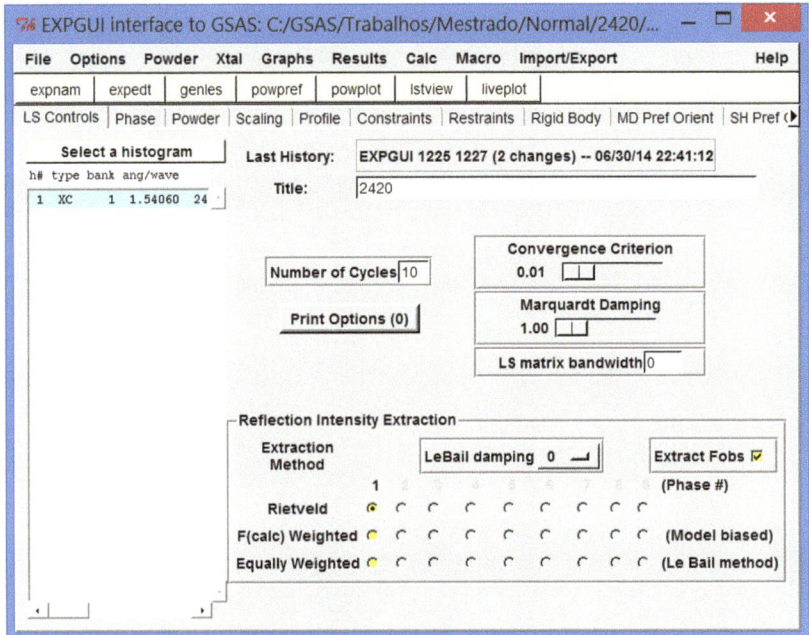

Fig. 6 Marquardt damping use equals 1 in cycles of 10

the preferred orientation were corrected by the use of the March–Dollase model for the planes h k l (2 1 1), (1 2 1), and (1 1 2) (Fig. 7).

6 Results and Discussion

6.1 Ultrasonometry Analysis

Information from the cadaver was collected from the city morgue regarding the age, height, and weight and with the equipment Achilles InSight (GE Medical Systems Lunar), quantitative data in measures of three was obtained, where the average T-score is represented according to Table 4.

These values of T-score represent an indication of the clinical condition of the individuals between 64 and 86 years of age, serving as a reference to separate and correlate the groups, since the T-score relates only to the BMD. However, with further analysis the ultrasonometry prediction was confirmed.

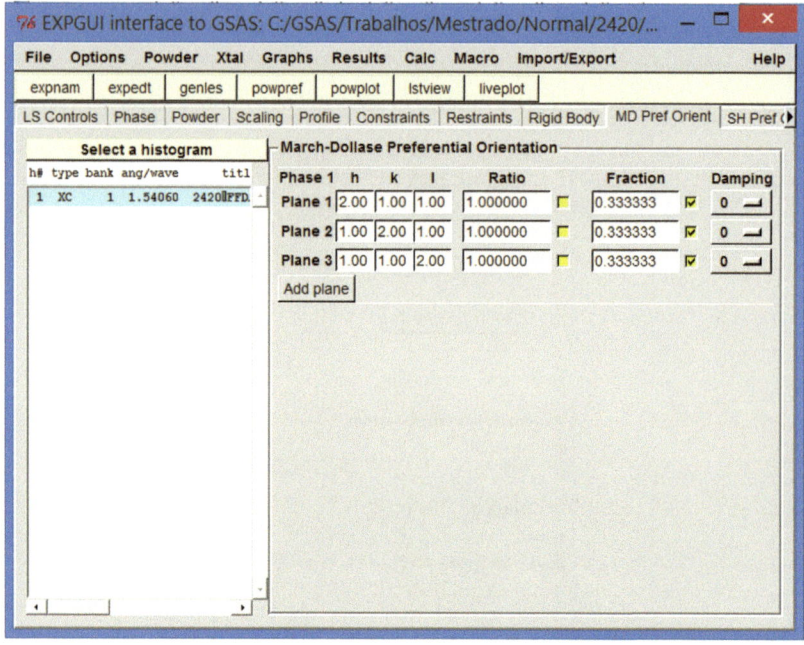

Fig. 7 Correction of the effects caused by preferred orientation using the March–Dollase method in the planes (hkl) equals to (2 1 1), (1 2 1), and (1 1 2)

Table 4 Group classification by the analysis of ultrasonometry of the calcaneus [70]

Group	Age	Height (M)	Weight (Kg)	T-score
Normal 1	67	1.69	46	−0.067
Normal 2	64	1.73	72.2	0.383
Normal 3	81	1.54	45	−0.183
Osteopenic 1	68	1.63	42	−1.383
Osteopenic 2	85	1.75	56	−1.517
Osteopenic 3	82	1.78	70	−1.333
Osteoporotic 1	86	1.61	42	−2.767
Osteoporotic 2	79	1.39	25.4	−2.883
Osteoporotic 3	68	1.7	84	−2.750

This table is reproduced with permissions from Elsevier

6.2 Scanning Electron Microscopy (SEM)

A relation between cracks, fractures, trabecular density with or without plaque formation, connectivity, and the sizes of the pores in the three groups was observed. One normal, one osteopenic, and one osteoporotic were selected. For normal bones

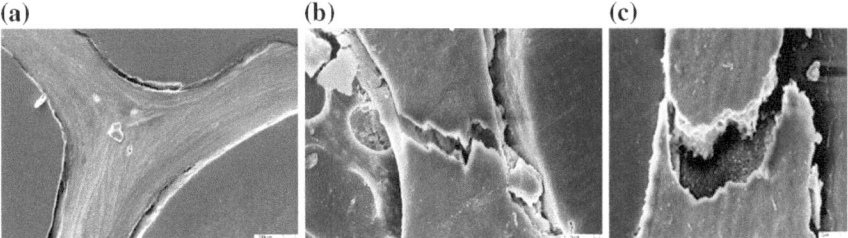

Fig. 8 Scanning electron microscopy of the trabecular bone in human vertebral fixed by epoxy resin and polished: **a** Normal (700 times); **b** Osteopenic (3000 times); **c** Osteoporotic (3000 times)

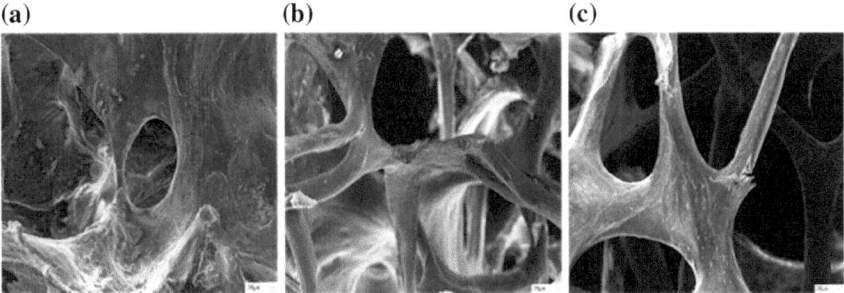

Fig. 9 Scanning electron microscopy of human vertebrae trabecular bone (300 times): **a** Normal; **b** Osteopenic; **c** Osteoporotic

(Figs. 8a and 9a), the microarchitecture was formed by rounded regular pores, with higher connectivity, trabecular number, and the presence of thick and well-organized plaques.

In osteopenic bones (Figs. 8b and 9b), the existence of fragile fractures caused by axial or shear loads, the thinning of the trabeculae and the inexistence of plaque connections were observed. Finally, in bones considered osteoporotic (Figs. 8c and 9c), cracks and fractures with higher irregularity, low connectivity, thinning of the trabeculae and microarchitecture deterioration were observed.

6.3 Dispersive Energy Spectroscopy (DES)

The values found for the phosphorous and calcium proportions are represented in Table 5 and also in Fig. 10.

The Ca/P proportion is higher for normal bones, and the higher the Ca/P proportion, until a certain critical point, the lower the tendency for rupture to take place as seen by Fountos et al. [61] and Kourkoumelis et al. [62] due to probable calcium ions replacements [63], lowering its quantity and increasing the disorganization of

Table 5 Calcium and phosphorous proportions in the 3 stages of the bone

	Normal	Osteopenic	Osteoporotic
Sample 1	1.91	1.80	1.68
Sample 2	2.13	1.80	1.73
Sample 3	2.03	1.82	1.78
Average	2.02	1.81	1.73
Standard deviation	0.11	0.01	0.05

Fig. 10 Calcium and phosphorous proportions in each group

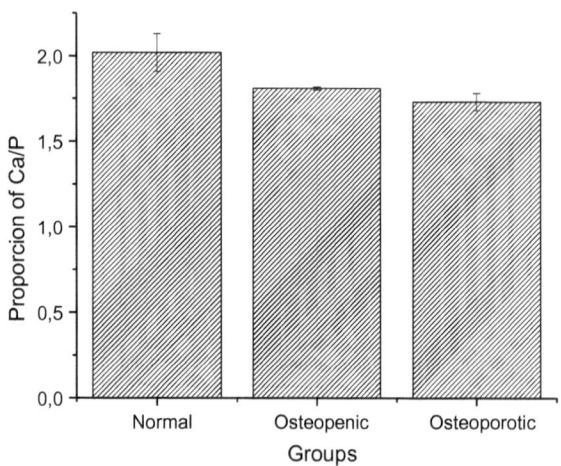

the unit cells in osteopenic and osteoporotic bones. This loss makes the boneless rigid [64, 65], which was confirmed by our microhardness tests, and the disorganization probably caused an increase in microdeformations [61–65].

6.4 Microhardness (HK)

The values analyzed for Knoop microhardness (HK) are presented in Tables 6, 7, and 8 and also in Fig. 11.

Normal bone samples had an average result of 30.27 and standard deviation of 0.36, resulting in 30.27 ± 0.36 HK.

Table 6 Normal sample measurements

Sample	Measure 1	Measure 2	Measure 3	Average
1	30.7	30.7	30.6	30.67
2	28.2	28.8	32.9	29.97
3	30.1	33.9	26.5	30.17

Table 7 Osteopenic samples measurements

Sample	Measure 1	Measure 2	Measure 3	Average
1	26.8	27.9	24	26.23
2	27.3	26.9	23.3	25.83
3	28.9	26.6	24.4	26.63

Table 8 Osteoporotic samples measurements

Samples	Measure 1	Measure 2	Measure 3	Average
1	22	23.2	20.4	21.87
2	23.4	22	19.4	21.6
3	20.3	20.9	19.4	20.2

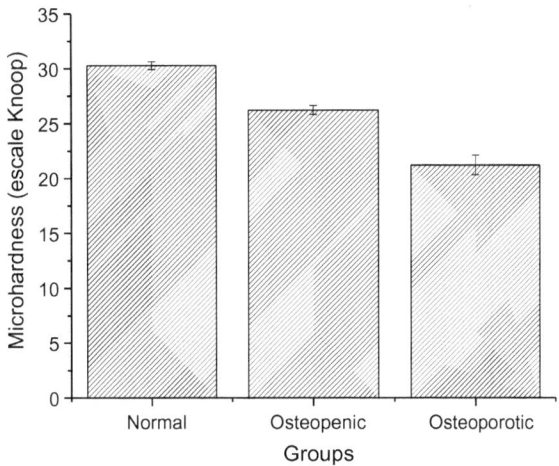

Fig. 11 Microhardness in the Knoop scale for each group

The osteopenic samples in number of three resulted in an average of 26.23 with a standard deviation of 0.440, resulting in 26.23 ± 0.40 HK.

The osteoporotic samples resulted in an average of 21.22 with a standard deviation of 0.89 = 21.22 ± 0.89 HK.

The averages and their respective standard deviations are presented in Fig. 11.

It was observed that osteoporotic bones correspond to lower values of microhardness when compared to the normal ones, corroborating the studies of Li et al. [66], Moran et al. [67], and Boivin et al. [68]. The higher the microhardness value, the bigger the resistance to deformation, according to Ferrante [69], which agrees with the reality for osteoporotic bones, that suffer fractures more easily [1, 66–69].

6.5   X-Ray Diffractometry (XRD) and the Rietveld Method

The X-ray diffractometry spectrum of normal, osteopenic, and osteoporotic bones is represented in Fig. 12. It is possible to observe the hydroxyapatite peaks, with no other phase present and hexagonal group P6$_3$/m, typical of hydroxyapatite crystals.

Even though the visual identification is not possible, it is known that normal peaks are larger than osteopenic and osteoporotic ones, once the crystallite sizes are bigger, as a result of Eqs. (3) and (8) [48, 45].

The Bravais solid was produced from the hexagonal three-dimensional structure of the hydroxyapatite through the software Crystal Maker, as shown in Fig. 13. Apparently, the O–H groups are located in the corners of the crystal unit cell, while

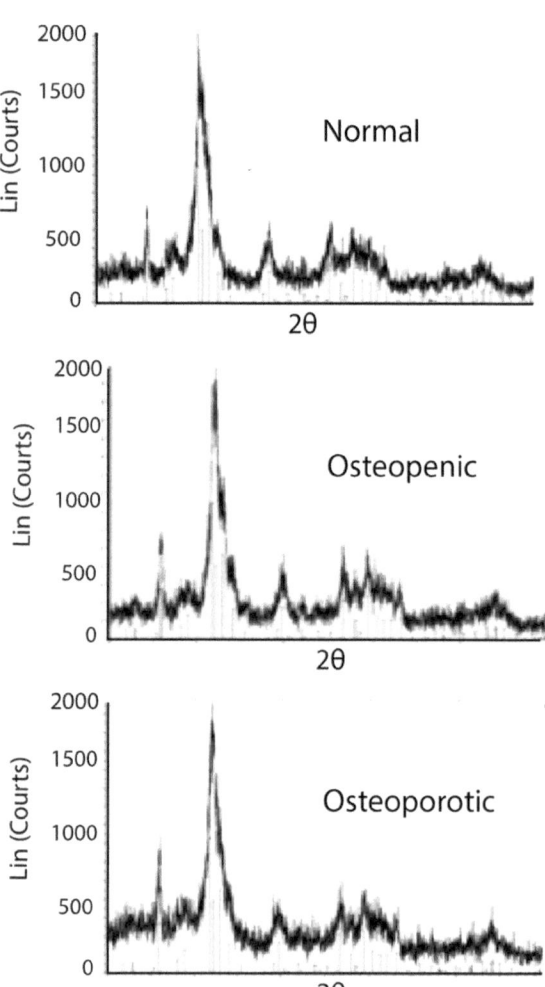

Fig. 12 X-ray diffractometry for each type of bone, from top to bottom; normal, osteopenic, and osteoporotic

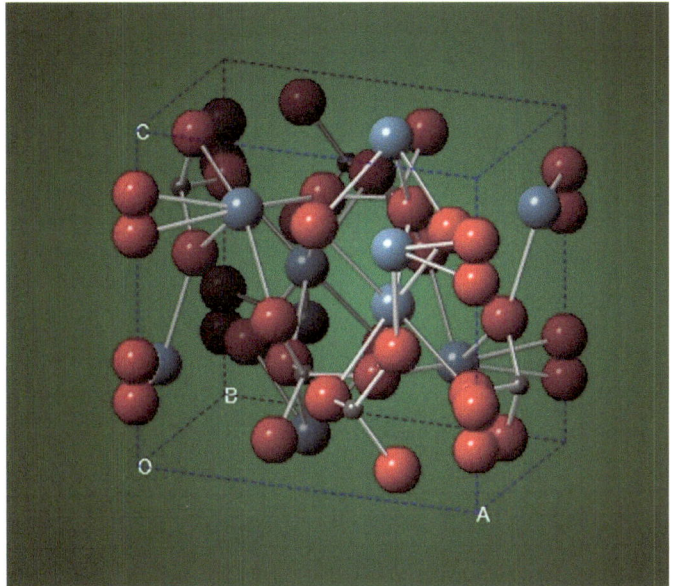

Fig. 13 Crystalline structure of the bone, where O = red, Ca = blue, and P = gray

the Ca, O, and P atoms are located inside the volume, where $a = b \neq c$, with angles $\alpha = \beta = 90°$ and $\gamma = 120°$.

Through the Rietveld method, it was possible to find the sizes of the crystallites, as shown in Table 9 and in Fig. 14.

Microdeformations were also analyzed, as shown in Table 10 and Fig. 15.

Normal bones were found with crystallite sizes bigger than the osteopenic and osteoporotic ones, and the relative variation of the net parameter or interplanar distance caused by the defects (microdeformation), different from the crystallite size, has a lower average for normal bones, when compared to osteopenic and osteoporotic ones. A possible explanation is the increase in microdeformation (disorder) takes place because for the osteoporotic bones there will be a lower exchange of random ions; for the same way, the unbalancing of bone remodeling takes place due to fractures inherent to the aging process, and there is also the addition of different ions to the unit cell. Due to the increase of the disorder in the crystallite and knowing from the literature that a preference for ions with a smaller atomic ratio, it is understood that there will be a decrease in crystallite size. Once it decreases, there is the

Table 9 Crystallite size for each group

Group	Crystallite size (Å)	Standard deviation
Normal	669.34	27.70
Osteopenic	467.38	65.99
Osteoporotic	213.01	86.00

Fig. 14 Crystallite size for the different groups

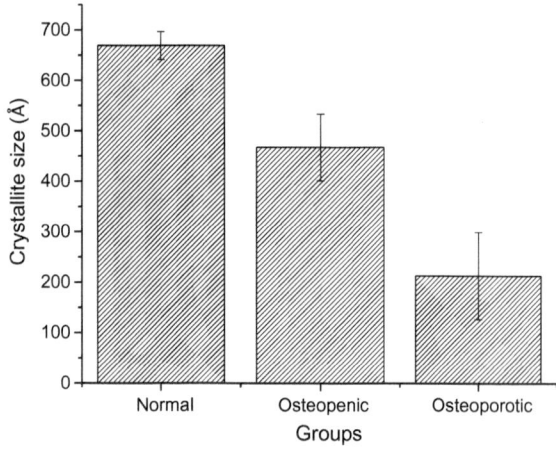

Table 10 Microdeformation for each group

Group	Microdeformation	Standard Deviation
Normal	5.41	1.58
Osteopenic	11.48	1.57
Osteoporotic	16.88	1.42

Fig. 15 Microdeformation for each group

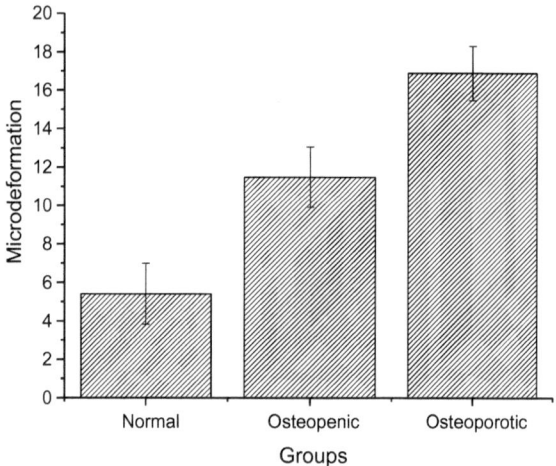

possibility of crystallites becoming more compact, which will, once more, increase microdeformation, that also measures the internal tension or the residual tension in the crystallite.

7 Conclusion

The bone microarchitecture of dried trabecular bones vertebrae could be evaluated by the methods: scanning electron microscopy, EDS, microhardness and X-ray diffractometry with the Rietveld refinement method. The microstructural characterization of hydroxyapatite crystals in dried trabecular bones allowed for the identification of the three types of bones (normal, osteopenic, and osteoporotic) and to complement the evaluation and detection of osteoporosis with emphasis on bone quality.

References

1. N.I.H. (2013) National Institute of Arthritis and Muscoloskeletal and Skin Deseases. Consensus statement—osteoporosis prevention, diagnosis, and therapy
2. Bouxsein M (2003) Bone quality: where do we go from here? Osteoporosis Int 14(5):118–127. 9 Jan 2003. ISSN 0937-941X. Available at: http://dx.doi.org/10.1007/s00198-003-1489-x
3. Deborah M, Olof J, Hans W (1996) Meta-analysis of how well measures of bone mineral density predict occurrence of osteoporotic fractures. BMJ 312
4. Kleerekoper M, Balena R (1991) Fluorides and Osteoporosis. Ann Rev Nutr 11(1):309–324. 01 July 1991. ISSN 0199-9885. Available at: http://dx.doi.org/10.1146/annurev.nu.11.070191. 001521. Consulted on: 16 July 1991
5. Miller P (2006) Guidelines for the diagnosis of osteoporosis: T-scores vs fractures. Rev Endocr Metab Disord 7(1–2):75–89. 01 June 2006. ISSN 1389-9155. Available at: http://dx.doi.org/10.1007/s11154-006-9006-0
6. Qu Y et al (2005) The effect of raloxifene therapy on the risk of new clinical vertebral fractures at three and six months: a secondary analysis of the MORE trial. Curr Med Res Opin 21(12):1955–1959. 01 Dec 2005. ISSN 0300-7995. Available at: http://informahealthcare.com/doi/abs/10.1185/030079905X75032. Consulted on: 16 July 2013
7. Genant HK, Jiang Y (2006) Advanced imaging assessment of bone quality. Ann NY Acad Sci 1068(1):410–428. ISSN 1749-6632. Available at: http://dx.doi.org/10.1196/annals.1346.038
8. Gourion-Arsiquaud S et al (2009) Use of FTIR spectroscopic imaging to identify parameters associated with fragility fracture. J Bone Miner Res 24(9):1565–1571. ISSN 1523-4681. Available at: http://dx.doi.org/10.1359/jbmr.090414
9. Licata A (2009) Bone density vs bone quality: what's a clinician to do? Clevel Clin J Med 76(6):331–336. Available at: http://www.ccjm.org/content/76/6/331.abstract
10. N.O.F. (2013) National Osteoporosis Foundation—Physicians' guide to prevention and treatment of osteoporosis
11. Parfitt AM (2002) Targeted and nontargeted bone remodeling: relationship to basic multicellular unit origination and progression. Bone 30(1):5–7. ISSN 8756-3282. Available at: http://www.sciencedirect.com/science/article/pii/S8756328201006421
12. Qiu S et al (2005) The morphological association between microcracks and osteocyte lacunae in human cortical bone. Bone 37(1):10–15. ISSN 8756-3282. Available at: http://www.sciencedirect.com/science/article/pii/S8756328205000086
13. Van Der Linden JC et al (2001) Mechanical consequences of bone loss in cancellous bone. J Bone Miner Res 16(3):457–465. ISSN 1523-4681. Available at: http://dx.doi.org/10.1359/jbmr.2001.16.3.457
14. Van Der Meulen MCH, Jepsen KJ, Mikić B (2001) Understanding bone strength: size isn't everything. Bone 29(2):101–104. ISSN 8756-3282. Available at: http://www.sciencedirect.com/science/article/pii/S8756328201004914

15. Viguet-Carrin S, Garnero P, Delmas PD (2006) The role of collagen in bone strength. Osteo-porosis Int 17(3):319–336. 01 Mar 2006. ISSN 0937-941X. Available at: http://dx.doi.org/10.1007/s00198-005-2035-9

16. W.H.O. (2001) World Health Organization Study Group

17. Zebaze RMD et al (2005) Femoral neck shape and the spatial distribution of its mineral mass varies with its size: clinical and biomechanical implications. Bone 37(2):243–252. ISSN 8756-3282. Available at: http://www.sciencedirect.com/science/article/pii/S8756328205001006

18. Allen MR, Burr DB (2011) Bisphosphonate effects on bone turnover, microdamage, and mechanical properties: what we think we know and what we know that we don't know. Bone 49(1):56–65. ISSN 8756-3282. Available at: http://www.sciencedirect.com/science/article/pii/S8756328210018636

19. Boskey AL, Marks SC. Mineral and matrix alterations in the bones of incisors-absent (ia/ia) osteopetrotic rats. Calcified Tissue International 37(3):287–292. 01 May 1985. ISSN 0171-967X. Available at: http://dx.doi.org/10.1007/BF02554876

20. Cundy T, Reid IR (2012) Paget's disease of bone. Clin Biochem 45(1–2):43–48. ISSN 0009-9120. Available at: http://www.sciencedirect.com/science/article/pii/S0009912011026890

21. Dilworth L et al (2008 Bone and faecal minerals and scanning electron microscopic assess-ments of femur in rats fed phytic acid extract from sweet potato (Ipomoea batatas). BioMetals 21(2):133–141. 01 Apr 2008. ISSN 0966-0844. Available at: http://dx.doi.org/10.1007/s10534-007-9101-z

22. Fratzl P et al (1996) Effects of sodium fluoride and alendronate on the bone mineral in minipigs: A small-angle X-ray scattering and backscattered electron imaging study. J Bone Miner Res 11(2):248–253. ISSN 1523-4681. Available at: http://dx.doi.org/10.1002/jbmr.5650110214

23. Leventouri T (2006) Synthetic and biological hydroxyapatites: crystal structure questions. Bio-materials 27(18):3339–3342. ISSN 0142-9612. Available at: http://www.sciencedirect.com/science/article/pii/S0142961206001761

24. Marcus R et al (2009) Fundamentals of osteoporosis, 1ª edn. Elsevier, UK. eBook ISBN: 9780123751089

25. Noor A et al (2011) Assessment of microarchitecture and crystal structure of hydroxyapatite in osteoporosis. Univ Med 31(1)

26. Ou-Yang H et al (2001) Infrared microscopic imaging of bone: spatial distribution of CO_3^{2-}. J Bone Miner Res 16(5):893–900. ISSN 1523-4681. Available at: http://dx.doi.org/10.1359/jbmr.2001.16.5.893

27. Saito M, Marumo K (2010) Collagen cross-links as a determinant of bone quality: a possi-ble explanation for bone fragility in aging, osteoporosis, and diabetes mellitus. Osteoporos Int 21(2):195–214. 01 Feb 2010. ISSN 0937-941X. Available at: http://dx.doi.org/10.1007/s00198-009-1066-z

28. Shen Y et al (2009) Postmenopausal women with osteoarthritis and osteoporosis show different ultrastructural characteristics of trabecular bone of the femoral head. BMC Musculoskelet Disord 10(1):35. ISSN 1471-2474. Available at: http://www.biomedcentral.com/1471-2474/10/35

29. Ginebra M-P et al (1999) Modeling of the hydrolysis of α-tricalcium phosphate. J Am Ceram Soc 82(10):2808–2812. ISSN 1551-2916. Available at: http://dx.doi.org/10.1111/j.1151-2916.1999.tb02160.x

30. Klein CPAT et al (1990) Studies of the solubility of different calcium phosphate ceramic particles in vitro. Biomaterials 11(7):509–512. ISSN 0142-9612. Available at: http://www.sciencedirect.com/science/article/pii/014296129090067Z

31. Barrere F et al (2002) Influence of ionic strength and carbonate on the Ca-P coating formation from SBF × 5 solution. Biomaterials 23(9):1921–1930. ISSN 0142-9612. Available at: http://www.sciencedirect.com/science/article/pii/S0142961201003180

32. Raynaud S et al (2002) Calcium phosphate apatite with variable Ca/P atomic ratio I. Synthesis, characterisation and thermal stability of powders. Biomaterials 23(4):1065–1072. ISSN 0142-9612. Available at: http://www.sciencedirect.com/science/article/pii/S0142961201002186

33. Dekker RJ et al (2005) Bone tissue engineering on amorphous carbonated apatite and crystalline octacalcium phosphate-coated titanium discs. Biomaterials 26(25):5231–5239. ISSN 0142-9612. Available at: http://www.sciencedirect.com/science/article/pii/S0142961205001079

34. Hill RJ, Howard CJ (1987) Quantitative phase analysis from neutron powder diffraction data using the Rietveld method. J Appl Crystallogr 20(6):467–474. ISSN 0021-8898. Available at: http://dx.doi.org/10.1107/S0021889887086199

35. Langford JI, Louer D, Scardi P (2000) Effect of a crystallite size distribution on X-ray diffraction line profiles and whole-powder-pattern fitting. J Appl Crystallogr 33(3):964–974. ISSN 0021-8898. Available at: http://dx.doi.org/10.1107/S002188980000460X

36. Pecharsky VK (2009) Fundamentals of powder diffraction and structural characterization of materials, 2nd edn. Springer, Berlin, 744 pp. ISBN 978-0-387-09579-0

37. Rietveld H (1967) Line profiles of neutron powder-diffraction peaks for structure refinement. Acta Crystallogr 22(1):151–152. 01 Oct 1967. ISSN 0365-110X. Available at: http://dx.doi.org/10.1107/S0365110X67000234

38. Rietveld HM (1969) A profile refinement method for nuclear and magnetic structures. J Appl Crystallogr 2(2):65–71. 06 Feb 1969. ISSN 0021-8898. Available at: http://dx.doi.org/10.1107/S0021889869006558

39. Young RA, Mackie PE, Von Dreele RB (1977) Application of the pattern-fitting structure-refinement method of X-ray powder diffractometer patterns. J Appl Crystallogr 10(4):262–269. 08 Jan 1977. ISSN 0021-8898. Available at: http://dx.doi.org/10.1107/S0021889877013466

40. Enzo S et al (1988) A profile-fitting procedure for analysis of broadened X-ray diffraction peaks. I. Methodology. J Appl Crystallogr 21(5):536–542. 10 Jan 1988. ISSN 0021-8898. Available at: http://dx.doi.org/10.1107/S0021889888006612

41. Louer D, Langford JI (1988) Peak shape and resolution in conventional diffractometry with monochromatic X-rays. J Appl Crystallogr 21(5):430–437. 10 Jan 1988. ISSN 0021-8898. Available at: http://dx.doi.org/10.1107/S002188988800411X

42. Madsen IC, Hill RJ (1988) Effect of divergence and receiving slit dimensions on peak profile parameters in Rietveld analysis of X-ray diffractometer data. J Appl Crystallogr 21(5):398–405. 10 Jan 1988. ISSN 0021-8898. Available at: http://dx.doi.org/10.1107/S0021889888003474

43. Press WH, Teukolsky SA (2007) Numerical recipes in C++, 3rd edn. The Art of Scientific Programming Cambridge University Press, 1256 pp. ISBN: 978-0521880688

44. Teixeira EM (2013) Particle size refinement and microstrain polycrystalline samples by X-ray diffraction profiles using kinetic and dynamic theories. 46 (Physics Bachelor). Department of Physics, Federal University of Ceará, Fortaleza

45. Caglioti G, Paoletti A, Ricci FP (1958) Choice of collimators for a crystal spectrometer for neutron diffraction. Nucl Instrum 3(4):223–228. ISSN 0369-643X. Available at: http://www.sciencedirect.com/science/article/pii/0369643X5890029X

46. Dehlinger U, Kochendörfer A (1939) Linienverbreiterung von verformten Metallen. Zeitschrift Für Kristallographie. Cryst Mater 101(1–6). https://doi.org/10.1524/zkri.1939.101.1.134

47. Azàroff LV (1968) Elements of x-ray crystallography. McGraw-Hill Book Company, New York

48. Williamson GK, Hall WH (1953) X-ray line broadening from filed aluminium and wolfram. Acta Metallurgica 1(1):22–31. ISSN 0001-6160. Available at: http://www.sciencedirect.com/science/article/pii/0001616053900066

49. Aguado F et al (1996) Behavior of bone mass measurements—dual energy X-ray absorptiometry total body bone mineral content, ultrasound bone velocity, and computed metacarpal radiogrammetry, with age, gonadal status, and weight in healthy women. Investigative Radiol 31(4):218-222. ISSN 0020-9996. Available at: Go to ISI: WOS:A1996UE33000006

50. Faulkner KG et al (1994) Quantitative ultrasound of the heel—correlation with densitometric measurements at different skeletal sites. Osteoporos Int 4(1):42–47. ISSN 0937-941X. Available at: Go to ISI: WOS:A1994MT96000008

51. Hans D et al (1996) Ultrasonographic heel measurements to predict hip fracture in elderly women: the EPIDOS prospective study. Lancet 348(9026):511–514. 24 Aug 1996. ISSN 0140-6736. Available at: Go to ISI : WOS:A1996VD42700011

52. Kwok T et al (2012) Predictive values of calcaneal quantitative ultrasound and dual energy X-ray absorptiometry for non-vertebral fracture in older men: results from the MrOS study (Hong Kong). Osteoporos Int 23(3):1001–1006. ISSN 0937-941X. Available at: Go to ISI: WOS:000300251200023

53. Ross P et al (1995) Predicting vertebral deformity using bone densitometry at various skeletal sites and calcaneus ultrasound. Bone 16(3):325–332. ISSN 8756-3282. Available at: Go to ISI: WOS:A1995RB63800007

54. Salamone LM et al (1994) Comparison of broad-band ultrasound attenuation to single x-ray absorptiometry measurements at the calcaneus in postmenopausal women. Calcif Tissue Int 54(2):87–90. ISSN 0171-967X. Available at: Go to ISI: WOS:A1994MT40200002

55. Turner CH et al (1995) Calcaneal ultrasonic measurements discriminate hip fracture independently of bone mass. Osteoporos Int 5(2):130–135. ISSN 0937-941X. Available at: Go to ISI: WOS:A1995QM68000010

56. Waud CE, Lew R, Baran DT (1992) The relationship between ultrasound and densitometric measurements of bone mass at the calcaneus in women. Calcif Tissue Int 51(6):415–418. ISSN 0171-967X. Available at: Go to ISI: WOS:A1992JY74200004

57. Yeap SS et al (1998) The relationship between bone mineral density and ultrasound in postmenopausal and osteoporotic women. Osteoporos Int 8(2):141–146. ISSN 0937-941X. Available at: Go to ISI: WOS:000078768900008

58. Cortet B et al (2004) Does quantitative ultrasound of bone reflect more bone mineral density than bone microarchitecture? Calcif Tissue Int 74(1):60–67. 01 Jan 2004. ISSN 0171-967X. Available at: http://dx.doi.org/10.1007/s00223-002-2113-3

59. Webb S (2012) The physics of medical imaging, 2 edn. CRC Press is an imprint of Taylor & Francis Group, Boca Raton. ISBN-13: 978-1-4665-6895-2 (eBook—PDF)

60. Njeh CF et al (1999) Quantitative ultrasound assessment of osteoporosis and bone status. London Martin Dunitz. ISBN: 1-85317-679-6, 420 pp. doi: https://doi.org/10.1016/S0301-5629(00)00280-5

61. Fountos G et al (1998) The effects of inflammation-mediated osteoporosis (IMO) on the skeletal Ca/P ratio and on the structure of rabbit bone and skin collagen. Appl Radiat Isotopes 49(5–6):657–659. ISSN 0969-8043. Available at: http://www.biomedsearch.com/nih/effects-inflammation-mediated-osteoporosis-IMO/9569570.html

62. Kourkoumelis N, Balatsoukas I, Tzaphlidou M (2012) Ca/P concentration ratio at different sites of normal and osteoporotic rabbit bones evaluated by Auger and energy dispersive X-ray spectroscopy. J Biol Phys 38(2):279–291. ISSN 0092-0606. Available at: http://dx.doi.org/10.1007/s10867-011-9247-3

63. Costa ACFM et al (2009) Hydroxyapatite: collection, characterization and applications. Eletr J Mater Proces 4(3):10

64. Garnet LP, Hiatt JL (2003) Treaty of histology. 2. Rio de Janeiro: Guanabara Koogan. ISBN: 8527708132

65. Junqueira LC, Carneiro J (2008) In: de Janeiro R (ed) Basic histology, 11th edn. Guanabara Koogan. ISBN: 9788527731812

66. Li B, Aspden RM (1997) Mechanical and material properties of the subchondral bone plate from the femoral head of patients with osteoarthritis or osteoporosis. Ann Rheum Dis 56(4):247–254, 1997. Available at: http://ard.bmj.com/content/56/4/247.abstract

67. Moran P et al (2007) Preliminary work on the development of a novel detection method for osteoporosis. J Mater Sci Mater Med 18(6):969–974. 01 Jun 2007. ISSN 0957-4530. Available at: http://dx.doi.org/10.1007/s10856-006-0037-6

68. Boivin G et al (2008) The role of mineralization and organic matrix in the microhardness of bone tissue from controls and osteoporotic patients. Bone 43(3):532–538. ISSN 8756-3282. Available at: http://www.sciencedirect.com/science/article/pii/S8756328208002834

69. Ferrante M (1996) Material selection. EDUFSCar, São Carlos

70. Rollo JMDA et al (2015) Assessment of trabecular bones microarchitectures and crystal structure of hydroxyapatite in bone osteoporosis with application of the rietveld method. Procedia Eng 110:8–14. ISSN 1877-7058. https://doi.org/10.1016/j.proeng.2015.07.003

Bone Remodelling Algorithms

Meshless, Bone Remodelling and Bone Regeneration Modelling

M. C. Marques, Jorge Belinha, R. Natal Jorge and A. F. Oliveira

Abstract In this chapter, it is presented an extensive bibliographic survey about meshless methods, bone remodelling and bone regeneration modelling. Here, the regeneration and remodelling processes are shown with detail and, in addition, it is presented a description of the mathematical models approaching both regeneration and remodelling processes. Three different classifications of models are presented, the mechanoregulated models, the bioregulated models and the mechanobioregulated models. The literature shows that the combination of remodelling models with meshless techniques allows to numerically achieve more realistic trabecular distributions. Thus, in this chapter, an introduction to meshless methods is presented, with a special focus on radial point interpolation meshless methods, such as the radial point interpolation method (RPIM) and the natural neighbour RPIM (NNRPIM).

M. C. Marques
Institute of Mechanical Engineering and Industrial Management (INEGI), Rua Dr. Roberto Frias, s/n, 4600-465 Porto, Portugal
e-mail: mcmarques@inegi.up.pt

J. Belinha (✉)
Department of Mechanical Engineering, School of Engineering, Polytechnic of Porto (ISEP), Rua Dr. António Bernardino de Almeida, 431, 4200-072 Porto, Portugal
e-mail: job@isep.ipp.pt

R. Natal Jorge
Faculty of Engineering, Mechanical Engineering Department, University of Porto (FEUP), Rua Dr. Roberto Frias, s/n, 4600-465 Porto, Portugal
e-mail: rnatal@fe.up.pt

A. F. Oliveira
Medical Teaching Department—CHP/HSA, Instituto de Ciencias Biomédicas Abel Salazar (ICBAS), Rua de Jorge Viterbo Ferreira, 228, 4050-313 Porto, Portugal
e-mail: afoliveira@netc.pt

© Springer Nature Switzerland AG 2020
J. Belinha et al. (eds.), *The Computational Mechanics of Bone Tissue*,
Lecture Notes in Computational Vision and Biomechanics 35,
https://doi.org/10.1007/978-3-030-37541-6_3

1 Introduction

Bone is a structure mainly defined by bone matrix and by bone cells, being these bone cells responsible to produce the bone matrix, in which they became entrapped. The bone cells that create the bone matrix are the same that reabsorb it to allow the replacement of old bone matrix by a newer one. These bone cells also act in the bone regeneration and remodelling processes. All the bones in the human body form the skeletal system, which possesses several vital functions, such as: support, protection, movement, storage and blood cell production [1–4].

Bone possesses the intrinsic capacity for regeneration and remodelling, being bone remodelling the biological process whereby living bone tissue renews itself in the course of life. Regeneration, besides being the repair process in response to injury, is as well a process that starts with the skeletal development in early life and continues in the form of bone remodelling throughout adult life [5].

One of the most significant activities performed by engineers and scientists is to model natural phenomena in order to study and simulate them. Much of the physical phenomena studied, particularly those related to transient continuum mechanics theory (involving both time and space changes), can be formulated in terms of algebraic, ordinary or partial differential and/or integral equations. These conceptual and mathematical models simulate physical events, whether they are biological, chemical, geological, or mechanical, they are based on scientific laws of physics. Thus, it is possible to formulate constitutive equations and laws ruling distinct phenomena, such as heat transfer, stress/strain relations, solid deformations, fluid flow, etc. [6].

Depending on the importance of the constitutive equations in the studied phenomenon, they may or not be considered, and their combination results in the differential equations that govern the phenomenon. There are many phenomena associated with different fields, that can be obtained using the same mathematical model and consequently the same differential equations [6]. Most real engineering problems cases are characterized by very complex equations modelled on geometrically complex regions and defined by equations with nonlinear nature, making it virtually impossible to achieve an analytical exact solution. For these cases, in order to obtain the solutions, numerical approximation methods are preferable. The development of computers in recent decades allowed to use these methods more efficiently and so, to obtain the solution of many problems heretofore unsolvable. Additionally, computer progress as allowed the development of new formulations and new algorithms for solving many other complex problems [6]. With massive use of computers, it has been an increase in the development and use of numerical methods and simulates in silico several biomechanical scenarios. The finite difference method, the finite element method, meshless methods and their variants, are the most used numerical methods in the analysis of practical engineering problems. These methods are created on the idea that the complete system could be analysed as the integrated sum of its different parts.

1.1 Bone Regeneration

Bone regeneration is a well-arranged series of biological events where the damaged bone, with deficiencies or discontinuities, is regenerated to a newly formed bone, rescuing its original biomechanical properties. This process is affected by a gathering of genetic, environmental, mechanical, cellular and endocrine factors, which in the end allows to obtain a newly formed bone properties possessing almost indistinguishable properties from the properties of the adjacent old bone. This process results in a bone geometry, in the healing site, that usually is the same to its initial shape and with the peculiarity of the non-formation of a scar tissue [7, 8]. The most common case of bone regeneration is the fracture healing that is divided in three main stages, the inflammatory phase, bone repair and bone remodelling, that are represented in Fig. 1. During these three stages, two types of ossification occur, intramembranous and the endochondral ossification [9, 10].

The inflammatory phase, the first stage, starts immediately after the injury, that is characterized by the haemorrhage, occurring from the damaged blood vessels and leading to the supply suspension of nutrient and oxygen to the bone cells. This first inflammatory reaction leads to the fracture immobilization in a direct and indirect way. Indirectly, since the pain caused leads the individual to protect the injury, and directly by swelling hydrostatically in the fracture zone, that keeps it from moving. Just after the start of the haemorrhage, the blood began to clot and the platelets (trapped in the clot) become activated and start to release growth factors, which regulate the processes of bone healing. The supply interruption of nutrients and oxygen to the bone cells leads to its dead. Thus, there occurs a necrotic process that starts an inflammatory response, which is characterized by the migration of macrophages, phagocytes and leucocytes to the wound site aiming to remove the necrotic tissue.

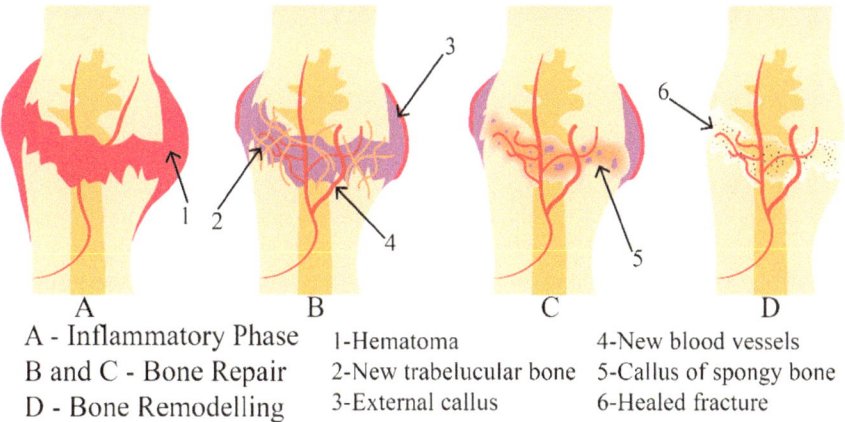

A - Inflammatory Phase
B and C - Bone Repair
D - Bone Remodelling

1-Hematoma
2-New trabelucular bone
3-External callus

4-New blood vessels
5-Callus of spongy bone
6-Healed fracture

Fig. 1 Bone regeneration three main stages

The reparative phase, the second stage, occurs before the inflammatory phase subsides, and lasts for several weeks. In this phase, it is developed the reparative callus in the fracture site, which in the last phase—the remodelling phase—will be replaced by bone. The composition of repair tissue and rate of repair may differ depending on where the fracture occurs in the bone, the extent of soft tissue damage, and mechanical stability of the fracture site [11].

In this phase, the fibroblasts, mesenchymal stem cells (MSC's) and endothelial cells, from the periosteum and from the soft tissues surrounding the bone, migrate into the healing site. This migration is regulated by growth factors released by the activated platelets and by the above-mentioned cells. Under the effect of different stimuli, MSC's differentiate into different cells, as well as osteoblasts, chondrocytes and fibroblasts.

In the case that the MSC's differentiate into osteoblast, they start to synthesize bone directly on a pre-existing surface without the mediation of the cartilage phase, occurring the intramembranous ossification [10]. With this apposition of new bone matrix on a solid surface, it is observed the creation of a bone-forming surface that is often referred to as the hard callus [12–14].

In the case that the MSC's differentiate to chondrocytes, it occurs the endochondral ossification, that usually occurs in the middle of the fracture area. With the proliferation of chondrocytes, it follows the growth of cartilage tissue, and with the maturation of chondrocytes is followed by the calcification of cartilage. Chondrocytes are also responsible by the apoptosis and blood vessels grow into the cavities that were initially occupied by the chondrocytes. The mineralized extracellular matrix of the cartilage tissue acts as a scaffold where osteoblasts create woven bone [15].

The remodelling phase, the final phase, is characterized by the replacement of the woven bone by lamellar bone and by the resorption of excess callus. The woven bone is replaced by the osteoclastic resorption and by the osteoblastic deposition. This remodelling is a gradual modification of the fracture site that adapts itself to the new configuration to get the optimal stability under the new mechanical load scenario.

1.2 Bone Remodelling

Bone remodelling term is usually used for the phase of general growth, reinforcement and resorption processes. This remodelling is progressive and it is induced in order to adapt the bone morphology to any new external load. Bone remodelling is a complex process performed by the coordinated activities of osteoclasts (that resorb bone), osteoblasts (that replace bone), osteocytes (within the bone matrix), bone-lining cells (covering the bone surface) and the capillary blood supply. Together, osteoclasts and the osteoblasts cells form temporary anatomical structures, called basic multicellular unite (BMU's), which execute bone remodelling. The interactions between osteoblasts and osteoclasts assuring a proper balance between bone gain and loss, is known as coupling [16].

All the millions BMU present in the skeleton are in different stages, being the life span of individual cells that create a BMU much shorter than that BMU itself [17–19].

These BMUs are constantly remodelling bone tissue during life, adult and senescent skeleton, preventing its premature deterioration, and maintaining its overall strength. If there is an interruption of this bone remodelling process due to a biochemical or cellular link cut, such as osteoporosis or hyperparathyroidism, BMUs' activity might be disrupted by a metabolic bone disease.

Normal bone remodelling occurs in discrete bone locations, taking 2–5 years for a discrete location to complete one bone remodelling cycle [20]. Bone remodelling can be classified as targeted remodelling, or by random remodelling. For instances, if a specific region of the bone is induced to remodel due to a structural microdamage, it is said to occur a targeted remodelling [21, 22]. In this case, remodelling permits the restore of the microdamage caused by fatigue and/or by shock.

Bone remodelling, is a main biological process (possessing a relevant role in the mineral homeostasis) , which provides calcium and phosphate. Thus, for example, if the calcium and phosphate are removed from random locations from the skeleton, it occurs a random remodelling [21, 22].

Bone balance is defined as the relation between the amount of bone removed and of new bone restored in the bone remodelling process. Bone balance can be easily modified by diseases, hormonal factors and even by external mechanical stimulus. Being bones, a major reservoir of body calcium, bone are under the hormonal control of parathyroid hormone (PTH), the most important hormone regulating calcium homeostasis and bone remodelling [23]. The modification of the bone remodelling behaviour, possessing a direct influence in the bone mass, is affected by the effect of PTH, in which a continuous increase of the PTH levels decreases bone mass and discontinuous PTH administration leads to a increases of bone mass [24–27].

Also long-term physical activity on a regular basis plays a particularly important role in bone remodelling. The mechanical stimulus induced by physical exercise can maintain or increase bone strength by increasing bone mass or by changing bone structures at micro and macro levels. Two main types of exercise are beneficial to bone health: weight-bearing, exercise performed while a person is standing so that gravity is exerting a force, and resistance exercise, exercises involving lifting weights with arms or legs. Long-term physical activity on a regular basis plays a particularly important role in maintaining healthy bones. Exercise can maintain and increase bone strength by increasing bone mass or by changing bone structures at micro and macro levels [28]. More recently was reported that, as the presence of external loads leads to the bone remodelling, the absence of load occurring in conditions of disuse, such as during immobility, space flight and long-term bed rest lead to the same processes, leads to the opposite result: bone loss and mineral changes [3, 29, 30].

Bone remodelling can be divided in five distinct phases occurring in a coordinate and sequential way. These five phases are the activation, resorption, reversal, formation and the termination.

The activation phase is characterized by a continuum process occurring in the boundaries of the BMU in which it is detected the presence of an inducing remodelling signal, that can be of mechanical or hormonal nature [20]. Bone cells are exposed to a dynamic environment of biophysical stimuli that includes strain, stress, shear, pressure, fluid flow, streaming potentials and acceleration, having these parameters, the ability to regulate independently the cellular responses and influence the bone remodelling. The osteocytes located in the cortical bone possess the ability of sensing this biophysical stimulus and, using the canaliculi network, they are capable to activate the regulation of the proteins sclerostin and RANKL, which possess a major role in bone remodelling, [28, 31].

The resorption phase is characterized by the formation and activity of osteoclasts that create a sealed section where the resorption process occurs. This formation and activity of osteoclasts are controlled by osteoblast cells that active the movement of mature osteoclasts into a bone remodelling site with the expression of CSF-1, RANKL, OPG and by PTH [1, 3, 4, 20, 32, 33].

The reversal phase lasts around nine days and occurs when the maximum eroded depth is achieved, between 60 and 40 μm. This phase is characterized by transition of the activity from osteoclast to osteoblast [18]. The osteoclasts start the process of apoptosis and at the same time, the bone-lining cells enter the lacuna and clean bone matrix remains. This clean up allows the deposition of a proteins (collagenous) layer in the resorption pits that form a cement line (glycoprotein), which helps the attachment of preosteoblasts that begin to differentiate [34–36].

The formation phase is the characterized by the formation of new bone. Once osteoclasts resorbed a cavity of bone, they are replaced by cells of the osteoblast lineage that initiate the bone formation, the preosteoblast. The preosteoblast is attracted by the growth factors liberated from the matrix that act as chemotactics and, in addition, stimulate its proliferation.

The preosteoblasts synthesize a cementing substance upon which the new tissue is attached and express bone morphogenic proteins (BMP) responsible for differentiation.

Bone resorption liberates TGF-β from the matrix, which is a key protein for recruiting mesenchymal stem cells to sites of bone resorption. This recruitment of mesenchymal stem cells and the presence of the bone morphogenic proteins (BMP) lead to the differentiation of preosteoblasts to osteoblasts.

The already differentiated osteoblasts synthesize the osteoid matrix and also secrete collagen, whose accumulation contributes to the cessation of cell growth.

The osteoid, the non-mineralized organic portion of the bone matrix, has to be mineralized with hydroxyapatite to create a mature bone tissue [28, 37].

The termination phase is characterized by the terminal differentiation of the osteoblast. Some of the osteoblast transform to lining cells, covering the newly formed bone surface, while other osteoblasts differentiate into osteocytes and remain in the matrix [28, 31, 37].

1.3 Meshless Methods

Nowadays, one of the most popular discrete numerical tools is the finite element method (FEM), but in the last few years, meshless methods, such as Smooth particle hydrodynamics (SPH) method [38], radial point interpolation method (RPIM) [39–41] and as natural neighbour radial point interpolation method (NNRPIM) [42], came into focus of interest.

The first meshless method to be developed dates from 1977, when Gingold and co-workers proposed the smooth particle hydrodynamics (SPH) method [38]. The SPH possesses a kernel approximation for a single function $u(x)$ in a domain Ω. This method was used for modelling astrophysical phenomena without boundaries, such as exploding stars and dust clouds [38]. From 1997 to now, many different methods were developed using different approaches: generalized finite difference method [43], diffuse element method (DEM) [44], element-free Galerkin method (EFGM) [45], meshless local Petrov–Galerkin (MLPG) method [46], point interpolation method (PIM) [39–41] and radial point interpolation method (RPIM) [47, 48].

The point interpolation method (PIM) was developed using the Galerkin weak form and shape functions that are constructed based only on a group of nodes arbitrarily distributed in a local support domain by means of polynomial interpolation [39–41].

The major advantage of PIM is that the shape functions created possess the Kronecker delta function property, which allows to enforce essential boundary conditions using simple numerical techniques, as in the conventional finite element method. PIM can use two types of shape functions: polynomial basis functions and radial basis functions (RBFs), being termed RPIM when using the RBFs [47, 48].

More recently, new methods were developed, such as the natural neighbour radial point interpolation method (NNRPIM) [42], which is based on the combination of the natural neighbour finite element method with the radial point interpolation method, and natural radial element method (NREM) [49], combining the simplicity of low-order finite elements connectivity with the geometric flexibility of meshless methods. The main advantage of the meshless methods is that they do not require elements to discretize the problem domain [1].

In meshless methods, the problem domain is discretized using an unstructured nodal mesh, as can be seen in Fig. 2b, opposing with the domain discretization using elements used in FEM, represented in Fig. 2a. In biomechanics, this discretization flexibility is advantageous, since it permits to discretize the problem domain using directly medical images.

Meshless methods can be divided in approximation meshless methods [42, 50–52] and interpolation meshless methods [7, 8, 24, 34, 53, 54]. The major advantage of using interpolator meshless methods is the possibility to impose directly the essential and natural boundary conditions, since the constructed test functions possess the delta Kronecker property.

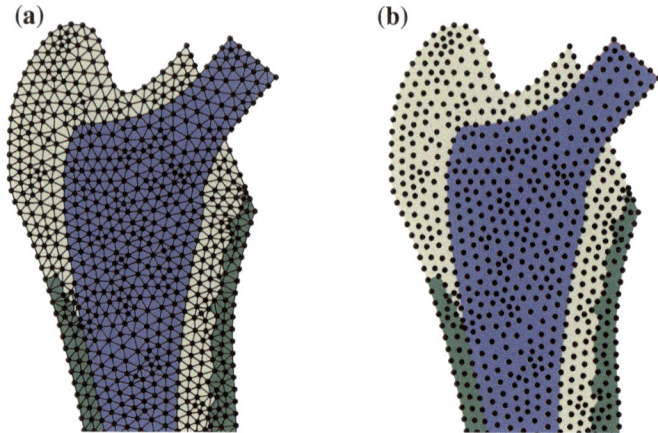

Fig. 2 **a** Domain discretization in FEM using elements; **b** Domains discretization in meshless using unstructured nodes

They can also be separated in meshless methods that use the strong form solution, using directly the partial differential equations governing the studied physical phenomenon, and others that uses the weak form solution. The weak form stands for a formulation that uses the variational principle to minimize the residual weight of the differential equations ruling a phenomenon, where the residual is obtained replacing the exact solution by an approximated one, affected by a test function.

The meshless methods start with the problem domain being discretized by a regular or irregular node set. This nodal set is not considered a "mesh", since the meshless method does not require any previous information about the nodes' vicinity in order to create the approximation or interpolation functions to the unknown variable field function.

In meshless methods, the nodal spatial distribution and nodal density discretization affect the performance of the method. Usually, having a denser nodal distributions results in more accurate solutions; however, the increase of the density of the dense distributions increases the computational costs. Furthermore, unbalanced nodal distribution leads to a lower accuracy, where location with predictable concentration of stress should have a higher nodal density when compared with locations with expected smoother stress distributions.

After discretizing the problem domain with the nodal set, a background integration mesh is created, which can be nodal dependent or nodal independent. This background mesh allows to numerically integrate the weak form governing the problem. The most popular integration schemes use the Gauss-Legendre quadrature technique. However, the majority of the integration schemes is nodal dependent, which hinders the "meshless" concept. Thus, other techniques comprise integration schemes that are completely nodal independent, as the ones using the Voronoï diagram or the natural neighbour mathematical concept [42, 49].

After the domain discretization and the construction of the background integration mesh, the meshless method has to assure the nodal connectivity, which is not predefined by elements as in FEM. Thus, in meshless methods, for each point of interest, one must defined areas or volumes acting as influence-domains.

Meshless methods and the FEM are discrete numerical methods and so, both require the discretization of the problem domain. The FEM discretize the problem domain using nodes and elements, where the finite element concept assures the nodal connectivity. In the meshless methods, the problem's domain is discretized using only nodes, and consequently, there are no elements, as in the FEM assuring that the nodes belonging to the same element interact directly between each other and with the boundary nodes of neighbour finite elements. Herewith, in the meshless methods, it is necessary to define the nodal interaction. Such interaction is enforced by means of "influence-domains". A given node i, or interest point, searches for the closest nodes. The area or volume in which those closest nodes are contained is called the influence domain of node i. In the literature, it is possible to find several techniques to define influence domains [1].

2 Bone Remodelling and Regeneration Modelling

The models used to simulate the behaviour of the bone regeneration and remodelling are based on various concepts and approaches. The models can be related to various forms of bone regeneration and remodelling, such as models for fracture healing, models for bone regeneration around endosseous implants and chemical models predicting the pharmacological effect on the remodelling process. As said before, the bone remodelling is the last phase of bone regeneration, even so there are models for both processes. This occurs due to the distinct duration of both processes. Bone regeneration is a process that lasts for some mouths, but the regeneration (occurring as the last phase of the regeneration or as a natural process on a healthy bone) can last the entire bone life.

It is usually to separate the models into three different classes with respect to the essential mechanisms regulating the bone regeneration/regeneration process. The classification in mechanical, biological and mechanobiological model allows to characterize the model according to the nature of the factures influencing the bone regeneration/remodelling process. Additionally, this classification permits to create a chronological perspective of the mathematical modelling of bone.

Mechanoregulatory models are governed by laws where the local mechanical environment is assumed as the only factor that influences the bone regeneration/remodelling. In bioregulatory models, only biochemical factors are considered in the model, and in the mechanobioregulatory models, both mechanical and biochemical factors are assumed. Despite mechanobioregulatory models are more realistic, when compared to mechanoregulatory and bioregulatory approaches, they also are more complex, due to the larger number of assumptions involved in their formulation.

The mechanoregulatory was the first to appear in 1960 by Pauwels [55]. In 1999, Adam et al. [56, 57] created one of the firsts bioregulatory modes to characterize the bone behaviour. Later, the mechanobioregulatory models appeared as a combination of the mechanoregulatory with the bioregulatory. Curiously, the generalization of the bioregulatory models did not stop the developments of the mechanoregulatory models and the development of the mechanobioregulatory models did not stop the developments of the bioregulatory models.

2.1 Mechanoregulatory Models

The first mathematical model related to mechanoregulation of bone remodelling describing "Wolff's law" was developed in 1960 by Pauwels [55], and later applied in 1965 [58]. This law defines that a tissue differentiation depends on local stresses and strains. Pauwels assumed that deviatoric strains stimulate the formation of fibrous tissue and that hydrostatic pressure stimulates formation of cartilage.

In 1964, Frost developed the "Curvature Model" describing the remodelling process controlled by strain actions [59].

Later in 1976, Cowin and co-workers presented the adaptive elasticity model, defining bone internal remodelling as the sum of chemical reactions between bone matrix and the extracellular fluids, where the rates of these chemical reactions depend upon the strain values [60–63]. In 1979, Perren presented a model for bone regeneration where was considered the influence of the local mechanical environment on the morphology of fracture healing, suggesting that a repair tissue can only be formed if the tissue tolerates the local mechanical strain. This approach sustains that different bone repair patterns occur when different physical influences are present, including strain tolerance [64].

In 1986, Carter and co-workers introduced the "self-optimization" concept, a remodelling model that assumed that mechanical stimulus is proportional to effective stress field. Thus, stress can be related with trabecular orientation and apparent density based on the idea that bone is a self-optimizing material [65–68].

In 1987, Huiskes and co-workers developed an adaptive remodelling model that used the strain energy density (SED) as a feedback control variable to determine shape or bone density adaptations [69].

Later in 1990, Reiter and co-workers modified Huiskes' model, introducing the effects of overstrain necrosis [70]. Harrigan and Hamilton, in 1992, introduced the SED for stress-induced remodelling as mechanical stimuli, using "Adaptive Elasticity" concept [71–73].

In 1997, Pettermann and co-workers (also using a modified Huiskes' model) introduced the anisotropic material behaviour in a regeneration model [74]. Prendergast and co-workers considered the biological tissues based on a poroelastic (biphasic) material. In this model, the differentiation process assumes as feedback controller the maximal distortional strain and relative fluid velocity [75].

In 2000, similar to the model developed by Prendergast et al. [75], Kuiper and co-workers created a regeneration model to study the fracture healing process as a function of applied movement. In this model, the shear strain and fluid shear stress regulate bone fracture healing, by adding or resorbing tissue and by modifying tissue properties. The model was able to control the callus development with the proliferation of granulation tissue [76]. Also in 2000, Ament and Hofer [77] developed a fuzzy logic formulation that using strain energy density as mechanical stimulus modified the behaviour of the tissue differentiation in the process of bone regeneration [77].

Later on 2002, Lacroix, Prendergast and co-workers extended Prendergast's work (1997) to include mesenchymal cell migration in a regeneration model [78, 79]. McNamara and Prendergast developed in 2007 four mechanoregulated regeneration models where the mechanical stimulus was strain, damage, combined strain/damage, and either strain or damage with damage-adaptive remodelling prioritised when damage was above a critical level. Each model was implemented with both bone-lining cell (surface) sensors and osteocyte cell (internal) sensors, and it also was applied to predict the BMU remodelling on the surface of a bone trabecula [80].

Also in 2002, Doblaré and García-Aznar proposed a remodelling model based on the principles of damage mechanics [81]. This model identifies bone voids with the cavities or micro cracks of other material damage models, but changes some of the standard assumptions in damage-mechanics theory to adapt it to the special requirements of living adaptive materials. A remodelling tensor, defined in terms of the apparent density and analogous to the standard damage tensor, was proposed to characterize the state of the homogenized bone microstructure.

In 2008, Liu and Niebur developed and modified the version of Lacroix's regeneration algorithm. This version enforces tissue differentiation pathway by transitioning from differentiation to bone adaptation [82]. In 2010, Mulvihill and Prendergast using the concepts of the algorithms developed by McNamara and Prendergast, 2007, and the concept proposed by Frost [59], developed a new regeneration algorithm where both strain and microdamage work as stimuli for BMU activity and that ON/OFF thresholds operate to control osteoclast and osteoblast activation [83].

Later in 2012, Belinha et al. [84, 85] developed a material law correlating the bone apparent density and the obtained level of stress. Using this new material law, it was developed a biomechanical remodelling model, an adaptation of Carter's models for predicting bone density distribution, based on the assumption that the bone structure is a gradually self-optimizing anisotropic biological material that maximizes its own structural stiffness [66–68, 86]. This model assumes that mechanical stimulus acts as the principal driving force in the bone tissue remodelling process [84, 85].

2.2 Bioregulatory Models

In 1999, Adam and co-workers developed a regeneration model where the critical size defect in the fracture healing site was defined by a set of a partial differential equation (PDE) governing a growth factor concentration [56, 57].

By 2001, Bailon-Plaza and Vander Meulen developed a more extended biological regeneration model that was defined by a system of PDE's [15]. In this approach, the PDE's system was defined by seven variables: densities of MSC's, of osteoblasts, of chondrocytes, concentrations of chondrogenic and osteogenic growth factors, and densities of connective/cartilage extracellular matrix and bone extracellular matrix. This model allowed the authors to represent the processes of cell differentiation, proliferation, migration and death, synthesis and resorption of tissues. This model is rather popular due to its generality. This model can be modified, allowing the development of several other bioregulatory models [15].

Later in 2003, Komarova et al. [87] developed a remodelling model describing the population dynamics of bone cells accordingly with the number of osteoclasts and osteoblasts at a single BMU. The interactions occurring among osteoblasts and osteoclasts encouraged by autocrine and paracrine allowed the authors to calculate cell population dynamics and changes in bone mass at a discrete site of bone remodelling [87]. To condense the net effect of local factors on the rates of cell production, this model uses a power-law approximation that was developed by Herries [88], as effective tools for analysis of highly nonlinear biochemical systems. In this model, all factors leading to a cell response are taken together in a single exponential parameter [87].

In 2004, Martin and Buckland-Wright developed a mathematical remodelling model to predict the depth of erosion and the duration of the resorption phases in healthy adult cancellous bone-based wholly on biological information. This model uses Michaelis-Menten-like feedback mechanics to affect bone resorption [89].

Also in 2004, Lemaire and co-workers developed the first mathematical remodelling model that includes the RANKL/RANK/OPG pathway, which affects the bone remodelling process. This model is based on the idea that the relative proportions of immature and mature osteoblasts control the degree of osteoclastic activity [90].

In 2008, Geris and co-workers extended the work of Bailon-Plaza and Meulen [15] by creating a regeneration model that considers angiogenesis. This was achieved by separating fibrous tissue and cartilage densities, and by defining additional chemotactic and haptotaxis terms for cell migration in the governing equations [91, 92].

Also in 2008, Pivonka and co-workers, using as base the Lemaire's model [90], suggested a new remodelling model where were added four new parameters, a rate equation (describing changes in bone volume with time), a rate equation (describing release of TGF-β from the bone matrix), the expression of OPG and RANKL on osteoblastic cell lines, and a modified activator/repressor functions [93].

In 2009, Ryser co-workers using as base Komarova's work developed a mathematical remodelling model of BMU describing changes in time and space of the concentrations of proresorptive cytokine RANKL and its inhibitor osteoprotegerin (OPG), in osteoclast and osteoblast numbers, and in bone mass [94].

In 2011, Amor and co-workers adapted the Bailon-Plaza and Vander Meulen model [15] for peri-implant osseointegration [95, 96]. This regeneration model postulates that intermediate cartilaginous phase is not observed experimentally in bone healing occurring near implants, [12, 13, 97, 98] and so, the model disregard chondrogenic growth factors, chondrocytes and cartilage. The big difference from the original Bailon-Plaza and Vander Meulen model is that the chemotaxis of MSC's is represented in equations and that the density of activated platelets was included.

2.3 Mechanobioregulatory Models

In 2002, Lacroix and co-workers using the models developed by Prendergast [75] and by Huiskes [99] developed a new regeneration model considering cellular processes together with mechanical factors by incorporating the random walk of MSC's [78, 79].

Later in 2003, Bailon-Plaza and Meulen, extending their previous work [15], created a new regeneration model in which it was predicted the beneficial effects of moderate, early loading and adverse effects of delayed or excessive loading on bone healing. The mechanical factors were modelled by stimulating and inhibiting the effects of dilatational and deviatoric strains [100].

In 2005, Gómez-Benito and co-workers and García-Aznar and co-workers developed a mechanobioregulatory regeneration model as a system of PDE's model that had as main variables cell concentrations of the four basic skeletal cells: MSC's, osteoblasts, fibroblasts and cartilage cells, and the four types of skeletal tissues, namely granulation tissue, fibrous tissue, cartilage and bone [101–103]. In this model, it was assumed that the fracture site is invaded by proliferating and migrating MSCs in the first stage of the bone healing process and that these may differentiate into cartilage cells, bone cells or fibroblasts, depending on the value of a mechanical stimulus. It was also assumed that very high mechanical stimulus causes MSCs to die. In this model, the process of bone healing was simulated as a process driven by a mechanical stimulus, the second invariant of the deviatoric strain tensor [103].

In 2008, Andreykiv et al. [104] and Isaksson et al. [105] developed a regeneration model where the biological part allowed the simulation of cell migration, proliferation, differentiation, tissue deposition and replacement. The mechanical component of the model calculates the mechanical stimuli influencing the cellular processes. Cell differentiation, proliferation and tissue production in this model are regulated by tissue shear strain and interstitial fluid velocity, as proposed by Prendergast et al. [75].

In 2014, Hambli [106] developed a remodelling model where the cellular behaviour was based on Komarova et al. [87] dynamic law. In this model, the mechanical stimuli are related with the biological one using strain–damage stimulus function controlled by the level of autocrine and paracrine. In 2010, Geris et al. [107] extended their previous bioregulatory model [108] by defining dependencies for the model parameters based on mechanical stimuli, according to Prendergast et al. [75]. This model was applied to study impaired fracture healing. The authors checked predictions of the model for various mechanoregulatory relations. In this model, both angiogenesis and osteogenesis were assumed to be affected by the mechanical loading.

Later in 2015, Yi and co-workers developed a microscale bone remodelling model using the equivalent strain as the mechanical stimuli. In this model, the mechanical stimuli affect the proliferation and differentiation of the osteoblast cells. The influence of the different physiological conditions, restricted by Denosumab (a used drug for treating osteoporosis), on the formation/resorption rate was also considered [109].

In 2016, Lerebours et al. [110] developed a multi-scale mechanobiological remodelling model describing the evolution of a human mid-shaft femur scan subjected in osteoporosis and mechanical disuse. In this model, hormonal regulation and biochemical coupling of bone cell populations are included. This models also include a mechanical adaptation of the tissue and factors that influence the microstructure on bone turnover rate [110].

3   Meshless Methods Applications

3.1   Meshless Methods and Mechanics

Many studies were performed using meshless method in order to prove it's efficiency in many different fields of mechanics. Belinha and Dinis, 2006, extended the usage of element-free Galerkin method (EFGM) in order to perform the elasto-plastic analysis of isotropic plates [111]. Later in 2007, Belinha and Dinis extended the EFGM to the analysis of anisotropic plates and laminates considering a Reissner–Mindlin laminate theory (FSDT) [112]. Also in 2007, Dinis and co-workers proposed a new meshless method, the natural neighbour radial point interpolation method (NNRPIM) based on the combination of the natural neighbour finite element method with the radial point interpolation method. In order to prove the high accuracy and convergence rate of the proposed method, well-known 2D and 3D problems were solved and compared with other studies and methods [42]. Later in 2008, NNRPIM was extended for the analysis of thick plates and laminates [51] being later in 2009 NNRPIM extended to dynamic analysis (free vibrations and forced vibrations) of 2D, 3D and bending plate problems [113]. In 2010, NNRPIM was combined with an Unconstrained Third-Order Plate Theory to analyse functionally graded plates [114], being later in 2011, using also Unconstrained Third-Order Plate Theory applied to laminates [50]. Also in 2010, Dinis and co-workers developed a unique NNRPIM approach when 3D-thin structures are considered. In order to demonstrate the effectiveness of the method, several isotropic and orthotropic thin plates and shells problems were solved [115]. Later in 2013, NNRPIM was extended to the analysis of composite laminated plates [116] and to the elastostatic analysis of thick plates [117]. In 2014, NNRPIM was extended to the analysis of laminated plates using Timoshenko theory [118]. In 2015, NNRPIM was extended to the field of fracture mechanics [119], and the radial point interpolation method (RPIM) to the elastostatic analysis of circular plates assuming the 2D axisymmetric deformation theory [54]. Later in 2016 was analysed and compared the performance of distinct meshless techniques assuming first-order shear deformation theory (FSDT) using FEM, and also an approximation meshless method (EFGM) and three interpolation meshless methods RPIM, NNR-PIM and natural radial element method (NREM) [120]. Also in 2016, NNRPIM was used to simulate the crack growth phenomenon in brittle materials [121].

3.2   Meshless Methods and Biomechanics

Meshless methods are used in different fields, but are getting an increasing interest on the biomechanical field, since the discretization flexibility in meshless is advantageous, allowing to discretize the problem domains using directly medical images.

These biomechanical studies focus in the bone tissue. Belinha and co-workers extended the NNRPIM to bone remodelling analysis, developing a biomechanical model to predict the bone density distribution [85, 122]. Later, some studies using meshless approach were performed in order to evaluate the bone response after the insertion of implants. Most of these studies focus in dental implants, [52, 123, 124]. More recent studies analysed the bone remodelling behaviour after the insertion of a femoral stem [125].

3.3   Meshless Methods and Bone Remodelling

In the last few years, meshless methods gradually become to enter into bone remodelling simulation field [53].

García and co-workers in 2000 developed one of the first works using bone structures and the meshless methods, studying the bone internal remodelling by means of α-shape-based natural element method (α-NEM). The considered remodelling model is based on the principles of continuum damage mechanics [126].

Liew and co-workers in 2002 explored stress distribution phenomena in the human proximal femur, having in consideration the detrimental effects of infarction as well as ageing. Liew and co-workers in 2002 concluded that meshless methods were efficient numerical techniques suitable for biomechanics [127].

James and co-workers in 2007 developed a model to analyse osteoporosis process. In this model, the CT imaging data was linked to the meshless method in order to analyse the mechanical properties of the porous, heterogeneous trabecular bone and the property–microstructure relationship [128].

Buti and co-workers in 2010 developed a uniform, particle-based, space- and geometry-oriented approach for bone remodelling. The developed methodology uses a multi-scale approach, at tissue scale and at cell scale. This methodology uses a meshless method approach developed by Taddei and co-workers in 2008 [129], which is a numerical approach based on a direct discrete formulation of physical laws—the Cell Method [130].

Recently Belinha and co-workers started to use meshless methods applied to bone tissue analysis. Firstly, using a microscale analysis of the bone remodelling phenomenon [85], and later using a macroscale analysis to study innumerable structures as the calcaneus, femur, mandible and the maxillary bone [84, 123, 131].

In 2016, Belinha and co-workers using the developed bone remodelling algorithm, the developed material law and the developed meshless methods, analysed the bone tissue remodelling of the femur due to the insertion of a stem [125].

4   Conclusion

Bone remodelling and bone regeneration are natural processes that are controlled very efficiently by the human body, by a combination of mechanical and biological factors. Along the years, many mathematical models were developed to simulate these two processes, evolving from simpler models (that considered only one type of factors) to the more recent models (that combine both mechanical and biological factors)—the mechanobiological models. These models, in many cases, are evolutions of previous models by adding more variables. The objective of these more complex models is to simulate the natural events in a more precise way. In the case of the biological models, it was found in the literature survey that they usually are well supported on laboratory experiments using bone tissue.

As said before, the last phase of the regeneration process is the remodelling phase. The models concerning the regeneration process consider the remodelling phase, although with a much smaller temporal windows when compared with the pure remodelling models. Thus, tell apart both model phases might not be immediately.

Being FEM, a mature and widely used method, it is natural that the first remodelling and regeneration models where firstly combined with the FEM.

Meshless methods are discretization techniques similar to FEM. Thus, the same fields and studies that have been explored using the FEM can now be explored using meshless methods (and also some topics where the FEM cannot be used).

For the particular case of bone regeneration, in the bibliographic survey, the authors did not found any study in which meshless methods where applied, which creates a research opportunity for meshless methods.

Taking into account some of the vantages of the meshless methods, such as the fact that they do not require elements to discretize the problem domain, and that in biomechanics domains with high complexity are very common, it is very likely that meshless methods became a standard in biomechanical analysis.

In the particular case of bone remodelling, the literature shows some relevant recent studies combining meshless methods with bone remodelling models. This recent interest could be explained by the meshing flexibility of meshless methods and by the fact that meshless methods can discretize numerically the problem domain using directly the medical images. In addition, meshless methods allow to adapt and update the discretization along the simulation, for example, by adding or removing nodes (domain) at any time-step during the simulation.

Acknowledgements The authors truly acknowledge the funding provided by Ministério da Educação e Ciência—Fundação para a Ciência e a Tecnologia (Portugal), under grants SFRH/BD/110047/2015, and by project funding MIT-EXPL/ISF/0084/2017. Additionally, the authors gratefully acknowledge the funding of Project NORTE-01-0145-FEDER-000022— SciTech—Science and Technology for Competitive and Sustainable Industries, co-financed by Programa Operacional Regional do Norte (NORTE2020), through Fundo Europeu de Desenvolvimento Regional (FEDER).

References

1. Belinha J (2014) Meshless methods in biomechanics: lecture notes in computational vision and biomechanics, vol 16. Springer International Publishing, Switzerland. ISBN: 9783319063997
2. Feng X, McDonald JM (2011) Disorders of bone remodeling. Annu Rev Pathol Mech Dis 6:121–145. https://doi.org/10.1146/annurev-pathol-011110-130203
3. Qin Q-H (2012) Mechanics of cellular bone remodeling: coupled thermal, electrical, and mechanical field effects, 1st edn. CRC Press, Boca Raton. ISBN: 9781466564176
4. Van Putte CL, Regan J, Russo A, Seeley RR, Stephens T, Tate P, Regan J, Russo A, Seeley RR, Stephens T, Tate P (2014) Seeley's anatomy & physiology, 10th edn. McGraw-Hill, New York. ISBN: 9780073403632
5. Wnek GE, Bowlin GL (2008) Encyclopedia of biomaterials and biomedical engineering, 2nd edn. CRC Press, Boca Raton. ISBN: 9781420078022
6. Reddy JN (2004) An introduction to nonlinear finite element analysis, 1st edn. Oxford University Press, Oxford. ISBN: 9780198525295
7. Dimitriou R, Jones E, McGonagle D, Giannoudis PV (2011) Bone regeneration: current concepts and future directions. BMC Med 9:66. https://doi.org/10.1186/1741-7015-9-66
8. Epari DR, Duda GN, Thompson MS (2010) Mechanobiology of bone healing and regeneration: In vivo models. Proc Inst Mech Eng Part H J Eng Med 224:1543–1553. https://doi.org/10.1243/09544119JEIM808
9. Geris L, Vander Sloten J, Van Oosterwyck H (2009) In silico biology of bone modelling and remodelling: regeneration. Philos Trans R Soc A Math Phys Eng Sci 367:2031–2053. https://doi.org/10.1098/rsta.2008.0293
10. Shapiro F (2008) Bone development and its relation to fracture repair. The role of mesenchymal osteoblasts and surface osteoblasts. Eur Cells Mater 15:53–76. https://doi.org/10.22203/eCM.v015a05 [pii]
11. Lieberman JR, Friedlaender GE (2005) Bone regeneration and repair: in biology and clinical applications, 1st edn. Humana Press, Totowa. ISBN: 9780896038479
12. Abrahamsson I, Berglundh T, Linder E, Lang NP, Lindhe J (2004) Early bone formation adjacent to rough and turned endosseous implant surfaces. An experimental study in the dog. Clin Oral Implants Res 15:381–392. https://doi.org/10.1111/j.1600-0501.2004.01082.x
13. Berglundh T, Abrahamsson I, Lang NP, Lindhe J (2003) De novo alveolar bone formation adjacent to endosseous implants. A model study in the dog. Clin Oral Implants Res 14:251–262. https://doi.org/10.1034/j.1600-0501.2003.00972.x
14. Davies J (2003) Understanding peri-implant endosseous healing. J Dent Educ 67:932–949
15. Bailón-Plaza A, Van Der Meulen MCH (2001) A mathematical framework to study the effects of growth factor influences on fracture healing. J Theor Biol 212:191–209. https://doi.org/10.1006/jtbi.2001.2372

16. Rodan GA, Martin TJ (1981) Role of osteoblasts in hormonal control of bone resorption—a hypothesis. Calcif Tissue Int 33:349–351. https://doi.org/10.1007/BF02409454

17. Hauge EM, Qvesel D, Eriksen EF, Mosekilde L, Melsen F (2001) Cancellous bone remodeling occurs in specialized compartments lined by cells expressing osteoblastic markers. J Bone Miner Res 16:1575–1582. https://doi.org/10.1359/jbmr.2001.16.9.1575

18. Kular J, Tickner J, Chim SM, Xu J (2012) An overview of the regulation of bone remodelling at the cellular level. Clin Biochem 45:863–873. https://doi.org/10.1016/j.clinbiochem.2012.03.021

19. Smit TH, Burger EH (2010) Is BMU-coupling a strain-regulated phenomenon? A finite element analysis. J Bone Miner Res 15:301–307. https://doi.org/10.1359/jbmr.2000.15.2.301

20. Raggatt LJ, Partridge NC (2010) Cellular and molecular mechanisms of bone remodeling. J Biol Chem 285:25103–25108. https://doi.org/10.1074/jbc.R109.041087

21. Burr DB, Robling AG, Turner CH (2002) Effects of biomechanical stress on bones in animals. Bone 30:781–786. https://doi.org/10.1016/S8756-3282(02)00707-X

22. Parfitt AM (2002) Targeted and nontargeted bone remodeling: Relationship to basic multicellular unit origination and progression. Bone 30:5–7. https://doi.org/10.1016/S8756-3282(01)00642-1

23. Kroll M (2000) Parathyroid hormone temporal effects on bone formation and resorption. Bull Math Biol 62:163–188. https://doi.org/10.1006/bulm.1999.0146

24. Dobnig H, Turner RT (1997) The effects of programmed administration of human parathyroid hormone fragment (1–34) on bone histomorphometry and serum chemistry in rats. Endocrinology 138:4607–4612. https://doi.org/10.1210/endo.138.11.5505

25. Karaplis AC, Goltzman D (2000) PTH and PTHrP effects on the skeleton. Rev Endocr Metab Disord 1:331–341. https://doi.org/10.1023/A:1026526703898

26. Locklin RM, Khosla S, Turner RT, Riggs BL (2003) Mediators of the biphasic responses of bone to intermittent and continuously administered parathyroid hormone. J Cell Biochem 89:180–190. https://doi.org/10.1002/jcb.10490

27. Rubin MR, Bilezikian JP (2003) New anabolic therapies in osteoporosis. Endocrinol Metab Clin North Am 32:285–307. https://doi.org/10.1016/S0889-8529(02)00056-7

28. Valds-Flores M, Orozco L, Velzquez-Cruz R (2013) Molecular aspects of bone remodeling. In: Valds-Flores M (ed) Topics in osteoporosis. InTech, pp 1–28

29. Bauman WA, Spungen AM, Wang J, Pierson RN, Schwartz E (1999) Continuous loss of bone during chronic immobilization: a monozygotic twin study. Osteoporos Int 10:123–127. https://doi.org/10.1007/s001980050206

30. Zerwekh JE, Ruml LA, Gottschalk F, Pak CYC (2009) The effects of twelve weeks of bed rest on bone histology, biochemical markers of bone turnover, and calcium homeostasis in eleven normal subjects. J Bone Miner Res 13:1594–1601. https://doi.org/10.1359/jbmr.1998.13.10.1594

31. Bonewald LF (2011) The amazing osteocyte. J Bone Miner Res 26:229–238. https://doi.org/10.1002/jbmr.320

32. Tang Y, Wu X, Lei W, Pang L, Wan C, Shi Z, Zhao L, Nagy TR, Peng X, Hu J, Feng X, Van Hul W, Wan M, Cao X (2009) TGF-β1–induced migration of bone mesenchymal stem cells couples bone resorption with formation. Nat Med 15:757–765. https://doi.org/10.1038/nm.1979

33. Teti A (2011) Bone development: overview of bone cells and signaling. Curr Osteoporos Rep 9:264–273. https://doi.org/10.1007/s11914-011-0078-8

34. Everts V, Delaissié JM, Korper W, Jansen DC, Tigchelaar-Gutter W, Saftig P, Beertsen W (2002) The bone lining cell: its role in cleaning Howship's lacunae and initiating bone formation. J Bone Miner Res 17:77–90. https://doi.org/10.1359/jbmr.2002.17.1.77

35. Gallagher JC, Sai AJ (2010) Molecular biology of bone remodeling: implications for new therapeutic targets for osteoporosis. Maturitas 65:301–307. https://doi.org/10.1016/j.maturitas.2010.01.002

36. Matsuo K, Irie N (2008) Osteoclast-osteoblast communication. Arch Biochem Biophys 473:201–209. https://doi.org/10.1016/j.abb.2008.03.027

37. Kini U, Nandeesh BN (2012) Physiology of bone formation, remodeling, and metabolism. Radionuclide and hybrid bone imaging. Springer Berlin Heidelberg, Berlin, Heidelberg, pp 29–57

38. Gingold RA, Monaghan JJ (1977) Smoothed particle hydrodynamics: theory and application to non-spherical stars. Mon Not R Astron Soc 181:375–389. https://doi.org/10.1093/mnras/181.3.375

39. Liu GR, Gu YT (2001) Local point interpolation method for stress analysis of two-dimensional solids. Struct Eng Mech 11:221–236. https://doi.org/10.12989/sem.2001.11.2.221

40. Liu GR, Gu YT (2001) A local radial point interpolation method (LRPIM) for free vibration analyses of 2-D solids. J Sound Vib 246:29–46. https://doi.org/10.1006/jsvi.2000.3626

41. Liu GR, Gu YT (2001) A point interpolation method for two-dimensional solids. Int J Numer Methods Eng 50:937–951. https://doi.org/10.1002/1097-0207(20010210)50:4%3c937:AID-NME62%3e3.0.CO;2-X

42. Dinis LMJS, Jorge RM Natal, Belinha J (2007) Analysis of 3D solids using the natural neighbour radial point interpolation method. Comput Methods Appl Mech Eng 196:2009–2028. https://doi.org/10.1016/j.cma.2006.11.002

43. Liszka T, Orkisz J (1980) The finite difference method at arbitrary irregular grids and its application in applied mechanics. Comput Struct 11:83–95. https://doi.org/10.1016/0045-7949(80)90149-2

44. Nayroles B, Touzot G, Villon P (1992) Generalizing the finite element method: diffuse approximation and diffuse elements. Comput Mech 10:307–318. https://doi.org/10.1007/BF00364252

45. Belytschko T, Lu YY, Gu L (1994) Element-free Galerkin methods. Int J Numer Methods Eng 37:229–256. https://doi.org/10.1002/nme.1620370205

46. Atluri SN, Zhu T (1998) A new Meshless Local Petrov-Galerkin (MLPG) approach in computational mechanics. Comput Mech 22:117–127. https://doi.org/10.1007/s004660050346

47. Wang JG, Liu GR (2002) A point interpolation meshless method based on radial basis functions. Int J Numer Methods Eng 54:1623–1648. https://doi.org/10.1002/nme.489

48. Wang JG, Liu GR (2002) On the optimal shape parameters of radial basis functions used for 2-D meshless methods. Comput Methods Appl Mech Eng 191:2611–2630. https://doi.org/10.1016/S0045-7825(01)00419-4

49. Belinha J, Dinis LMJS, Jorge RM Natal (2013) The natural radial element method. Int J Numer Methods Eng 93:1286–1313. https://doi.org/10.1002/nme.4427

50. Dinis LMJS, Jorge RMN, Belinha J (2011) Static and dynamic analysis of laminated plates based on an unconstrained third order theory and using a radial point interpolator meshless method. Comput Struct 89:1771–1784. https://doi.org/10.1016/j.compstruc.2010.10.015

51. Dinis LMJS, Jorge RM Natal, Belinha J (2008) Analysis of plates and laminates using the natural neighbour radial point interpolation method. Eng Anal Bound Elem 32:267–279. https://doi.org/10.1016/j.enganabound.2007.08.006

52. Duarte HMS, Andrade JR, Dinis LMJS, Jorge RMN, Belinha J (2016) Numerical analysis of dental implants using a new advanced discretization technique. Mech Adv Mater Struct 23:467–479. https://doi.org/10.1080/15376494.2014.987410

53. Doblaré M, Cueto E, Calvo B, Martínez MA, Garcia JM, Cegoñino J (2005) On the employ of meshless methods in biomechanics. Comput Methods Appl Mech Eng 194:801–821. https://doi.org/10.1016/j.cma.2004.06.031

54. Farahani BV, Berardo JMV, Drgas R, de Sá JC, Ferreira A, Belinha J (2015) The axisymmetric analysis of circular plates using the radial point interpolation method. J Comput Methods Eng Sci Mech 16:336–353. https://doi.org/10.1080/15502287.2015.1103819

55. Pauwels F (1960) Eine neue Theorie über den Einfluß mechanischer Reize auf die Differenzierung der Stützgewebe. Brain Struct Funct 121:478–515. https://doi.org/10.1007/BF00523401

56. Adam JA (1999) A simplified model of wound healing (with particular reference to the critical size defect). Math Comput Model 30:23–32. https://doi.org/10.1016/S0895-7177(99)00145-4

57. Arnold JS, Adam JA (1999) A simplified model of wound healing II: the critical size defect in two dimensions. Math Comput Model 30:47–60. https://doi.org/10.1016/S0895-7177(99)00197-1

58. Pauwels F (1965) Gesammelte Abhandlungen zur funktionellen Anatomie des Bewegungsapparates. Springer. ISBN: 978-3-642-86842-9

59. Frost HM (1964) The laws of bone structure, 1st edn. Springfield, Ill, Thomas. ISSN: 0002-9629

60. Cowin SC, Hegedus DH (1976) Bone remodeling I: theory of adaptive elasticity. J Elast 6:313–326. https://doi.org/10.1007/BF00041724

61. Cowin SC, Nachlinger RR (1978) Bone remodeling III: uniqueness and stability in adaptive elasticity theory. J Elast 8:285–295. https://doi.org/10.1007/BF00130467

62. Cowin SC, Sadegh AM, Luo GM (1992) An evolutionary Wolff's law for trabecular architecture. J Biomech Eng 114:129. https://doi.org/10.1115/1.2895436

63. Hegedus DH, Cowin SC (1976) Bone remodeling II: small strain adaptive elasticity. J Elast 6:337–352. https://doi.org/10.1007/BF00040896

64. Perren SM (1979) Physical and biological aspects of fracture healing with special reference to internal fixation. Clin Orthop Relat Res 175–96

65. Carter DR (1987) Mechanical loading history and skeletal biology. J Biomech 20:1095–1109. https://doi.org/10.1016/0021-9290(87)90027-3

66. Carter DR, Orr TE, Fyhrie DP (1989) Relationships between loading history and femoral cancellous bone architecture. J Biomech 22:231–244. https://doi.org/10.1016/0021-9290(89)90091-2

67. Fyhrie DP, Carter DR (1986) A unifying principle relating stress to trabecular bone morphology. J Orthop Res 4:304–317. https://doi.org/10.1002/jor.1100040307

68. Whalen RT, Carter DR, Steele CR (1988) Influence of physical activity on the regulation of bone density. J Biomech 21:825–837. https://doi.org/10.1016/0021-9290(88)90015-2

69. Huiskes R, Weinans H, Grootenboer HJ, Dalstra M, Fudala B, Slooff TJ (1987) Adaptive bone-remodeling theory applied to prosthetic-design analysis. J Biomech 20:1135–1150. https://doi.org/10.1016/0021-9290(87)90030-3

70. Reiter TJ, Rammerstorfer FG, Bohm HJ (1990) Numerical algorithm for the simulation of bone remodeling. Am Soc Mech Eng Bioeng Div BED 17:181–184

71. Harrigan TP, Hamilton JJ (1992) An analytical and numerical study of the stability of bone remodelling theories: dependence on microstructural stimulus. J Biomech 25:477–488. https://doi.org/10.1016/0021-9290(92)90088-I

72. Harrigan TP, Hamilton JJ (1992) Optimality conditions for finite element simulation of adaptive bone remodeling. Int J Solids Struct 29:2897–2906. https://doi.org/10.1016/0020-7683(92)90147-L

73. Harrigan TP, Hamilton JJ (1993) Finite element simulation of adaptive bone remodelling: a stability criterion and a time stepping method. Int J Numer Methods Eng 36:837–854. https://doi.org/10.1002/nme.1620360508

74. Pettermann HE, Reiter TJ, Rammerstorfer FG (1997) Computational simulation of internal bone remodeling. Arch Comput Methods Eng 4:295–323. https://doi.org/10.1007/BF02737117

75. Prendergast PJ, Huiskes R, Søballe K (1997) Biophysical stimuli on cells during tissue differentiation at implant interfaces. J Biomech 30:539–548. https://doi.org/10.1016/S0021-9290(96)00140-6

76. Kuiper JH, Richardson JB, Ashton BA (2000) Computer simulation to study the effect of fracture site movement on tissue formation and fracture stiffness restoration. Eur Congr Comput Methods Appl Sci Eng 1–6

77. Ament C, Hofer EP (2000) A fuzzy logic model of fracture healing. J Biomech 33:961–968. https://doi.org/10.1016/S0021-9290(00)00049-X

78. Lacroix D, Prendergast PJ, Li G, Marsh D (2002) Biomechanical model to simulate tissue differentiation and bone regeneration: application to fracture healing. Med Biol Eng Comput 40:14–21. https://doi.org/10.1007/BF02347690

79. Lacroix D, Prendergast PJJ (2002) A mechano-regulation model for tissue differentiation during fracture healing: analysis of gap size and loading. J Biomech 35:1163–1171. https://doi.org/10.1016/S0021-9290(02)00086-6

80. McNamara LM, Prendergast PJ (2007) Bone remodelling algorithms incorporating both strain and microdamage stimuli. J Biomech 40:1381–1391. https://doi.org/10.1016/j.jbiomech.2006.05.007

81. Doblaré M, García JM (2002) Anisotropic bone remodelling model based on a continuum damage-repair theory. J Biomech 35:1–17. https://doi.org/10.1016/S0021-9290(01)00178-6

82. Liu X, Niebur GL (2008) Bone ingrowth into a porous coated implant predicted by a mechano-regulatory tissue differentiation algorithm. Biomech Model Mechanobiol 7:335–344. https://doi.org/10.1007/s10237-007-0100-3

83. Mulvihill BM, Prendergast PJ (2010) Mechanobiological regulation of the remodelling cycle in trabecular bone and possible biomechanical pathways for osteoporosis. Clin Biomech 25:491–498. https://doi.org/10.1016/j.clinbiomech.2010.01.006

84. Belinha J, Jorge RMN, Dinis LMJSJS (2012) A meshless microscale bone tissue trabecular remodelling analysis considering a new anisotropic bone tissue material law. Comput Methods Biomech Biomed Engin 5842:1–15. https://doi.org/10.1080/10255842.2012.654783

85. Belinha J, Jorge RMM Natal, Dinis LMJS (2012) Bone tissue remodelling analysis considering a radial point interpolator meshless method. Eng Anal Bound Elem 36:1660–1670. https://doi.org/10.1016/j.enganabound.2012.05.009

86. Carter DR, Fyhrie DP, Whalen RT (1987) Trabecular bone density and loading history: regulation of connective tissue biology by mechanical energy. J Biomech 20:785–794. https://doi.org/10.1016/0021-9290(87)90058-3

87. Komarova SV, Smith RJ, Dixon SJ, Sims SM, Wahl LM (2003) Mathematical model predicts a critical role for osteoclast autocrine regulation in the control of bone remodeling. Bone 33:206–215. https://doi.org/10.1016/S8756-3282(03)00157-1

88. Herries DG (1977) Biochemical systems analysis. a study of function and design in molecular biology. Biochem Educ 5:84. https://doi.org/10.1016/0307-4412(77)90075-9

89. Martin MJ, Buckland-Wright JC (2004) Sensitivity analysis of a novel mathematical model identifies factors determining bone resorption rates. Bone 35:918–928. https://doi.org/10.1016/j.bone.2004.06.010

90. Lemaire V, Tobin FL, Greller LD, Cho CR, Suva LJ (2004) Modeling the interactions between osteoblast and osteoclast activities in bone remodeling. J Theor Biol 229:293–309. https://doi.org/10.1016/j.jtbi.2004.03.023

91. Geris L, Reed AAC, Vander Sloten J, Simpson AHRW, van Oosterwyck H (2010) Occurrence and treatment of bone atrophic non-unions investigated by an integrative approach. PLoS Comput Biol 6:e1000915. https://doi.org/10.1371/journal.pcbi.1000915

92. Geris L, Sloten J Vander, Van Oosterwyck H (2008) Mathematical modelling of bone regeneration including angiogenesis : design of treatment strategies for atrophic non-union. In: 54th annual meeting of the orthopaedic research society, p 2005

93. Pivonka P, Zimak J, Smith DW, Gardiner BS, Dunstan CR, Sims NA, John Martin T, Mundy GR, Martin TJ, Mundy GR, John Martin T, Mundy GR (2008) Model structure and control of bone remodeling: a theoretical study. Bone 43:249–263. https://doi.org/10.1016/j.bone.2008.03.025

94. Ryser MD, Nigam N, Komarova SV (2009) Mathematical modeling of spatio-temporal dynamics of a single bone multicellular unit. J Bone Miner Res 24:860–870. https://doi.org/10.1359/jbmr.081229

95. Amor N, Geris L, Vander Sloten J, Van Oosterwyck H (2009) Modelling the early phases of bone regeneration around an endosseous oral implant. Comput Methods Biomech Biomed Engin 12:459–468. https://doi.org/10.1080/10255840802687392

96. Amor N, Geris L, Vander Sloten J, Van Oosterwyck H (2011) Computational modelling of biomaterial surface interactions with blood platelets and osteoblastic cells for the prediction of contact osteogenesis. Acta Biomater 7:779–790. https://doi.org/10.1016/j.actbio.2010.09.025

97. Buser D, Broggini N, Wieland M, Schenk RK, Denzer AJ, Cochran DL, Hoffmann B, Lussi A, Steinemann SG (2004) Enhanced bone apposition to a chemically modified SLA titanium surface. J Dent Res 83:529–533. https://doi.org/10.1177/154405910408300704

98. Colnot C, Romero DM, Huang S, Rahman J, Currey JA, Nanci A, Brunski JB, Helms JA (2007) Molecular analysis of healing at a bone-implant interface. J Dent Res 86:862–867. https://doi.org/10.1177/154405910708600911

99. Huiskes R, Van Driel WD, Prendergast PJ, Søballe K (1997) A biomechanical regulatory model for periprosthetic fibrous-tissue differentiation. J Mater Sci Mater Med 8:785–788. https://doi.org/10.1023/A:1018520914512

100. Bailón-Plaza A, Van Der Meulen MCH (2003) Beneficial effects of moderate, early loading and adverse effects of delayed or excessive loading on bone healing. J Biomech 36:1069–1077. https://doi.org/10.1016/S0021-9290(03)00117-9

101. García-Aznar JM, Kuiper JH, Gómez-Benito MJ, Doblaré M, Richardson JB (2007) Computational simulation of fracture healing: influence of interfragmentary movement on the callus growth. J Biomech 40:1467–1476. https://doi.org/10.1016/j.jbiomech.2006.06.013

102. García JM, Doblaré M, Cueto E (2000) Simulation of bone internal remodeling by means of the α -shape-based natural element method. In: European congress on computational methods in applied sciences and engineering. Barcelona

103. Gómez-Benito MJ, García-Aznar JM, Kuiper JH, Doblaré M (2005) Influence of fracture gap size on the pattern of long bone healing: a computational study. J Theor Biol 235:105–119. https://doi.org/10.1016/j.jtbi.2004.12.023

104. Andreykiv A, Van Keulen F, Prendergast PJ (2008) Simulation of fracture healing incorporating mechanoregulation of tissue differentiation and dispersal/proliferation of cells. Biomech Model Mechanobiol 7:443–461. https://doi.org/10.1007/s10237-007-0108-8

105. Isaksson H, van Donkelaar CC, Huiskes R, Ito K (2008) A mechano-regulatory bone-healing model incorporating cell-phenotype specific activity. J Theor Biol 252:230–246. https://doi.org/10.1016/j.jtbi.2008.01.030

106. Hambli R (2014) Connecting mechanics and bone cell activities in the bone remodeling process: an integrated finite element modeling. Front Bioeng Biotechnol 2:6. https://doi.org/10.3389/fbioe.2014.00006

107. Geris L, Vander Sloten J, Van Oosterwyck H (2010) Connecting biology and mechanics in fracture healing: an integrated mathematical modeling framework for the study of nonunions. Biomech Model Mechanobiol 9:713–724. https://doi.org/10.1007/s10237-010-0208-8

108. Geris L, Gerisch A, Vander Sloten J, Weiner R, Van Oosterwyck H (2008) Angiogenesis in bone fracture healing: a bioregulatory model. J Theor Biol 251:137–158. https://doi.org/10.1016/j.jtbi.2007.11.008

109. Yi W, Wang C, Liu X (2015) A microscale bone remodeling simulation method considering the influence of medicine and the impact of strain on osteoblast cells. Finite Elem Anal Des 104:16–25. https://doi.org/10.1016/j.finel.2015.04.007

110. Lerebours C, Buenzli PR, Scheiner S, Pivonka P (2016) A multiscale mechanobiological model of bone remodelling predicts site-specific bone loss in the femur during osteoporosis and mechanical disuse. Biomech Model Mechanobiol 15:43–67. https://doi.org/10.1007/s10237-015-0705-x

111. Belinha J, Dinis LMJS (2006) Elasto-plastic analysis of plates by the element free Galerkin method. Eng Comput 23:525–551. https://doi.org/10.1108/02644400610671126

112. Belinha J, Dinis LMJS (2007) Nonlinear analysis of plates and laminates using the element free Galerkin method. Compos Struct 78:337–350. https://doi.org/10.1016/j.compstruct.2005.10.007

113. Belinha J, Dinis LMJS, Jorge RMN (2009) The natural neighbour radial point interpolation method: dynamic applications. Eng Comput 26:911–949. https://doi.org/10.1108/02644400910996835

114. Dinis LMJS, Jorge RMN, Belinha J (2010) An unconstrained third-order plate theory applied to functionally graded plates using a meshless method. Mech Adv Mater Struct 17:108–133. https://doi.org/10.1080/15376490903249925

115. Dinis LMJS, Jorge RMN, Belinha J (2010) A 3D shell-like approach using a natural neighbour meshless method: isotropic and orthotropic thin structures. Compos Struct 92:1132–1142. https://doi.org/10.1016/j.compstruct.2009.10.014

116. Belinha J, Dinis LMJS, Jorge RMN (2013) Composite laminated plate analysis using the natural radial element method. Compos Struct 103:50–67. https://doi.org/10.1016/j.compstruct.2013.03.018

117. Belinha J, Dinis LMJS, Jorge RM Natal (2013) Analysis of thick plates by the natural radial element method. Int J Mech Sci 76:33–48. https://doi.org/10.1016/j.ijmecsci.2013.08.011

118. Moreira S, Belinha J, Dinis LMJSJS, Jorge RMNN (2014) Análise de vigas laminadas utilizando o natural neighbour radial point interpolation method. Rev Int Metod Numer para Calc y Disen en Ing 30:108–120. https://doi.org/10.1016/j.rimni.2013.02.002

119. Azevedo JMC, Belinha J, Dinis LMJS, Jorge RM Natal (2015) Crack path prediction using the natural neighbour radial point interpolation method. Eng Anal Bound Elem 59:144–158. https://doi.org/10.1016/j.enganabound.2015.06.001

120. Belinha J, Araújo AL, Ferreira AJM, Dinis LMJS, Jorge RM Natal (2016) The analysis of laminated plates using distinct advanced discretization meshless techniques. Compos Struct 143:165–179. https://doi.org/10.1016/j.compstruct.2016.02.021

121. Belinha J, Azevedo JMC, Dinis LMJS, Jorge RM Natal (2016) The natural neighbor radial point interpolation method extended to the crack growth simulation. Int J Appl Mech 08:1650006. https://doi.org/10.1142/S175882511650006X

122. Belinha J, Dinis LMJS, Jorge RMN (2015) The meshless methods in the bone tissue remodelling analysis. Procedia Eng 110:51–58. https://doi.org/10.1016/j.proeng.2015.07.009

123. Belinha J, Dinis LMJS, Jorge RMN (2015) The mandible remodeling induced by dental implants: a meshless approach. J Mech Med Biol 15:1550059. https://doi.org/10.1142/S0219519415500591

124. Tavares CSS, Belinha J, Dinis LMJS, Jorge RMN (2015) The elasto-plastic response of the bone tissue due to the insertion of dental implants. Procedia Eng 110:37–44. https://doi.org/10.1016/j.proeng.2015.07.007

125. Belinha J, Dinis LMJSMJS, Jorge RMM Natal (2016) The analysis of the bone remodelling around femoral stems: A meshless approach. Math Comput Simul 121:64–94. https://doi.org/10.1016/j.matcom.2015.09.002

126. García-Aznar JM, Rueberg T, Doblare M (2005) A bone remodelling model coupling microdamage growth and repair by 3D BMU-activity. Biomech Model Mechanobiol 4:147–167. https://doi.org/10.1007/s10237-005-0067-x

127. Liew KM, Wu HY, Ng TY (2002) Meshless method for modeling of human proximal femur: treatment of nonconvex boundaries and stress analysis. Comput Mech 28:390–400. https://doi.org/10.1007/s00466-002-0303-5

128. Lee JD, Chen Y, Zeng X, Eskandarian A, Oskard M (2007) Modeling and simulation of osteoporosis and fracture of trabecular bone by meshless method. Int J Eng Sci 45:329–338. https://doi.org/10.1016/j.ijengsci.2007.03.007

129. Taddei F, Pani M, Zovatto L, Tonti E, Viceconti M (2008) A new meshless approach for subject-specific strain prediction in long bones: evaluation of accuracy. Clin Biomech 23:1192–1199. https://doi.org/10.1016/j.clinbiomech.2008.06.009

130. Buti F, Cacciagrano D, Corradini F, Merelli E, Tesei L, Pani M (2010) Bone remodelling in bioshape. Electron Notes Theor Comput Sci 268:17–29. https://doi.org/10.1016/j.entcs.2010.12.003

131. Moreira SF, Belinha J, Dinis LMJS, Natal Jorge RM (2014) A global numerical analysis of the "central incisor/local maxillary bone" system using a meshless method. MCB Mol Cell Biomech 11:151–184. https://doi.org/10.3970/mcb.2014.011.151

Dynamic Biochemical and Cellular Models of Bone Physiology: Integrating Remodeling Processes, Tumor Growth, and Therapy

Rui M. Coelho, Joana P. Neto, Duarte Valério and Susana Vinga

Abstract Bone is an active connective tissue composed of different types of cells. The dynamic behavior of bone remodeling processes is typically represented through differential equations, which represent the physiological phenomena occurring in this organ. These models take into account the tight biochemical regulation between osteoclasts and osteoblasts and have also been enriched with variables and parameters related to bone pathologies and treatment. This chapter reviews some of the more recent models describing bone physiology, focusing on those that include the main cellular processes, along the biochemical control, and also the pharmacokinetic/pharmacodynamic (PK/PD) of the most common treatments for diseases such as cancer. These models are then compared in terms of the simulations obtained and, finally, some highlights on integrating them with the biomechanical component of the system which will be given. These models are expected to provide a valuable insight into this complex system and to support the development of clinical decision systems for bone pathologies.

R. M. Coelho · D. Valério
IDMEC, Instituto Superior Técnico, Universidade de Lisboa, Av. Rovisco Pais 1,
1049-001 Lisbon, Portugal
e-mail: rui.coelho@tecnico.ulisboa.pt

D. Valério
e-mail: duarte.valerio@tecnico.ulisboa.pt

J. P. Neto
Instituto Superior Técnico, Universidade de Lisboa, Av. Rovisco Pais 1,
1049-001 Lisbon, Portugal
e-mail: joana.neto@tecnico.ulisboa.pt

S. Vinga (✉)
INESC-ID, Instituto Superior Técnico, Universidade de Lisboa, Av. Rovisco Pais 1,
1049-001 Lisbon, Portugal
e-mail: susanavinga@tecnico.ulisboa.pt

© Springer Nature Switzerland AG 2020   95
J. Belinha et al. (eds.), *The Computational Mechanics of Bone Tissue*,
Lecture Notes in Computational Vision and Biomechanics 35,
https://doi.org/10.1007/978-3-030-37541-6_4

Overview

This chapter reviews some of the most recent models of bone remodeling processes, focusing on those that include known biochemical phenomena. These models are then complemented with variables related to healthy and pathological states, namely to study oncological processes and other bone pathologies. Additionally, they have been enriched with therapy, by including the pharmacokinetics (PK) and pharmacodynamics (PD) of several bone treatments.

The main goal of the present review is to compare these approaches and highlight future directions, in particular, how these models can be integrated with biomechanical knowledge to build integrated models for bone physiology.

To better understand the mentioned models, this chapter follows a growing complexity hierarchy. Firstly, basic physiology concepts are introduced in Sect. 1, ranging from healthy bone dynamics to existing treatments for tumors in its microenvironment. Local mathematical and computational models are then presented (Sect. 2), followed by their non-local adaptations (Sect. 3). Different models are reviewed and pertinent simulations results are presented, from the simplest one involving only three state variables to more complex ones that include other physiological processes and therapy.

1 Bone Remodeling Physiology

Modeling complex physiological systems are gaining an increasing interest in engineering due to the expected impact in clinical sciences. These multidisciplinary fields are promoting the 4P approach to medicine, which is becoming more predictive, personalized, preventive, and participatory [15].

In particular, there are now many efforts to model bone physiology, taking into account all the biochemical, cellular, and environmental processes. However, for an accurate modeling process, the inherent physiology must be known.

1.1 Healthy Bone Remodeling

Living tissue is constantly being renovated, and this process is also true for bone tissue. Static as it may seem to the naked eye, bone undergoes a constant remodeling process, being resorbed by cells termed *osteoclasts* and formed by cells called *osteoblasts*. They constitute part of a BMU, a basic multicellular unit.

- **Osteoclasts**—Originated from the fusion of mononucleated cells and progenitor cells that express receptor activator of NF-κB (RANK) and macrophage colony-stimulating factor receptor (c-FMS), and osteoclasts differentiate into multinucleated cells when colony-stimulating factor 1 (CSF1) and RANK-ligand (RANKL) are present in the bone microenvironment. Being capable of bone resorption, their generation rate determines the BMU extension; whereas, the life span determines the depth of resorption [3, 35].

- **Osteoblasts**—Osteoblasts are mononucleated cells able to form bone. They can regulate bone resorption and formation through parathyroid hormone (PTH) receptors and RANKL and osteoprotegerin (OPG) production. OPG is a soluble decoy receptor for RANKL that binds to RANK in osteoclasts precursors, consequently inhibiting osteoclast formation. Osteoblasts upregulate RANKL expression due to PTH, thus promoting their activation. PTH also contributes to osteoclastogenesis by decreasing the OPG production. The bone resorption and formation are regulated greatly by the RANK/RANKL/OPG pathway and PTH [10, 35].

- **Basic Multicellular Unit**—Basic multicellular units (BMU) are temporary anatomical structures, where autocrine and paracrine factors produced by osteoblasts and osteoclasts regulate the formation and activation of these cells. Bone resorbing osteoclasts lead an active BMU, removing old and damaged tissue. Osteoblasts then follow, occupying the tail portion of the BMU, secreting and depositing bone [33, 38].

Bone remodeling cycles can be activated by either mechanical *stimuli* on the bone, or through the production of estrogen or PTH [33] due to changes in homeostasis. The resorption phase is triggered by this hormone, which acts on cells of the osteoblastic lineage and leads to the differentiation and activation of osteoclasts. Active osteoclasts are then capable of degrading bone. At the resorbed site, the undigested demineralized collagen matrix is removed during the reversal phase, in preparation for bone formation. Then, osteoblasts form bone and replace the resorbed bone by the same amount, ending the bone remodeling cycle. The processes that link bone resorption to the initiation of bone formation may involve the release of coupling factors from bone during resorption phase, such as insulin growth factor I and II (IGF I and II) and transforming growth factor β (TGF-β), which attract osteoblasts to the sites of bone resorption [17]. Some of these processes are illustrated in Fig. 1.

1.2   *Tumor in the Bone Remodeling Cycle*

The previous section describes the main biochemical processes in healthy bone. When some pathology is present, the dynamic behavior of bone remodeling is severely affected, which usually disrupts its biochemical regulation and has a high impact on bone integrity. For example, in diseases such as cancer, either in bone tumors or through metastization of other primary tumors (e.g., breast or prostate cancer), the bone microenvironment is changed, affecting the osteoclasts/osteoblasts equilibrium. This leads, consequently, to bone lesions. Such pathophysiology is mostly explained by the theory of the vicious cycle proposed by Mundy and Guise in [14], according to which cancer cells resident in the bone can cause its destruction by stimulating osteoclast activity and receiving, in return, positive feedback from humoral factors released by the bone microenvironment during bone destruction [5].

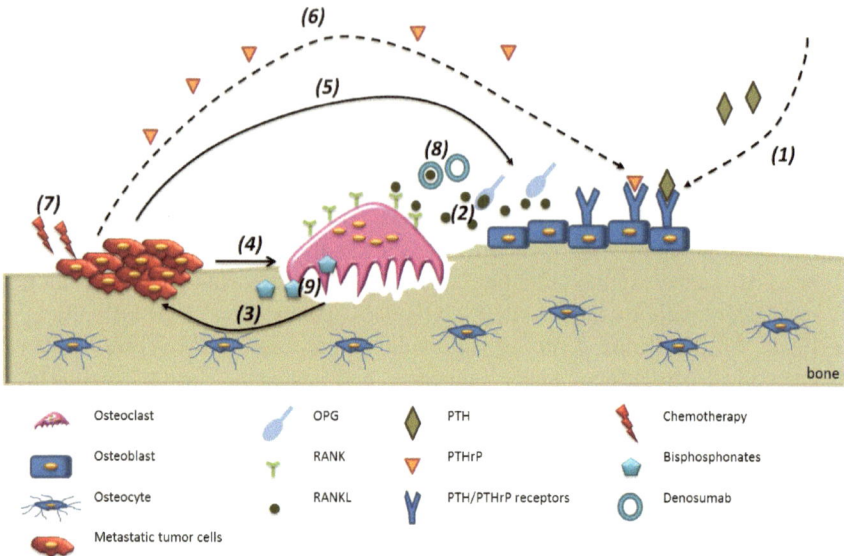

Fig. 1 Biochemical processes of bone remodeling, progression of bone metastases and treatment. Bone remodeling: (1) PTH stimulates RANKL production by osteoblasts; (2) RANK/RANKL/OPG pathway plays an important role in bone resorption and formation; bone metastases vicious cycle: (3) Bone-derived tumor growth factors (IGFs,TGF-β, bone morphogenetic protein (BMP), among others); (4) Tumor-derived factors stimulate bone resorption (parathyroid hormone-related protein (PTHrP), TGF-β, IL-8,11, among others); (5) Tumor-derived factors affect bone formation (DKK1, BMP, IGFs, among others); (6) PTHrP stimulates RANKL production by osteoblasts treatment: (7) Chemotherapy directly targets cancer cells; (8) Denosumab binds to RANKL, inhibiting osteoclast formation; (9) Bisphosphonates promote osteoclast apoptosis. Reprinted from [9]: Journal of Theoretical Biology, Vol 391, Coelho et al., Dynamic modeling of bone metastasis, microenvironment and therapy. Integrating parathyroid hormone (PTH) effect, anti-resorptive and anti-cancer therapy, Pages No. 1–12.Copyright 2016, with permission from Elsevier

Bone metastases can be osteolytic, in case bone resorption is increased, or osteoblastic, when bone formation is stimulated in an unstructured way. Both bone resorption and formation are still present in any case, although out of balance, resulting in loss of bone resistance and integrity. Breast cancer metastases are prone to develop osteolytic metastasis, and prostate cancer ones are usually osteoblastic [44].

For the bone remodeling deregulation resulting from *osteolytic* metastases, metastatic cells stimulate bone resorption [7], and TGF-β is released from the bone matrix, during bone resorption. TGF-β stimulates tumor growth and parathyroid hormone-related protein (PTHrP) production in metastatic cells that binds to PTH receptors on cells of osteoblastic lineage. RANKL levels are then enhanced, and, subsequently, osteoclasts are activated, leading to increased bone resorption [5]. Osteoclasts activity, in turn, will result in the release of TGF-β from the degraded bone, which further stimulates tumor growth and PTHrP secretion, giving rise to a vicious cycle.

In *osteoblastic* metastases, tumorous cells grow, as bone expresses endothelin-1 (ET-1) that stimulates osteoblasts through the endothelin A receptor (ETR), activating Wnt signaling. Tumor-derived proteases contribute to the release of osteoblastic factors from the extracellular matrix, including TGF-β and IGF I. RANKL is increased due to tumor-induced osteoblast activity, leading to the release of PTH and promoting osteoclast activity [5]. Thus, the tumor microenvironment leads to the accumulation of newly formed bone.

Although the reviewed models are mostly focused on cancer, these can be easily adapted to other bone pathologies. Such can be found in [41], where the model in [42] is extended to enable the simulation of postmenopausal osteoporosis effects on bone remodeling combined with the pharmacodynamical effects of drug treatment denosumab. Also, in [19], the original model in [17] is adapted to account for the effect of osteomyelitis, a bone pathology caused by bacteria infection (mostly *Staphylococcus aureus*), which alters the RANK/RANKL/OPG signaling dynamics that regulates osteoblasts and osteoclasts behavior in bone remodeling.

1.3 Bone Treatments

There are several possible targets for bone metastases treatment, as detailed in review [6], namely bone resorption, osteoblast, and tumor cells.

Anti-resorptive treatment targets osteoclasts, since these cells are essential to the vicious cycle of bone metastases. For osteolytic metastases, such as those from breast cancer, bisphosphonates or denosumab treatment is administrated as effective anti-resorptive therapies [5]. While bisphosphonates lodge in bone and poison osteoclasts as they degrade bone, denosumab, a fully human monoclonal antibody binds exclusively to RANKL, inhibiting osteoclast formation.

Although therapeutics that targets osteoblasts exists, such as PTH daily administration or endothelin, these are far less well-developed and used than anti-resorptive therapy, in order to decrease tumor burden on bone and recover bone mass.

Anti-cancer agents that target metastatic cells directly, such as chemotherapy and hormone therapy [22] should be used in combination with the other presented therapies.

2 Mathematical Bone Remodeling Local Models

Computational models of the dynamics of bone remodeling and its interaction with cancer cells are important to simulate the biochemical processes occurring in the bone microenvironment that potentiate the progression of such disease. The importance of understanding such a complex systems is highlighted in [26], which presents a review of mathematical modeling methodologies, applied to bone biology.

In all that follows, we will use systems of differential equations where D^1 is a first-order derivative in order to time, $\frac{d}{dt}$.

2.1 Healthy Remodeling Dynamics

The simplest model of bone remodeling involves a system of three ordinary differential equations representing osteoclast and osteoblast densities and normalized bone mass [17]. In this early proposal, bone remodeling is represented as an S-system [40], described by Eqs. (1a)–(1c), coupling the behavior of osteoclasts, $C(t)$, and osteoblasts, $B(t)$, through biochemical autocrine (g_{CC}, g_{BB}) and paracrine (g_{BC}, g_{CB}) factors expressed implicitly in the exponents.

$$D^1 C(t) = \alpha_c \, C(t)^{g_{CC}} \, B(t)^{g_{BC}} - \beta_c \, C(t) \tag{1a}$$

$$D^1 B(t) = \alpha_B \, C(t)^{g_{CB}} \, B(t)^{g_{BB}} - \beta_B \, B(t) \tag{1b}$$

$$D^1 z(t) = -k_c \, \max\{0,\, C(t) - C_{ss}\} + k_B \, \max\{0,\, B(t) - B_{ss}\} \tag{1c}$$

This is a local model that only takes into account temporal dynamics and whose variables and typical parameters values used in the simulations are described in the Table 3. It is worth noting that the RANK/RANKL/OPG pathway is implicitly encoded as their ratio is in the osteoblasts-derived osteoclasts paracrine regulator, g_{BC}. The constants $\alpha_{C,B}$ and $\beta_{C,B}$ represent the activation and apoptosis rate, respectively, of the corresponding cell type. Bone resorption from active osteoclasts and bone formation from active osteoblasts command the temporal evolution of the bone mass, z, through the rates k_C and k_B, respectively. The action of each cell type $C(t)$ and $B(t)$ is considered only when its number is above its steady-state values C_{ss} and B_{ss} (Eqs. 2a–2c).

$$C_{ss} = \left(\frac{\beta_c}{\alpha_c}\right)^{\frac{(1-g_{BB})}{\Gamma}} \left(\frac{\beta_B}{\alpha_B}\right)^{\frac{g_{BC}}{\Gamma}} \tag{2a}$$

$$B_{ss} = \left(\frac{\beta_c}{\alpha_c}\right)^{\frac{g_{CB}}{\Gamma}} \left(\frac{\beta_B}{\alpha_B}\right)^{\frac{(1-g_{CC})}{\Gamma}} \tag{2b}$$

$$\Gamma = g_{CB}\, g_{BC} - (1 - g_{CC})(1 - g_{BB}) \tag{2c}$$

By setting the values of the exponents appropriately, in particular g_{BC}, this model is capable of describing a single remodeling cycle or a periodic behavior, whose amplitude and frequency of oscillations depend on the initial conditions, as can be seen in Figs. 2 and 3, respectively. The parameters used in these simulations are those from the values in column A of Table 3, except those explicitly defined in the caption of the figures, in accordance with [17].

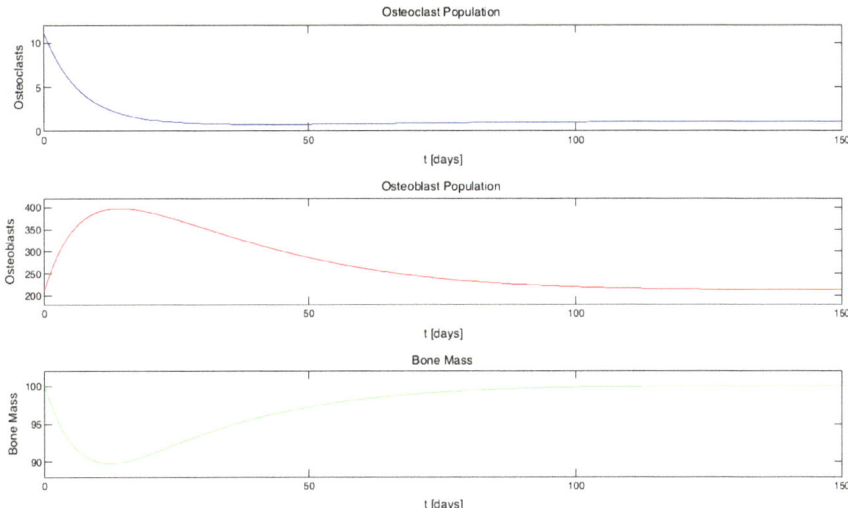

Fig. 2 Simulation of the model in Eqs. (1a)–(1c): number of osteoclasts, osteoblasts, and bone mass during normal bone modeling for a *single event*, triggered by an increase of the osteoclast population. The parameters are those in column A of Table 3 [2]

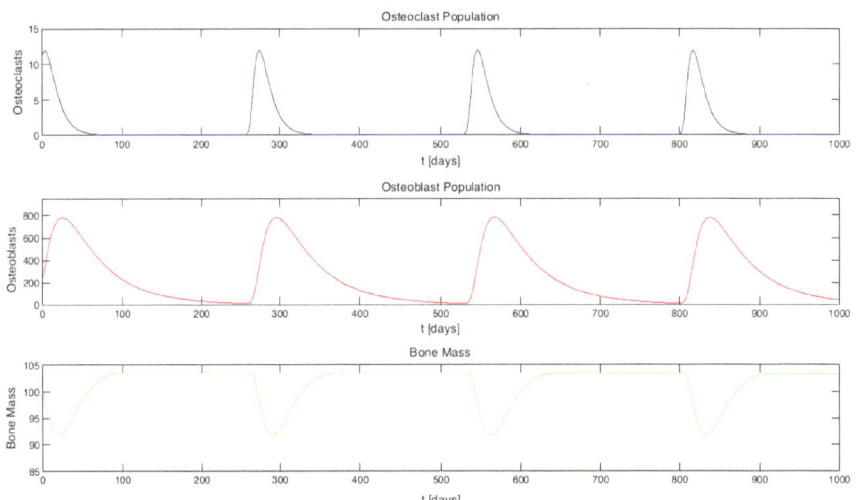

Fig. 3 Simulation of *oscillatory changes* in the number of osteoclasts, osteoblasts, and bone mass during normal bone remodeling for which the model presented in [17] has periodic solutions. Such behavior was triggered with an increase of ten units of the steady-state osteoclast population, $C_0 = 11.16$ and $C_{th} = 1.16$. The parameters were set according to [2]: $g_{CC} = 1.1$, being the responsible for the bone dynamic response, $\kappa_C = 0.0748$, $\kappa_B = 6.3952 \times 10^{-4}$ and $B_0 = B_{th} = 231.72$. Other parameters were set according to column A of Table 3

Due to its simplicity and versatility, this model serves as a basis for many other studies. In [16], the anabolic and catabolic effects of external administration of PTH on bone remodeling are studied, by using the single remodeling cycle behavior exhibited in [17]. The influence of PTH in the production of RANKL by osteoblasts is encoded in the exponent g_{BC}, adding a pulse of fixed duration at a specific time. The duration of application of PTH influences qualitatively the bone mass outcome and can be used to achieve different scenarios of bone modeling. It is possible to apply bifurcation analysis to generalized bone models [50], in particular to the model in [17], adding the osteoblasts precursors to the system. A stability analysis is hence achieved for this type of models.

More recently, the integration of PTH into the model was proposed [9], as given by Eqs. (3a)–(3e).

$$D^1 C(t) = \alpha_c C(t)^{g_{CC}} B(t)^{g_{BC} + K_{\text{PTH}_{\text{pool}} BC} \text{PTH}_{\text{pool}}(t)} - \beta_c C(t), \tag{3a}$$

$$D^1 B(t) = \alpha_B C(t)^{g_{CB}} B(t)^{g_{BB}} - \beta_B B(t), \tag{3b}$$

$$D^1 \text{PTH}_{\text{pool}}(t) = -\beta_{\text{PTH}} \text{PTH}_{\text{pool}}(t) + K_{\text{PTH}} \delta(t), \tag{3c}$$

$$P(\delta(t) = 1) = 1 - \exp\left(-\left(\frac{t}{\lambda_W}\right)^{k_W}\right), \tag{3d}$$

$$D^1 z(t) = -k_c \max\{0, C(t) - C_{ss}\} + k_B \max\{0, B(t) - B_{ss}\}, \tag{3e}$$

The PTH concentration is included in $\text{PTH}_{\text{pool}}(t)$, and $\delta(t)$ represents a train of Dirac deltas, occurring stochastically with a Weibull distribution with probability P given by Eq. (3d). K_{PTH} is the increase of $\text{PTH}_{\text{pool}}(t)$ with each Dirac delta. This increase in PTH concentration is responsible for initiating a single remodeling cycle, which increases the production of RANKL by osteoblasts, thus affecting g_{BC}, as in Eq. (3a). Variable $K_{\text{PTH}_{\text{pool}} BC}$ quantifies this influence on the RANKL/OPG ratio. [13] further extends the model in [17] to include osteocytes and pre-osteoblasts in the bone model, predicting an osteocyte-induced bone remodeling after apoptosis. The role of sclerostin is also included, enabling the study of anti-sclerostin drugs in bone turnover.

Lemaire et al. [18] present a different approach, incorporating explicitly the RANK/RANKL/OPG pathway, TGF-β, and PTH in the model. The kinetic reactions of these molecules can activate or repress mechanisms of differentiation and activation of bone cells, here considered to be the uncommitted osteoblast precursors B_u, osteoblast precursors B_p, active osteoblasts B_a, and active osteoclasts C_a. This model serves as the basis to [27], whose model is presented in Eqs. (4a)–(4d).

$$D^1 B_p(t) = \alpha_{B_u} \pi_{\text{act}, B_u}^{\text{TGF-}\beta} - \alpha_{B_p} B_p(t) \pi_{\text{rep}, B_p}^{\text{TGF-}\beta} \tag{4a}$$

$$D^1 B_a(t) = \alpha_{B_p} B_p(t) \pi_{\text{rep}, B_p}^{\text{TGF-}\beta} - \beta_{B_a} B_a(t) \tag{4b}$$

$$D^1 C_a(t) = \alpha_{C_p} C_p \pi_{\text{act}, C_p}^{\text{RANKL}} - \beta_{C_a} C_a(t) \pi_{\text{act}, C_p}^{\text{TGF-}\beta}, \tag{4c}$$

$$D^1 z = -k_C(C_a(t) - C_a(t_0)) + k_B(B_a(t) - B_a(t_0)) \tag{4d}$$

Functions $\pi_{\text{act/rep, cell}}^{\text{molecule}}$ represents activation (act) or repression (rep) of a given process in a cell type (cell) when binding to the molecule considered (molecule), and the corresponding coefficients are α_{cell}. It is usual to describe them as Hill functions, given by Eqs. (5a) and (5b), where X is the concentration of the molecule driving a cellular process, and K_1 and K_2 are activation or repression coefficients.

$$\pi_{\text{act}}^X = \frac{X}{K_1 + X} \tag{5a}$$

$$\pi_{\text{rep}}^X = \frac{1}{1 + X/K_2} \tag{5b}$$

Osteoblast precursors B_p differentiate from uncommitted osteoblast progenitors B_u at rate α_{B_u}, promoted by TGF-β. Osteoblast precursors differentiate into active osteoblasts B_a at rate α_{B_p}, inhibited by TGF-β. The number of active osteoblasts increases through this differentiation, but these cells have an apoptosis rate of β_{B_a}. The differentiation of osteoclasts precursors C_p into active osteoclasts C_a is promoted by RANKL, at rate α_{C_p}. TGF-β also promotes the apoptosis of active osteoclasts, occurring at rate β_{C_a}. It is assumed that the number of osteoclast precursors C_p and uncommitted osteoblast progenitors B_u is large and constant. The bone mass z is affected by bone resorption and formation, proportionally to the deviation from steady state of the number of active osteoclasts and osteoblasts, respectively. These cells are considered to be in their steady state at initial time t_0. Models for the concentrations of TGF-β and RANKL are proposed. The TGF-β release rate from the bone during resorption is proportional to the number of active osteoclasts. The RANK/RANKL/OPG pathway is explicitly modeled, including the influence of PTH on RANKL and OPG concentrations. The model is then studied for different scenarios, namely the importance of RANKL and OPG expression in the bone remodeling cycle. Moreover, the model parameter space is searched for physiologically sensible behaviors. In [28], the role of RANK/RANKL/OPG pathway is studied following the model in [27]. Furthermore, it was possible to simulate several bone diseases by changing the values of parameters, as well as different treatment strategies for these conditions, following the same approach.

2.2 Models Including Tumor Burden

Cancer interferes with the bone remodeling equilibrium, taking advantage of the dysregulated microenvironment to proliferate. Studying such alterations is important for cancers that begin in bone tissue, such as multiple myeloma, but also for bone metastases that often result from some kinds of cancer, such as breast or prostate cancer [44]. A modified version of the model in [17] which introduces multiple myeloma disease is presented in [2] and is described in Eqs. (6a)–(6d).

$$D^1 C(t) = \alpha_c C(t)^{\left(g_{cc}\left(1 + r_{cc}\frac{T(t)}{L_T}\right)\right)} B(t)^{\left(g_{BC}\left(1 + r_{BC}\frac{T(t)}{L_T}\right)\right)} - \beta_c C(t), \tag{6a}$$

$$D^1 B(t) = \alpha_B C(t)^{\left(\frac{g_{CB}}{1 + r_{CB}\frac{T(t)}{L_T}}\right)} B(t)^{\left(g_{BB} - r_{BB}\frac{T(t)}{L_T}\right)} - \beta_B B(t), \tag{6b}$$

$$D^1 T(t) = \gamma_T T(t) \log \frac{L_T}{T(t)}, \tag{6c}$$

$$D^1 z(t) = -k_c \max\{0, C(t) - C_{ss_T}\} - k_B \max\{0, B(t) - B_{ss_T}\}. \tag{6d}$$

In this model, $T(t)$ represents tumor concentration and its growth; $D^1 T(t)$ is made independent of the bone microenvironment, following a Gompertz law with growth rate γ_T and maximum size L_T, described by Eq. (6c). It affects the autocrine parameters in the exponents of the system through r_{cc} and r_{BB} and paracrine parameters through r_{BC} and r_{CB}, contributing to the increase in the number of osteoclasts and the decrease in osteoblasts when set to non-negative values. This results in the deregulation of the periodic remodeling cycles and, consequently, in the decrease of the overall bone mass. The equation for bone mass $z(t)$ keeps the same structure, although the thresholds to determine the number of active bone cells are no longer the steady states in healthy bone remodeling (C_{ss}, B_{ss}), but the steady states are (C_{ss_T}, B_{ss_T}) computed considering $T(t)$ to be at its maximum value L_T.

The model in [9], as presented in Eqs. (3a)–(3e), is extended in the same work to include the growth of bone metastasis and its influence on the bone microenvironment, described by Eqs. (7a)–(7g).

$$D^1 C(t) = \alpha_c C(t)^{\left(g_{cc} + r_{cc}\frac{T(t)}{L_T}\right)} B(t)^{\left(g_{BC} + K_{PTH_{pool\,BC}}PTH_{pool}(t)\right)} - \beta_c C(t), \tag{7a}$$

$$D^1 B(t) = \alpha_B C(t)^{\left(\frac{g_{CB}}{1 + r_{CB}\frac{T(t)}{L_T}}\right)} B(t)^{\left(g_{BB} - r_{BB}\frac{T(t)}{L_T}\right)} - \beta_B B(t), \tag{7b}$$

$$D^1 PTH_{pool}(t) = -\beta_{PTH} PTH_{pool}(t) + K_{PTH}\delta(t)$$
$$+ r_{PTHrP} \max\{0, C(t) - C_{th}(t)\}\frac{T(t)}{L_T}, \tag{7c}$$

$$P(\delta(t) = 1) = 1 - \exp\left(-\left(\frac{t}{\lambda_W}\right)^{k_W}\right), \tag{7d}$$

$$D^1 T(t) = k_T \max\{0, C(t) - C_{th}(t)\}\frac{T(t)}{\lambda_T + T(t)}, \tag{7e}$$

$$D^1 C_{th}(t) = \alpha_c C_{th}(t)^{\left(g_{cc} + r_{cc}\frac{T(t)}{L_T}\right)} B_{th}(t)^{g_{BC}} - \beta_c C_{th}(t), \tag{7f}$$

$$D^1 B_{th}(t) = \alpha_B C_{th}(t)^{g_{CB}} B_{th}(t)^{\left(g_{BB} + r_{BB}\frac{T(t)}{L_T}\right)} - \beta_B B_{th}(t). \tag{7g}$$

Metastatic cells produce PTHrP in the presence of TGF-β, which is released from bone after bone resorption, encoded by the term $r_{PTHrP} \max\{0, C(t) - C_{th}(t)\}\frac{T(t)}{L_T}$ in Eq. (7c). Osteoblasts have the same receptors for both PTH and PTHrP, which contribute to the production of RANKL. As such, PTHrP concentration is added to the

PTH$_{pool}$ dynamics, which now includes PTH and PTHrP concentration. Increased production of RANKL means the number of osteoclasts will rise, resulting in an increased bone resorption. Bone resorption releases tumor growth factors from bone, allowing the tumor to grow, as described by Eq. (7e), and to produce PTHrP, which will, in turn, indirectly stimulate the activation of osteoclasts and bone resorption. As such, this model is able to describe the vicious cycle of bone metastases through the action of PTHrP. The growth rate of metastases includes a saturation term on the size of the tumor, introduced by a sigmoid function. As bone metastases $T(t)$ depend on growth factors released during bone resorption to grow, the dynamics for tumor growth is given by Eq. (7e), in which its growth rate due to tumor size is limited by a sigmoid function. Besides affecting the number of osteoclasts indirectly through PTH$_{pool}$, the tumor also affects the dynamics of the system through the autocrine parameters, as in [2]. In the presence of tumor, the thresholds for active osteoclasts and osteoblasts are no longer static C_{th} and B_{th}, since the steady state is affected thereby. Dynamic thresholds $C_{th}(t)$ and $B_{th}(t)$ are then used to determine the number of active cells, as given by Eqs. (7f) and (7g), respectively.

The bone remodeling model of [27] is extended in [49] to include the influence of multiple myeloma in the bone microenvironment. The explicit inclusion of IL-6 and multiple myeloma bone marrow stromal cell adhesion in the model, along with the multiple myeloma disease, made it possible to predict a vicious cycle between bone and multiple myeloma cells.

2.3   Introducing Treatment

2.3.1   Pharmacokinetics and Pharmacodynamics

The **pharmacokinetics** (PK) of a drug describes its concentration evolution at the target tissue; whereas, the effect of such concentration is given by **pharmacodynamics** (PD). A PK one-compartment model with first-order absorption and elimination for subcutaneous administration is described by Eqs. (8a) and (8b), where C_g is the drug concentration yet to be absorbed and C_p the concentration in the plasma [12]. The drug is absorbed at rate k_a and eliminated at rate k_e.

$$D^1 C_g(t) = -k_a C_g(t) \tag{8a}$$

$$D^1 C_p(t) = k_a C_g(t) - k_e C_p(t) \tag{8b}$$

$$C^0 = \frac{D_0 F}{V_d} \tag{8c}$$

The initial drug concentration C^0 can be computed from the initial administered dose D_0, the bioavailability F, and volume distribution V_d, following Eq. (8c). The plasma concentration can be described in the Laplace domain, $C_p(s)$ as Eq. (9), where s is the Laplace transform variable of time.

$$C_p(s) = C^0 \frac{k_a}{(s + k_a)(s + k_e)} \tag{9}$$

From Eq. (9), the time response for single-dose administration is ruled by Eq. (10).

$$C_p(t) = C^0 \frac{k_a}{k_a - k_e} (e^{-k_d t} - e^{-k_a t}) \tag{10}$$

For multiple dose administration, assuming that every dose initial concentration is C^0 and they are administrated at a constant time interval τ, the drug concentration after the nth dose is described by

$$C_p(n, t') = C^0 \frac{k_a}{k_a - k_e} \left(\frac{1 - e^{-nk_e\tau}}{1 - e^{-k_e\tau}} e^{-k_e t'} - \frac{1 - e^{-nk_a\tau}}{1 - e^{-k_a\tau}} e^{-k_a t'} \right) \tag{11}$$

where $t' = t - (n - 1)\tau$ represents the time elapsed after the nth dose. For single intravenous (IV) administration, the PK is given by Eq. (12a), where the initial dose D_0 is included in the initial condition of C_p as Eq. (12b).

$$\frac{dC_p(t)}{dt} = -k_e C_p(t), \tag{12a}$$

$$C^0 = \frac{D_0}{V_d} \tag{12b}$$

In steady state, that is, after a large number of doses, the average concentration will be $\bar{C}_{p_{ss}} = \frac{1}{\tau} \frac{C^0}{k_e}$.

The effect of a drug $d(t)$ according to its concentration in plasma C_p can be given by a Hill function as in Eq. (13), where an effect of 50% is achieved at concentration C_{50}.

$$d(t) = \frac{C_p(t)}{C_{50} + C_p(t)} \tag{13}$$

The variables and parameters involved in the PK/PD model are summarized in Table 1.

2.3.2   Models for Tumor Treatment

The models proposed in [2] also allowed for tumor treatment. Since proteasome inhibitors are known to have direct anti-myeloma effects and to have direct effects on osteoblasts to stimulate osteoblast differentiation and bone formation, two time-dependent treatment step functions, $V_1(t)$ and $V_2(t)$, were introduced to model the inhibition of the osteoblasts apoptosis and tumor cells death, respectively. As such, Eqs. (6b) and (6c) are replaced by Eqs. (14a) and (14b), respectively, with $V_1(t)$ and $V_2(t)$ affecting the suitable parameters. This model including treatment is therefore

Table 1 Variables and parameters of the PK/PD model

Variable	Description	Units
$C_p(t)$	Effective drug concentration in the plasma	mg/L
$C_g(t)$	Concentration of drug remaining to be absorbed	mg/L
Parameter	Description	Units
C^0	Initial plasma concentration	mg/L
D_0	Drug dosage	mg
τ	Administration time interval	day
k_e	Drug elimination rate	day^{-1}
k_a	Drug absorption rate	day^{-1}
F	Bioavailability	–
V_d	Volume distribution	L
C_{50}	Drug concentration for 50% of maximum effect	mg/L

composed of Eqs. (6a), (14a), (14b), and (6d), resulting in an increase of bone mass and the elimination of tumor.

$$D^1 B(t) = \alpha_{_B} C(t)^{\left(\frac{g_{CB}}{1+r_{CB}\frac{T(t)}{L_T}}\right)} B(t)^{\left(g_{BB}-r_{BB}\frac{T(t)}{L_T}\right)} - \left(\beta_{_B} - V_1(t)\right) B(t) \quad (14a)$$

$$D^1 T(t) = \left(\gamma_{_T} - V_2(t)\right) T(t) \log\left(\frac{T(t)}{L_T}\right) \quad (14b)$$

This work [9] also proposes the treatment of bone metastases through anti-cancer therapy and anti-resorptive therapy corresponding to the administration of either bisphosphonates or denosumab. Bisphosphonates (zoledronic acid) promote osteoclast apoptosis, and denosumab acts as a decoy receptor for RANKL, indirectly inhibiting osteoclast formation. Since both therapies act on osteoclasts, the effects of bisphosphonates, d_1, and denosumab, d_2, are included in Eq. (7a) as described in Eq. (15a). Chemotherapy (paclitaxel), whose effect is given by d_3, is considered for anti-cancer therapy, targeting metastatic cells directly, and so it is included in Eq. (7e) as shown in Eq. (15b). The treatment also affects the threshold for active osteoclasts, C_{th}, described by Eq. (15c). Including treatment variables, this new model is composed of Eqs. (15a), (7b), (7c), (7d), (15b), (15c), and (7g).

$$D^1 C(t) = \alpha_{_c} C(t)^{g_{cc}+r_{cc}\frac{T(t)}{L_T}} B(t)^{g_{BC}+K_{\text{PTH}_{\text{pool}BC}}\text{PTH}_{\text{pool}}(t)-K_{d_1}d_1(t)}$$
$$- (\beta_{_c} + K_{d_2}d_2(t))C(t), \quad (15a)$$

$$D^1 T(t) = k_{_T}\max\{0, C(t) - C_{th}(t)\}\frac{T(t)}{\lambda_T + T(t)} - K_{d_3}d_3(t)T(t), \quad (15b)$$

$$D^1 C_{th}(t) = \alpha_{_c} C_{th}(t)^{g_{cc}+r_{cc}\frac{T(t)}{L_T}} B_{th}(t)^{g_{BC}-K_{d_1}d_1(t)}$$
$$- (\beta_{_c} + K_{d_2}d_2(t))C_{th}(t), \quad (15c)$$

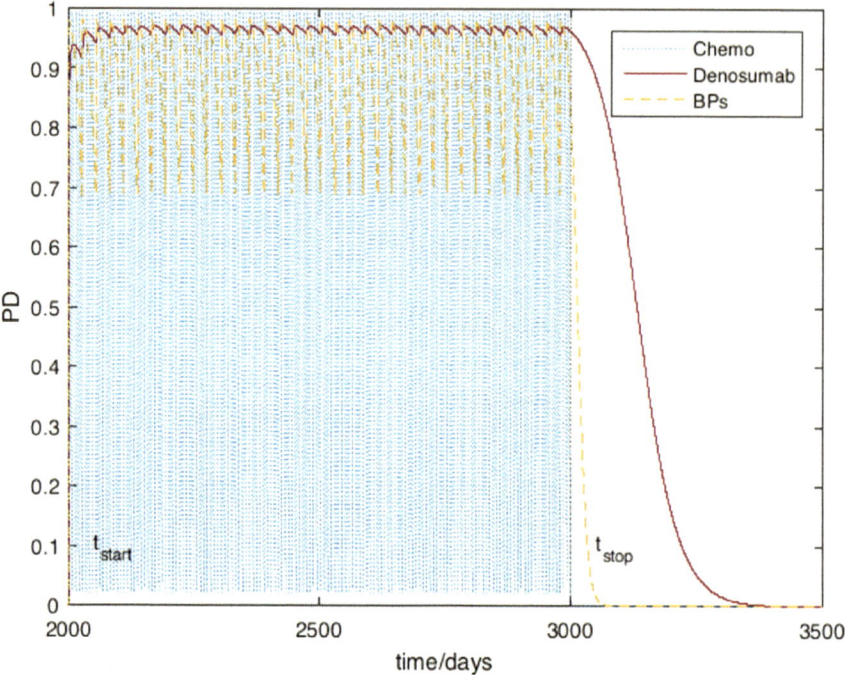

Fig. 4 PD effect of chemotherapy (paclitaxel), denosumab, and bisphosphonates (zoledronic acid), with administration parameters from Table 2 [9]

Table 2 Parameters for the therapy model (15), following [9]: denosumab, d_1; bisphosphonates, d_2 (zoledronic acid); anti-cancer therapy (paclitaxel), d_3

Parameter	d_1	d_2	d_3
D_0	120	4	176
τ	28	28	7
k_e	0.0248	0.1139	1.2797
k_a	0.2568	–	–
F	0.62	–	–
V_d	3.1508	536.3940	160.2570
C_{50}	1.2	0.0001	0.002
K_d	0.48	1.2	0.017

The PK/PD of denosumab (subcutaneous administration), zoledronic acid, and paclitaxel (both IV administration) is included in the model as described in Sect. 2.3.1. Figure 4 illustrates a simulation of the pharmacodynamics of these drugs, with administration starting at time $t_{\text{start}} = 2000$ days and interrupted at $t_{\text{stop}} = 3000$ days. The parameters for PK/PD are shown in Table 2. It is possible to observe that chemotherapy effect oscillates more than both drugs used for anti-resorptive therapy.

The treatment of bone metastases with chemotherapy alone and combined with either denosumab or bisphosphonates results in the dynamics shown in Figs. 5 and 6. The parameters used in the simulations are those from the values column B in Table 3, in accordance with [9].

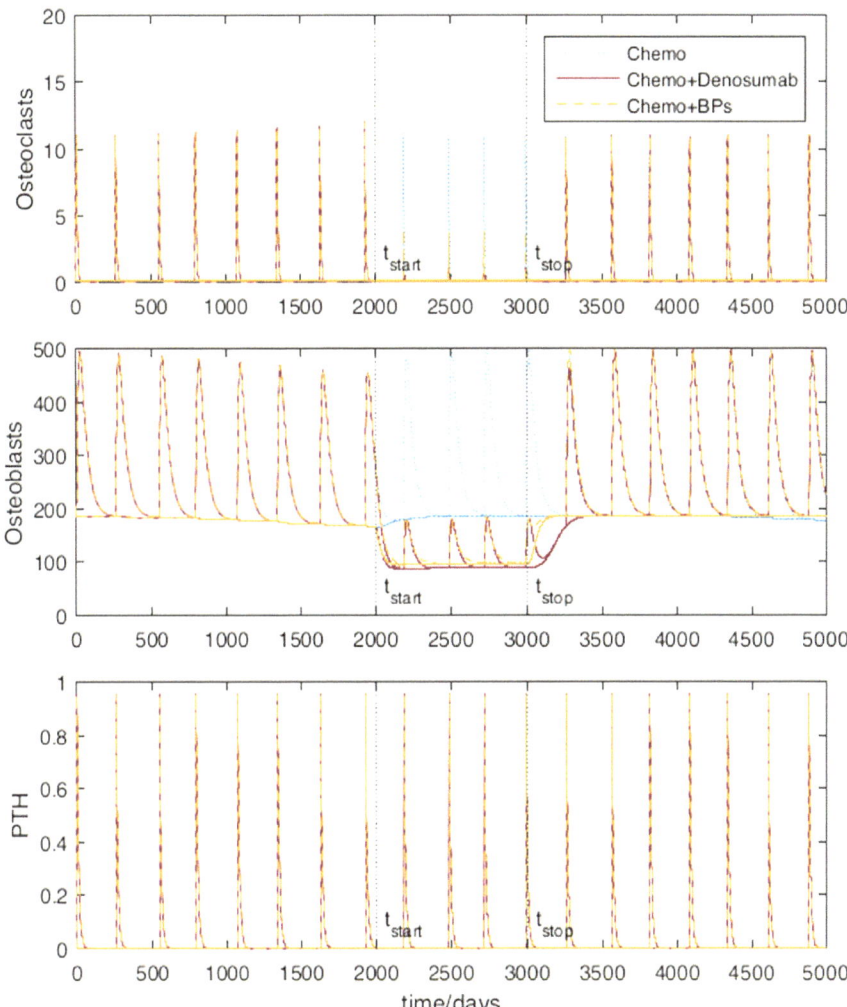

Fig. 5 Dynamics of bone microenvironment affected by metastases, with treatment through chemotherapy alone, chemotherapy combined with denosumab and combined with bisphosphonates, starting at time $t_{start} = 2000$ days and interrupted at $t_{stop} = 3000$ days. Following Eqs. (15a), (7b), (7c), (7d), (15b), (15c), and (7g), using the parameters from column B in Table 3, in accordance with [9]

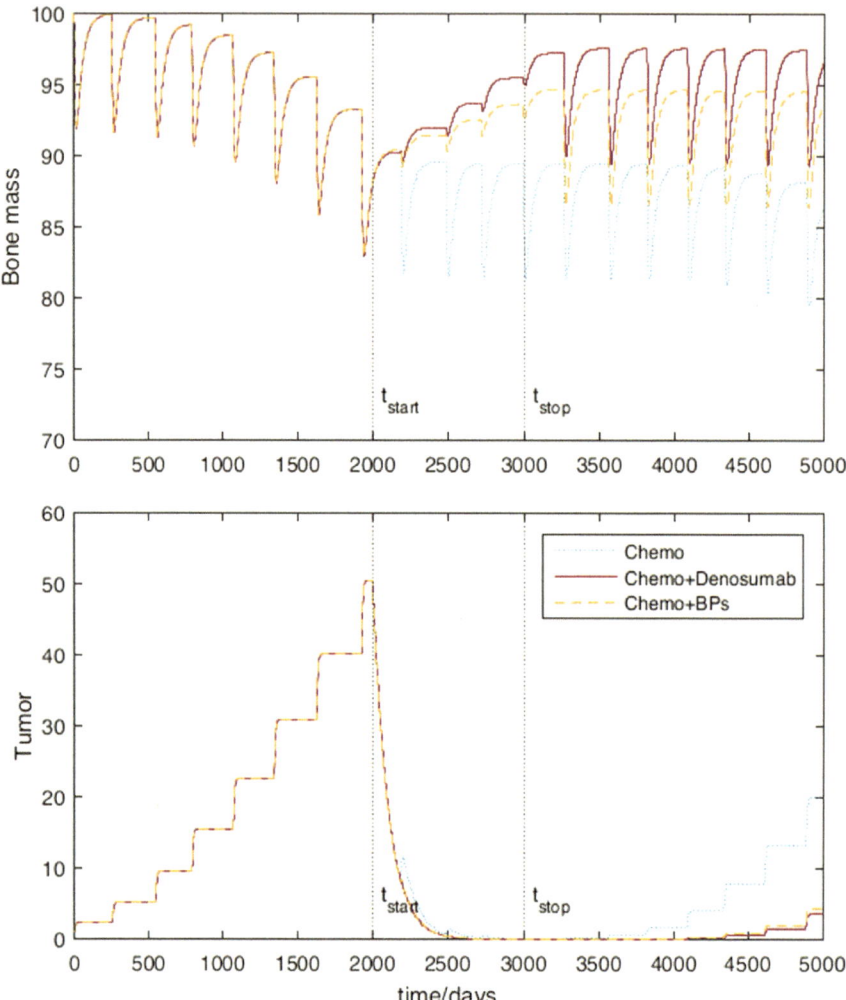

Fig. 6 Bone metastases and bone mass evolution, with treatment through chemotherapy alone, chemotherapy combined with denosumab and combined with bisphosphonates, starting at time $t_{start} = 2000$ days and interrupted at $t_{stop} = 3000$ days. Following Eqs. (15a), (7b), (7c), (7d), (15b), (15c), and (7g), using the parameters from column B in Table 3, in accordance with [9]

As it can be observed, the treatment with chemotherapy reduces the size of the tumor, but does not reduce bone loss during treatment. However, in combination with anti-resorptive therapy, it is possible to recover bone mass while eliminating tumor.

Table 3 Variables, parameters, and initial conditions of the simulated models: (*A*) Eqs. (1a)–(1c) [17], (*B*) Eqs. (15a), (7b), (7c), (7d), (15b), (15c), and (7g) [9], (*C*) Eqs. # [8]

Operator	Description				
D^1	First-order derivative, i.e., D^α with $\alpha = 1$				
D^α	Fractional order derivative				
$D^{\alpha(t)}$	Variable order derivative				
∇^2	Non-local, second-order, vector differential operator				
$\frac{\partial^2}{\partial x^2}$	Partial, second-order derivative for bone distance x				
Variable	Description	Units			
t	Time	day			
x	Adimensional distance along the bone	–			
$C(t, x)$	Number of osteoclasts	cells			
$B(t, x)$	Number of osteoblasts	cells			
$\text{PTH}_{\text{pool}}(t, x)$	PTH and PTHrP concentration variation	ng/L			
$T(t, x)$	Bone metastases size	%			
$z(t, x)$	Bone mass	%			
$d_1(t)$	Effect of denosumab	–			
$d_2(t)$	Effect of bisphosphonates	–			
$d_3(t)$	Effect of anti-cancer therapy	–			
$C_{th}(t, x)$	Threshold for active osteoclasts	cells			
$B_{th}(t, x)$	Threshold for active osteoblasts	cells			
$\alpha(t)$	Variable order for differential equations	cells			
Parameter	Description	Units	*Values A*	*Values B*	*Values C*
α	Fractional order of the equations	–	–	–	$\in\,]0, 2[$
σ_C	Diffusion coefficient of osteoclasts	–	–	–	10^{-6}
σ_B	Diffusion coefficient of osteoblasts	–	–	–	10^{-6}

(continued)

Table 3 (continued)

Parameter	Description	Units	Values A	Values B	Values C
σ_z	Diffusion coefficient of the bone	–	–	–	10^{-6}
σ_T	Diffusion coefficient of the tumor	–	–	–	10^{-6}
α_C	Osteoclasts activation rate	cell^{-1} day$^{-\alpha}$	3	3	3
α_B	Osteoblasts activation rate	cell^{-1} day$^{-\alpha}$	4	4	4
β_C	Osteoclasts apoptosis (programmed cell death) rate	day$^{-\alpha}$	0.2	0.2	0.2
β_B	Osteoblasts apoptosis rate	day$^{-\alpha}$	0.02	0.02	0.02
g_{CC}	Osteoclasts autocrine regulator	–	0.5	0.1	1.1
g_{BC}	Osteoblasts-derived osteoclasts paracrine regulator	–	-0.5	-1	-0.5
g_{CB}	Osteoclasts-derived osteoblasts paracrine regulator	–	1.0	0.8	1
g_{BB}	Osteoblasts autocrine regulator	–	0	0.2	0
k_C	Bone resorption rate	% cell^{-1} day$^{-\alpha}$	0.24	0.5	0.45
k_B	Bone formation rate	% cell^{-1} day$^{-\alpha}$	0.0017	0.00248723	0.0048
K_{PTH}	PTH growth rate	ng L^{-1} day$^{-\alpha}$	–	1	–
β_{PTH}	PTH and PTHrP degradation rate	day$^{-\alpha}$	–	0.1	–

(continued)

Table 3 (continued)

Parameter	Description	Units	*Values A*	*Values B*	*Values C*
$K_{\text{PTH}_{\text{pool}_{BC}}}$	Influence of PTH/PTHrP in RANKL/OPG ratio	ng^{-1} L	–	1.261	–
k_W	Shape parameter of Weibull distribution	–	–	15	–
λ_W	Scale parameter of Weibull distribution	–	–	300	–
k_T or γ_T	Bone metastases growth rate through bone resorption	% $cell^{-1}$ $day^{-\alpha}$	–	1	0.004
λ_T	Half-saturation constant for bone metastases size	%	–	10	–
L_T	Maximum size of bone metastases	%	–	100	100
r_{CC}	Effect of tumor in osteoclasts autocrine regulator	–	–	0.022	0.005
r_{CB}	Effect of tumor in osteoblasts-derived osteoclasts paracrine regulator	–	–	–	0
r_{BC}	Effect of tumor in Osteoclasts-derived osteoblasts paracrine regulator	–	–	–	0

(continued)

Table 3 (continued)

Parameter	Description	Units	Values A	Values B	Values C
r_{BB}	Effect of tumor in osteoblasts autocrine regulator	–	–	−0.198	0.2
r_{PTHrP}	Rate of PTHrP production by cancer cells	ng L^{-1} cell^{-1} day$^{-\alpha}$	–	0.0043	–
K_{d_1}	Maximum effect of denosumab	–	–	0.48	0.1
K_{d_2}	Maximum effect of bisphosphonates	day$^{-\alpha}$	–	1.2	0.1
K_{d_3}	Maximum effect of anti-cancer therapy	% day$^{-\alpha}$	–	0.017	2.3
C_{ss}	Steady-state value of $C(t, x)$	cells	1.06	–	
B_{ss}	Steady-state value of $B(t, x)$	cells	212.13	–	
Initial conditions	Description	Units	Values A	Values B	Values C
C_0 or $C(0, x)$	Initial number of osteoclasts	cells	$C_{ss} + 10$	C_{ss}	$C(0, x)$ [2]
B_0 or $B(0, x)$	Initial number of osteoblasts	cells	B_{ss}	B_{ss}	316
z_0 or $z(0, x)$	Initial bone mass percentage	%	100	100	100
PTH$_{\text{pool0}}$	Initial PTH and PTHrP concentration	ng/L	–	0	–
T_0 or $T(0, x)$	Initial size of bone metastases	%	–	1	

2.4 Models Based on Fractional and Variable Order Derivatives

2.4.1 Introduction to Fractional and Variable Order Calculus

Biological processes often present anomalous diffusion [21], and while it cannot be said with certainty that this is the case for bone remodeling, it is very likely that it be so. To account for this, it is necessary to introduce *fractional derivatives* in the existing mathematical models (such an adaptation has been done and studied in

[8]). Fractional derivatives are no more than a generalization of differentiation and integration of order $n \in \mathbb{N}$ to orders $\alpha \in \mathbb{R}$. The most natural and appealing example is that of the exponential function $f(x) = e^{ax}$, whose n*th* derivative is given simply by $a^n e^{ax}$. This suggests defining the derivative of order α that has not necessarily to be an integer, as $a^\alpha e^{ax}$. Since its idealization, the number of applications with fractional derivatives has been rapidly growing. Recently, the fractional order linear time invariant (FOLTI) systems have attracted significant attention in the control systems society [24], and fractional differential equations find a very wide range of applications: They are used to formulate and solve different physical models allowing a continuous transition from relaxation to oscillation phenomena; to predict the nonlinear survival and growth curves of foodborne pathogens; to adapt the viscoelasticity equations (Hooke's Law, Newtonian fluids Law), among other applications [34]. It also plays an important role in physics, thermodynamics, electrical circuits theory and fractances, mechatronics systems, signal processing, chemical mixing, chaos theory, to name a few [24]. However, many physical processes appear to exhibit a *fractional order behavior that varies with time or space*: In the field of viscoelasticity, the effect of temperature on the small amplitude creep behavior of certain materials is known to change its characteristics from elastic to viscoelastic or viscous, and real applications may require a time varying temperature to be analyzed; the relaxation processes and reaction kinetics of proteins that are described by fractional differential equations have been found to have a temperature dependence; the behavior of some diffusion processes in response to temperature changes can be better described using variable order elements rather than time varying coefficients; and more [20]. Such evidence allowed to consider that the fractional order of integrals and derivatives could be a function of time or some other variable, thus introducing the concept of variable order (or structure) operators, where the order of the operator is allowed to vary either as a function of the independent variable of integration or differentiation (t) or as a function of some other (perhaps spatial) variable (y) [20, 48]. More information on fractional and variable order calculus can be found in this chapter's Appendix.

2.4.2 Local Models with Variable Derivatives

A new model is proposed by the authors, where the model described in Eq. (6)—corresponding to local tumorous model, without treatment—was simplified by including variable order derivatives in the osteoclasts and osteoblasts equations, whose time-dependent order is given by $\alpha(t)$. Hence, by doing so, tumor itself would influence the order directly, and it is now the order responsible for the change in the behavior of the osteoclasts and osteoblasts, instead of inducing such behavior through the tumorous parameters (r_{ij}). The remaining parameters are to be set according to the healthy bone remodeling case, for periodic cycles.

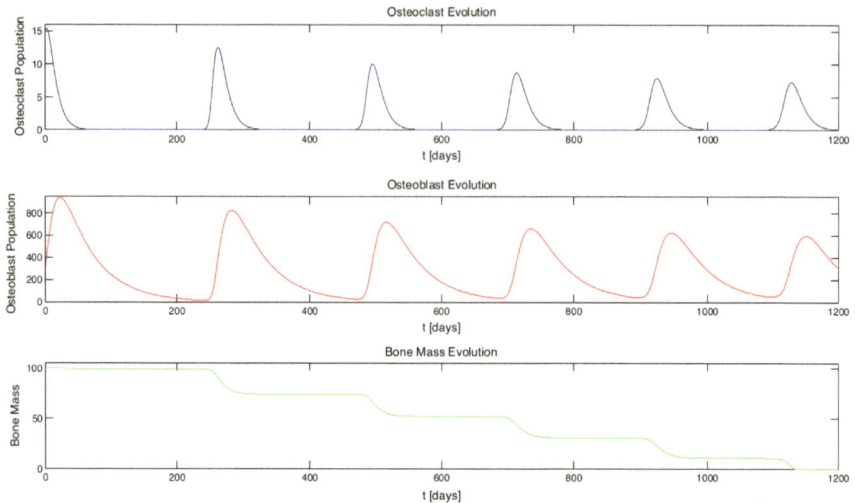

Fig. 7 Simulation results for the model in Sect. 2.4.2 simplifying the existing equations for the local tumorous influence in [2]. Parameters used for simulation are the same as the referred case

$$D^{\alpha(t)} C(t) = \alpha_c \, C(t)^{g_{CC}} \, B(t)^{g_{BC}} - \beta_c \, C(t) \tag{16a}$$

$$D^{\alpha(t)} B(t) = \alpha_B \, C(t)^{g_{CB}} \, B(t)^{g_{BB}} - \beta_B \, B(t) \tag{16b}$$

$$D^1 T(t) = \gamma_T \, T(t) \log \frac{L_T}{T(t)} \tag{16c}$$

$$\alpha(t) = 1 - T(t)\rho(t), \quad \text{where} \quad \rho(t) = \frac{0.00005}{T_{\text{simulation}}} t = 4.1667 \times 10^{-8} t \tag{16d}$$

Bone mass, like in previous cases, reflects the behavior of the osteoclasts and osteoblasts and is thus the same as that of the original model in [17]. This model consists, consequently, in Eqs. (16a)–(16d) and (1c).

The advantage of this model with variable order differential equations over all previous models is that the same phenomena are modeled with far less parameters (See Fig. 7).

3   Non-local Models

The previous section addressed local models for bone remodeling represented mostly by ordinary differential equations (ODEs). Under this framework, only the dynamic behavior of individual BMUs is taken into account. The present section extends these models to one-dimensional geometries, thus modeling diffusion processes in the bone. We will focus on the proposals based on partial differential equations (PDEs) although other approaches also exist, such as in [1] that uses hybrid cellular automata to describe spatiotemporal bone remodeling, both healthy and in the presence of bone metastases, and even the effect of anti-RANKL and bisphosphonates therapy.

3.1 Healthy Bone Remodeling

Ayati and colleagues [2] extend the model in [17], adding the effect in the bone dynamics and allowing for diffusion over one dimension (i.e., a one-dimensional bone is assumed), by introducing diffusion terms $\sigma \frac{\partial^2}{\partial x^2}$ into the osteoclasts, osteoblasts, and bone mass equations. Consequently, variables C, B, and z now depend on x as well as on t.

$$D^1 C(t, x) = \sigma_c \frac{d^2}{dx^2} C(t, x) + \alpha_c C(t, x)^{g_{CC}} B(t, x)^{g_{BC}} - \beta_c C(t, x) \quad (17a)$$

$$D^1 B(t, x) = \sigma_B \frac{d^2}{dx^2} B(t, x) + \alpha_B C(t, x)^{g_{CB}} B(t, x)^{g_{BB}} - \beta_B B(t, x) \quad (17b)$$

$$D^1 z(t, x) = \sigma_z \frac{\partial^2}{\partial x^2} z(t, x) - k_c \max\{0, C(t, x) - C_{th}(t, x)\}$$
$$+ k_B \max\{0, B(t, x) - B_{th}(t, x)\} \quad (17c)$$

In Eq. (17c), the diffusion term of $z(t, x)$ represents the stochastic nature of bone dynamics, but not cells migration. More information about variables and parameters can be found in Table 3. Other proposal by Ryser and colleagues [37] explicitly introduces OPG and RANKL as separate variables and includes the spatial evolution of a single BMU. The model is described by Eq. (18), assuming $g_{BB} = 0$ and, since OPG, ϕ_O, RANKL and ϕ_R are modeled explicitly, $g_{BC} = 0$.

$$D^1 C(t, x) = \alpha_c C(t, x)^{g_{CC}} - \beta_c C(t, x) + f_1 \frac{\phi_R(t, x)}{\lambda + \phi_R(t, x)} \theta(y_C(t, x)) C(t, x)$$
$$- \zeta \nabla \cdot (y_C(t, x) \nabla \phi_R(t, x)), \quad (18a)$$

$$D^1 B(t, x) = \alpha_B C(t, x)^{g_{CB}} - \beta_B B(t, x), \quad (18b)$$

$$D^1 \phi_R(t, x) = \alpha_R y_{B,t_R} + f_R \Delta(\phi_R^\epsilon(t, x)) - f_B \frac{\phi_R(t, x)}{\lambda + \phi_R(t, x)} \theta(y_C(t, x)) C(t, x)$$
$$- f_{RO} \phi_R(t, x) \phi_O(t, x), \quad (18c)$$

$$D^1 \phi_O(t, x) = \alpha_O y_{B,t_O} + f_O \Delta(\phi_O^\delta(t, x)) - f_{RO} \phi_R(t, x) \phi_O(t, x), \quad (18d)$$

$$D^1 z(t, x) = -k_c \max\{0, C(t, x) - C_{ss}\} + k_B \max\{0, B(t, x) - B_{ss}\}$$
$$= -k_c y_c(t, x) + k_B y_B(t, x) \quad (18e)$$

The activation of osteoclasts is now dependent on autocrine parameters and RANK/RANKL binding, encoded by the term $f_1 \frac{\phi_R}{\lambda + \phi_R} \theta(y_C) C$. RANK receptors saturation threshold is described by a Hill function, with half-saturation concentration λ, and θ is the Heaviside function (thus $\theta(y_C) = 1$ if the number of active osteoclasts y_C verifies

$y_C > 0$; otherwise, $\theta(y_C) = 0$). Osteoclasts movement follows RANKL gradient, as described by the term $\zeta \nabla \cdot (y_C \nabla \phi_R)$, at migration rate ζ. RANKL and OPG are produced by active osteoblasts, considering that after osteoblasts precursors become mature, there is a time delay t_R and t_O before they can produce RANKL and OPG, encoded by y_{B,t_R} and y_{B,t_O}, respectively. RANKL and OPG porous diffusion in the bone is given by $f_R \Delta(\phi_R^\epsilon)$ and $f_O \Delta(\phi_O^\delta)$, at rates f_R and f_O. The binding of OPG to RANKL is described by the term $f_{RO} \phi_R \phi_O$, which is present in both O and R dynamics. For RANKL, however, an additional term $f_2 \frac{\phi_R}{\lambda + \phi_R} \theta(y_C)C$ encodes the RANK/RANKL binding in osteoclasts precursors, at rate f_2, similarly to the term in (18a). Sensitivity analysis and parameter estimation are conducted for this model in [36]. The model of [27] is extended in [4] by adding spatial evolution to the different components of the model and by introducing appropriate fluxes in cells and regulatory agents. Fluxes represent transport processes, such as diffusion or chemotaxis. This model is able to capture the known organized structure of a BMU, presenting a resorption zone at the front, then a reversal zone and lastly a bone formation zone. The model also captures how PTH and the RANK/RANKL/OPG pathway affect the bone cells in the different stages of maturation and during bone remodeling.

3.2 Models Including the Tumor Burden

Ayati and colleagues [2] also extended the model proposed in [17] by including the diffusion process of multiple myeloma disease effects on bone dynamics:

$$D^1 C(t, x) = \sigma_c \frac{\partial^2}{\partial x^2} C(t, x) + \alpha_c C(t, x)^{g_{CC}\left(1 + r_{CC} \frac{T(t,x)}{L_T}\right)} B(t, x)^{g_{BC}\left(1 + r_{BC} \frac{T(t,x)}{L_T}\right)}$$
$$- \beta_c C(t, x) \tag{19a}$$

$$D^1 B(t, x) = \sigma_B \frac{\partial^2}{\partial x^2} B(t, x) + \alpha_B C(t, x)^{\frac{g_{CB}}{1 + r_{CB} \frac{T(t,x)}{L_T}}} B(t, x)^{g_{BB} - r_{BB} \frac{T(t,x)}{L_T}}$$
$$- \beta_B B(t, x) \tag{19b}$$

$$D^1 T(t, x) = \sigma_T \frac{\partial^2}{\partial x^2} T(t, x) + \gamma_T T(t, x) \log\left(\frac{T(t, x)}{L_T}\right) \tag{19c}$$

This model's bone mass equation is (17c). The tumorous variables and parameters now introduced have the same meaning as in the local case (see Sect. 2.2), except that $T(t)$ and $T(0)$ are now given by $T(t, x)$ and $T(0, x)$, respectively.

In [38], the effect of osteolytic bone metastases is included in the bone remodeling model in [36], accounting for spatial diffusion, as described by Eq. (20).

$$D^1 C(t, x) = \alpha_c C(t, x)^{g_{CC}} - \beta_c C(t, x) + f_1 \frac{\phi_R(t, x)}{\lambda + \phi_R(t, x)} y_C(t, x)$$
$$- \zeta \frac{\partial}{\partial x}\left(y_C(t, x) \frac{\partial \phi_R(t, x)}{\partial x}\right), \tag{20a}$$

$$D^1 \phi_R(t, x) = k_R \phi_P(t, x) z(t) + \tau_R T(t, x) + f_R \frac{\partial^2 \phi_R(t, x)}{\partial x^2}$$

$$- f_B \frac{\phi_R(t, x)}{\lambda + \phi_R(t, x)} y_C(t, x) - f_{RO} \phi_R(t, x) \phi_O(t, x), \quad (20b)$$

$$D^1 \phi_O(t, x) = \tau_O T(t, x) + S_o + f_o \frac{\partial^2 \phi_O(t, x)}{\partial x^2} - \beta_O \phi_O(t, x)$$

$$- f_{RO} \phi_R(t, x) \phi_O(t, x), \quad (20c)$$

$$D^1 \phi_P(t, x) = \tau_p T(t, x) + f_p \frac{\partial^2 \phi_P(t, x)}{\partial x^2} - \beta_p \phi_P(t, x), \quad (20d)$$

$$D^1 z(t, x) = -k_c \max\{0, C(t, x) - C_{th}\} = -k_c y_C(t, x), \quad (20e)$$

$$T(t, x) = 1 - z(t, x) \quad (20f)$$

This model assumes bone formation to be much slower than bone resorption and tumor growth in the presence of osteolytic bone metastases. As such, the osteoblast dynamics is not taken into account and is removed from the model. Hence, bone mass dynamics z in Eq. (20e) is dominated by bone resorption and tumor cells T fill in the space left in the bone after resorption. Tumor-derived PTHrP ϕ_P increases the production of RANKL ϕ_R by osteoblasts. Also, tumor cells can also produce RANKL. The conjugation of these factors leads to increased bone resorption in areas close to the tumor, which contributes to tumor growth, and thus the vicious cycle of bone metastases is captured by this model.

3.3 Non-local Treatment Approach

3.3.1 Normal Cell Diffusion

Following what was done in [2], a treatment possibility is applied accounting for the effects of proteasome inhibition ($V_1(t)$) and anti-cancer therapy ($V_2(t)$) in myeloma bone disease, similarly to what was done locally in the model (14). For the following equation, when $t < t_{\text{start}}$, treatment is yet to be started.

$$V_1(t) = v_1, \quad \text{for } t \geq t_{\text{start}} \quad (21a)$$

$$V_2(t) = v_2, \quad \text{for } t \geq t_{\text{start}} \quad (21b)$$

(a)

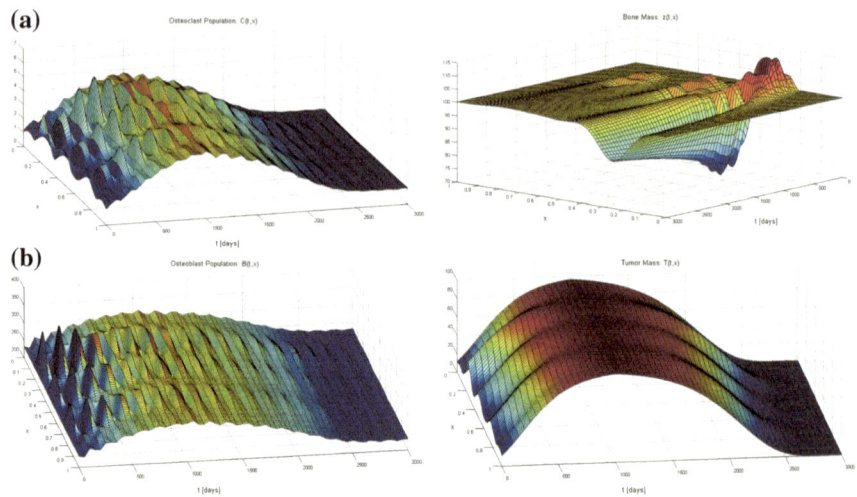

(b)

Fig. 8 Bone model simulation with an initial distribution of osteoclasts and tumor by $C(0, x) = T(0, x)$, as $C(0, x)$ presented in [2], where **a** represents the osteoclast evolution, and **b** the osteoblast evolution. Treatment was introduced at $t_{\text{start}} = 1000$ days, with parameters $V_1(t) = v_1 = 0.0005$ and $V_2(t) = v_2 = 0.007$. Non-spatial and bone mass parameters were set according to the local model presented for periodic remodeling cycles (see column A of Table 3) and as in the one-dimensional model without tumor (see [2] for more details), respectively; and all of the tumor-related parameters are the same as in the local case

$$D^1 C(t, x) = \sigma_c \frac{\partial^2}{\partial x^2} C(t, x) + \alpha_c C(t, x)^{g_{CC} \left(1 + r_{CC} \frac{T(t,x)}{L_T}\right)} B(t, x)^{g_{BC} \left(1 + r_{BC} \frac{T(t,x)}{L_T}\right)}$$
$$- \beta_c C(t, x) \tag{22a}$$

$$D^1 B(t, x) = \sigma_B \frac{\partial^2}{\partial x^2} B(t, x) + \alpha_B C(t, x)^{\left(\frac{g_{CB}}{1 + r_{CB} \frac{T(t,x)}{L_T}}\right)} B(t, x)^{\left(g_{BB} - r_{BB} \frac{T(t,x)}{L_T}\right)}$$
$$- \left(\beta_B - V_1(t)\right) B(t, x) \tag{22b}$$

$$D^1 T(t, x) = \sigma_T \frac{\partial^2}{\partial x^2} T(t, x) + (\gamma_T - V_2(t)) T(t, x) \log\left(\frac{T(t, x)}{L_T}\right) \tag{22c}$$

Again, the bone mass equation is (17c) (non-local healthy bone remodeling case). Simulations for such treatment, and its effect on the bone microenvironment, can be seen in Fig. 8, where the tumor is of initial small intensity, acting in several areas of the spatial bone environment x, but growing over time. Bone resorption and formation, both temporal and spatially, are disrupted with a delayed response from the osteoblast population [2]. Consequently, an increased bone mass loss is observed in areas where the tumor has a more intense effect until treatment is introduced at $t_{\text{start}} = 1000$ days. Osteoclast production is then promoted, and the tumor growth inhibited until its significant reduction and regular bone dynamics coupling are achieved. Bone mass only partially recovers, since areas of the bone where the tumor effect was stronger were unable to fully recover its original density.

The pharmacokinetics and pharmacodynamics presented in Sect. 2.3.1 can also be integrated with non-local bone remodeling models. This was done in [8], by considering that each drug's pharmacodynamic effect to be the same throughout the bone, for the same time instance. Again, the effect of a drug, $d_i(t)$, is given by Eq. (13), where $d_1(t)$ is the effect of bisphosphonates; $d_2(t)$ represents denosumab; and chemotherapy is encompassed in $d_{34}(t)$ (it is noted that, in [8], chemotherapy is considered to be a combination of two distinct drugs, hence the subscript 34). Simulation results, from [8], can be found in Fig. 9. This model's bone mass equation is once more (17c) (non-local healthy bone remodeling case), and the remaining equations are as follows.

$$D^1 C(t, x) = \sigma_c \frac{\partial^2}{\partial x^2} C(t, x)$$
$$+ \alpha_c C(t, x)^{g_{CC}\left(1 + r_{CC}\frac{T(t,x)}{L_T}\right)} B(t, x)^{g_{BC}\left(1 + r_{BC}\frac{T(t,x)}{L_T}\right)}\left(1 + K_{d_1} d_1(t)\right)$$
$$- \left(1 + K_{d_2} d_2(t)\right) \beta_c C(t, x) \tag{23a}$$

$$D^1 B(t, x) = \sigma_B \frac{\partial^2}{\partial x^2} B(t, x) + \alpha_B C(t, x)^{\left(\frac{g_{CB}}{1 + r_{CB}\frac{T(t,x)}{L_T}}\right)} B(t, x)^{\left(g_{BB} - r_{BB}\frac{T(t,x)}{L_T}\right)}$$
$$- \beta_B B(t, x) \tag{23b}$$

$$D^1 T(t, x) = \sigma_T \frac{\partial^2}{\partial x^2} T(t, x)$$

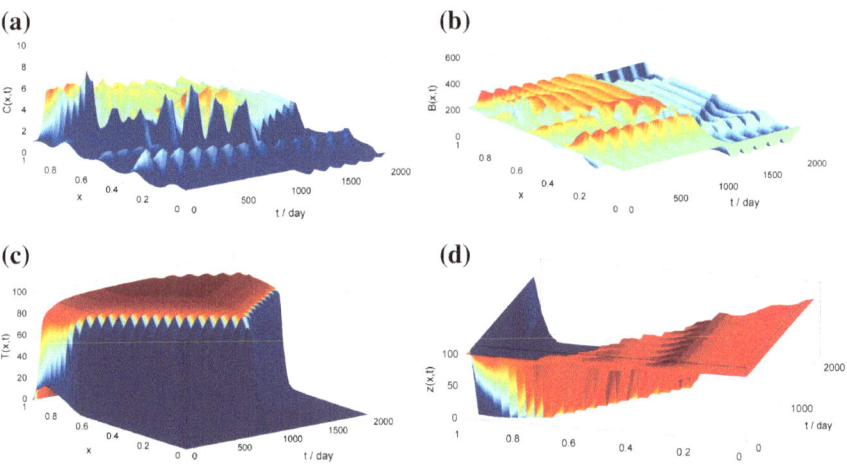

Fig. 9 Simulation of model of Eq. 23, where **a** is the osteoclasts evolution, **b** the osteoblasts evolution, **c** the tumorous influence, and **d** the consequences for the bone mass. Used parameters can be found in column C of Table 3

$$+ \left(1 - K_{d_{34}} d_{34}(t)\right) \gamma_{_T} T(t, x)^a \left(\frac{T(t, x)}{L_T}\right)^b \tag{23c}$$

Although different set of parameters were used, when compared to the local model in Eq. 15 and considering the different used expressions for the tumor evolution, the results behave similarly. The tumor is extinguished as osteoclasts and osteoblasts recover their steady-state value.

3.3.2   Working with Fractional Derivatives

Anomalous diffusion corresponds to partial differential equations with fractional derivatives in order to time and second-order derivatives in order to space [21]. In what pharmacokinetic and pharmacodynamic models are concerned, those presented in Sect. 2.3.1 were achieved for models with integer orders derivatives. Hence, they cannot be directly applied when one is considering fractional order: New PK equations are introduced and assumed to have the same fractional order as the model itself. Such adaptation is given by Eq. (24), where α is the fractional order of the system. The PD component, however, remains as in Eq. (13).

$$D^{\alpha} C_g(t) = -k_a C_g(t) \tag{24a}$$

$$D^{\alpha} C_p(t) = k_a C_g(t) - k_e C_p(t) \tag{24b}$$

$$C_P^0 = \frac{D_0 F}{V_d} \tag{24c}$$

$$d(t) = \frac{C_p(t)}{C_{50} + C_p(t)} \tag{24d}$$

The model from Eq. (23) is, but, a particular case of a general anomalous diffusion case, corresponding to setting the describing fractional order to $\alpha = 1$. The general model of order α (including for the PK/PD treatment) is as follows [8].

$$D^{\alpha} C(t, x) = \sigma_c \frac{\partial^2}{\partial x^2} C(t, x)$$
$$+ \alpha_c C(t, x)^{g_{CC}} \left(1 + r_{CC} \frac{T(t,x)}{L_T}\right) B(t, x)^{g_{BC}} \left(1 + r_{BC} \frac{T(t,x)}{L_T}\right) \left(1 + K_{d_1} d_1(t)\right)$$
$$- \left(1 + K_{d_2} d_2(t)\right) \beta_c C(t, x) \tag{25a}$$

$$D^{\alpha} B(t, x) = \sigma_{_B} \frac{\partial^2}{\partial x^2} B(t, x) + \alpha_{_B} C(t, x)^{\left(\frac{g_{CB}}{1 + r_{CB} \frac{T(t,x)}{L_T}}\right)} B(t, x)^{\left(g_{BB} - r_{BB} \frac{T(t,x)}{L_T}\right)}$$
$$- \beta_{_B} B(t, x) \tag{25b}$$

$$D^{\alpha} z(t, x) = \sigma_z \frac{\partial^2}{\partial x^2} z(t, x) - k_c \max\{0, C(t, x) - C_{th}(t, x)\}$$
$$+ k_{_B} \max\{0, B(t, x) - B_{th}(t, x)\} \tag{25c}$$

$$D^\alpha T(t, x) = \sigma_T \frac{\partial^2}{\partial x^2} T(t, x)$$

$$+ \left(1 - K_{d_{34}} d_{34}(t)\right) \gamma_T T(t, x)^a \left(\frac{T(t, x)}{L_T}\right)^b \tag{25d}$$

Also, the local PTH model in Eq. (15) was extended to a three-dimensional case in [45]. This was done replacing variable $\chi(t, x)$ (where χ can be C, B, T, z, C_{th} or B_{th}) with $\chi(t, \mathbf{x}) = \chi(t, x_1, x_2, x_3)$ and using the Laplacian operator $\nabla = \frac{\partial^2}{\partial x_1^2} + \frac{\partial^2}{\partial x_2^2} + \frac{\partial^2}{\partial x_3^2}$ instead of $\frac{\partial^2}{\partial x^2}$. However, for comparison sake, only the one-dimensional version is presented next.

$$D^\alpha C(t, x) = \sigma_C \frac{\partial^2}{\partial x^2} C(t, x)$$

$$+ \alpha_C C(t, x)^{g_{CC} + r_C \frac{T(t,x)}{L_T}} B(t, x)^{g_{BC} + K_{\text{PTH}_{\text{pool}21}} \text{PTHpool}(t,x) - K_{d_1} d_1(t)}$$

$$- (\beta_B + K_{d_2} d_2(t)) C(t, x) \tag{26a}$$

$$D^\alpha B(t, x) = \sigma_B \frac{\partial^2}{\partial x^2} B(t, x) + \alpha_B C(t, x)^{g_{CB}} B(t, x)^{g_{BB} + r_B \frac{T(t,x)}{L_T}} - \beta_B B(t, x) \tag{26b}$$

$$D^\alpha \text{PTH}_{\text{pool}}(t, x) = -\beta_{\text{PTH}} \text{PTH}_{\text{pool}}(t, x) + K_{\text{PTH}} \delta(t) + r_{\text{PTHrP}} \max\{0, C(t, x)$$

$$- C_{th}(t, x)\} \frac{T(t, x)}{L_T} \tag{26c}$$

$$P(\delta(t) = 1) = 1 - e^{-\left(\frac{t}{\lambda_w}\right)^{k_w}} \tag{26d}$$

$$D^\alpha T(t, x) = \sigma_T \frac{\partial^2}{\partial x^2} T(t, x) + k_T \max\{0, C(t, x) - C_{th}(t, x)\} \frac{T(t, x)}{\lambda_T + T(t, x)}$$

$$- K_{d_3} d_3(t) T(t, x) \tag{26e}$$

$$D^\alpha C_{th}(t, x) = \alpha_C C_{th}(t, x)^{g_{CC} + r_C \frac{T(t,x)}{L_T}} B_{th}(t, x)^{g_{BC} - K_{d_1} d_1(t)}$$

$$- (\beta_C + K_{d_2} d_2(t)) C_{th}(t, x) \tag{26f}$$

$$D^\alpha B_{th}(t, x) = \alpha_B C_{th}(t, x)^{g_{CB}} B_{th}(t, x)^{g_{BB} + r_B \frac{T(t,x)}{L_T}} - \beta_B B_{th}(t, x) \tag{26g}$$

The bone mass equation of this model is in Eq. (25c) for an order α. Simulations of this model can be found in Fig. 10.

Table 3 lists variables and parameters, used throughout presented simulations, giving their meaning and dimensions.

4 Conclusions and Future Work

This chapter overviewed some of the most recent models for bone remodeling dynamics that take into account the biochemical microenvironment and the main cellular

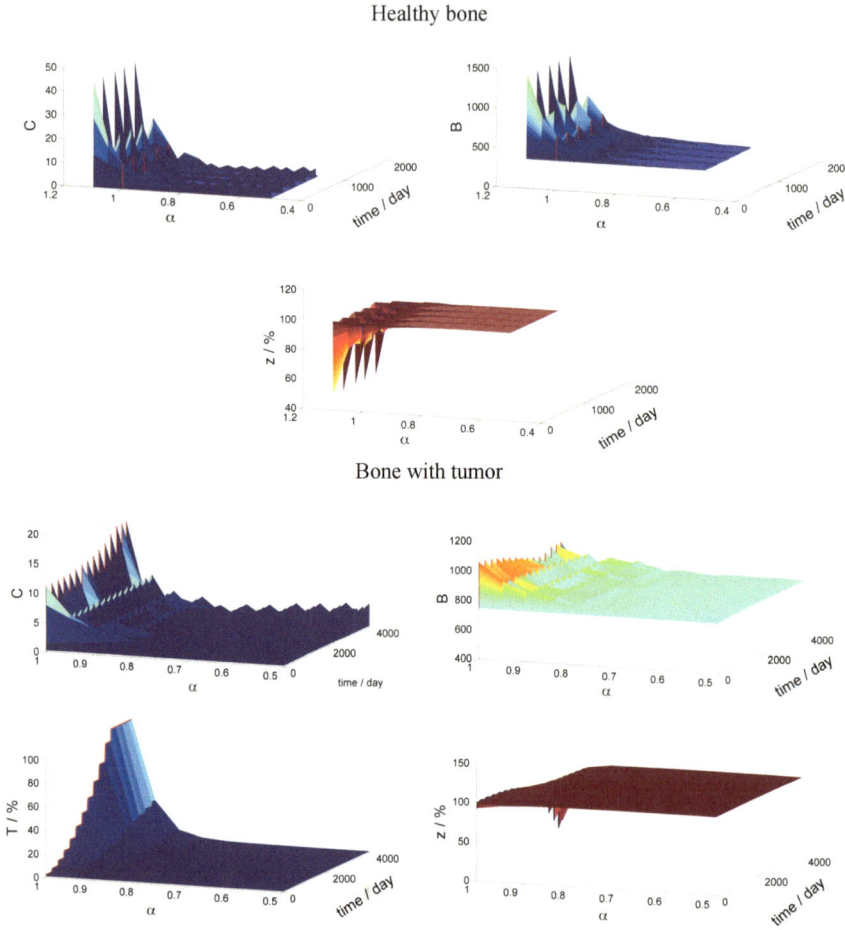

Fig. 10 Equation (26) model simulation, comparing the effect of several values of α on the model dynamics, where the $\alpha = 1$ case is represented by either a blue or red line. On the **left**, a healthy bone environment is represented and, on the **right**, a tumorous bone environment. From [45]

processes. These computational models allow for the integration of pathological phenomena such as cancer and also accommodate pharmacokinetic/pharmacodynamic (PK/PD) information.

There are many other models for the bone that are focused on the biomechanical aspects of the system, addressed in other chapters of this book. As future work, it would be crucial to integrate both approaches, as to improve the models in terms of accuracy to describe the real organ. In fact, these parallel lines of research are already being integrated in recent years. Scheiner, Pivonka, and colleagues [25, 42] proposed models that include the coupled regulations of geometrical, biomechanical, and biochemical factors. The ODE system uses biochemical factors already addressed such as TGFβ, RANK and RANKL, OPG and PTH with the f_{vas}, S_v and Ψ_{bm} of BMU. Although very complete, this model does not take into account one entire bone (e.g., femur) and the corresponding spatial effects, neither the inclusion of input functions representing PK/PD of the therapeutic drugs. These therapies were also analyzed separately, such as denosumab and bisphosphonates for osteoporosis [11, 41]. It is expected that all of these modeling and computational analysis of bone physiology will have an impact on the development of clinical decision support systems in the future.

Acknowledgements This work was supported by FCT, through IDMEC, under LAETA, projects UID/EMS/50022/2019 and BoneSys (Bone biochemical and biomechanic integrated modeling: addressing remodeling, disease and therapy dynamics), joint Polish–Portuguese project "Modeling and controlling cancer evolution using fractional calculus", and PERSEIDS (PTDC/EMS-SIS/0642/2014). R. Moura Coelho acknowledges the support by grant ZEUGMA-BiNOVA, number AD0075. S. Vinga acknowledges the support by Program Investigador FCT (IF/00653/2012) from FCT, co-funded by the European Social Fund (ESF) through the Operational Program Human Potential (POPH).

Appendix: Fractional and Variable Order Derivatives

To formally define fractional derivatives (*fractional* is the historical name, though order α can be irrational or have an imaginary part), one can present the Grünwald–Letnikoff construction, usually denoted by $D^\alpha f(t)$. The upper limit of the summations was set so that D^{-k}, $k \in \mathbb{N}$ to retrieve Riemann integrals: $_c D_t^{-1} f(t) = \int_c^t f(t)\,dt$, and in general

$$_c D_t^{-k} f(t) = \underbrace{\int_c^t \ldots \int_c^t f(t)\,dt \ldots dt}_{k \text{ integrations}}, \ k \in \mathbb{N} \tag{27}$$

Since it is not always reasonable to call c and t limits of integration, they are usually called terminals instead.

The Grünwald–Letnikoff construction is as follows.

$$_c D_t^\alpha f(t) = \lim_{h \to 0^+} \frac{\sum_{k=0}^{\left[\frac{t-c}{h}\right]} (-1)^k \binom{a}{k} f(t - kh)}{h^\alpha} \tag{28}$$

where $\begin{pmatrix} a \\ k \end{pmatrix}$ are the combinations of a things k at a time, given by

$$\binom{a}{k} = \begin{cases} \frac{\Gamma(\alpha+1)}{\Gamma(k+1)\Gamma(\alpha-k+1)}, & \text{if } \alpha, k, (\alpha-k) \in \mathbb{R} \setminus \mathbb{Z}^- \\ \frac{(-1)^k \Gamma(k-\alpha)}{\Gamma(k+1)\Gamma(-\alpha)}, & \text{if } \alpha \in \mathbb{Z}^- \wedge k \in \mathbb{Z}_0^+ \\ 0, & \text{if } (k \in \mathbb{Z}^- \vee (k-\alpha) \in \mathbb{N}) \wedge \alpha \notin \mathbb{Z}^- \end{cases} \quad (29)$$

Function $\Gamma(z)$, $z \in \mathbb{C} \setminus \mathbb{Z}_0^-$ generalizes the factorial since $\Gamma(k+1) = k!$, $k \in \mathbb{Z}_0^+$.

Since α can assume any real value, it is possible to make it change with time (continuously or not). There are different reasonable ways of accounting for a time varying order $\alpha(t)$, again considering the Grünwald–Letnikoff definition. The two most straightforward options are given below:

- in the first, the argument of the order is simply the current value of t, meaning that the result at a given instant will not depend on previous values of the order $\alpha(t)$:

$$_cD_t^{\alpha(t)} f(t) = \lim_{h \to 0^+} \frac{\sum_{k=0}^{\left|\frac{t-c}{h}\right|} (-1)^k \binom{\alpha(t)}{k} f(t-kh)}{h^{\alpha(t)}} \quad (30)$$

- in the second, the argument of α in the summation is the same argument of f, leading to a memory of previous values of the order [48]:

$$_cD_t^{\alpha(t)} f(t) = \lim_{h \to 0^+} \sum_{k=0}^{\left|\frac{t-c}{h}\right|} \frac{(-1)^k \binom{\alpha(t-kh)}{k} f(t-kh)}{h^{\alpha(t-kh)}} \quad (31)$$

Other possible definitions exist [43, 47]. Since it is reasonable to expect the previous evolution of a metastasis to have influence in the present time, variable order models presented presume this type of definition.

Readers interested in details can consult [21, 23, 39, 46, 48], which also address alternative definitions of fractional derivatives (namely those of Riemann–Liouville and of Caputo, that will, for functions well-behaved enough, lead to similar results).

Letting h have a small but finite value, approximations of fractional derivatives can be obtained. Values can be stored in matrices updated iteratively. This is the idea behind the numerical method of [29, 31, 32], termed matrix approach to discrete fractional calculus, used for the simulations in this chapter through MATLAB toolbox [30].

References

1. Araujo A, Cook LM, Lynch CC, Basanta D (2014) An integrated computational model of the bone microenvironment in bone-metastatic prostate cancer. Cancer Res 74(9):2391–401
2. Ayati BP, Edwards CM, Webb GF, Wikswo JP (2010) A mathematical model of bone remodeling dynamics for normal bone cell populations and myeloma bone disease. Biol Direct 5:28
3. Bellido T, Plotkin LI, Bruzzaniti A (2014) Bone Cells (Chap. 2). In: Basic and applied bone biology. Academic Press, Cambridge, pp 27 – 45

4. Buenzli PR, Pivonka P, Smith DW (2011) Spatio-temporal structure of cell distribution in cortical bone multicellular units: a mathematical model. Bone 48(4):918–26
5. Casimiro S, Ferreira AR, Mansinho A, Alho I, Costa L (2016) Molecular mechanisms of bone metastasis: which targets came from the bench to the bedside? Int J Mol Sci 17(9):1415
6. Casimiro S, Guise TA, Chirgwin J (2009) The critical role of the bone microenvironment in cancer metastases. Mol Cell Endocrinol 310(1–2):71–81
7. Chen Y-C, Sosnoski DM, Mastro AM (2010) Breast cancer metastasis to the bone: mechanisms of bone loss. Breast Cancer Res: BCR 12(6):215
8. Christ LF, Valério D, Coelho R, Vinga S (2018) Models of bone metastases and therapy using fractional derivatives. J Appl Nonlinear Dyn 7(1):81–94
9. Coelho RM, Lemos JM, Valério D, Alho I, Ferreira AR, Costa L, Vinga S (2016) Dynamic modeling of bone metastasis, microenvironment and therapy–integrating parathyroid hormone (PTH) effect, antiresorptive treatment and chemotherapy. J Theor Biol 391:1–12
10. Crockett JC, Rogers MJ, Coxon FP, Hocking LJ, Helfrich MH (2011) Bone remodelling at a glance. J Cell Sci 124(Pt 7):991–8
11. Cummings SR, San Martin J, McClung MR, Siris ES, Eastell R, Reid IR, Delmas P, Zoog HB, Austin M, Wang A, Kutilek S, Adami S, Zanchetta J, Libanati C, Siddhanti S, Christiansen C, Trial FREEDOM (2009) Denosumab for prevention of fractures in postmenopausal women with osteoporosis. New England J Med 361(8):756–765
12. Dhillon S, Kostrzewski A (2006) Basic pharmacokinetics (Chap. 1). In: Clinical pharmacokinetics, 1 edn. Pharmaceutical Press (2006)
13. Graham JM, Ayati BP, Holstein SA, Martin JA (2013) The role of osteocytes in targeted bone remodeling: a mathematical model. PloS one 8(5):10–14
14. Guise TA, Mundy GR (1998) Cancer and bone. Endocr Rev 19(1):18–54
15. Hood L, Friend SH (2011) Predictive, personalized, preventive, participatory (P4) cancer medicine. Nat Rev Clin Oncol 8(3):184–187
16. Komarova SV (2005) Mathematical model of paracrine interactions between osteoclasts and osteoblasts predicts anabolic action of parathyroid hormone on bone. Endocrinology 146(8):3589–95
17. Komarova SV, Smith RJ, Dixon SJ, Sims SM, Wahl LM (2003) Mathematical model predicts a critical role for osteoclast autocrine regulation in the control of bone remodeling. Bone 33(2):206–215
18. Lemaire V, Tobin FL, Greller LD, Cho CR, Suva LJ (2004) Modeling the interactions between osteoblast and osteoclast activities in bone remodeling. J Theor Biol 229(3):293–309
19. Liò P, Paoletti N, Moni MA, Atwell K, Merelli E, Viceconti M (2012) Modelling osteomyelitis. BMC Bioinf 13(Suppl 14):S12
20. Lorenzo CF, Hartley TT (2002) Variable order and distributed order fractional operators. Technical Report National Aeronautics and Space Administration (NASA)
21. Magin RL (2004) Fractional Calculus in Bioengineering. Begell House
22. Makatsoris T, Kalofonos HP (2009) The role of chemotherapy in the treatment of bone metastases. In: Kardamakis D, Vassiliou V, Chow E (eds) Bone metastases, volume 12 of Cancer metastasis biology and treatment. Springer Netherlands, Dordrecht, pp 287–297
23. Miller KS, Ross B (1993) An introduction to the fractional calculus and fractional differential equations. Wiley, New York
24. Petrás I (2009) Stability of fractional-order systems with rational orders: a survey. Int J Theory Appl 12(3)
25. Pivonka P, Buenzli PR, Scheiner S, Hellmich C, Dunstan CR (2013) The influence of bone surface availability in bone remodelling-a mathematical model including coupled geometrical and biomechanical regulations of bone cells. Eng Struct 47:134–147
26. Pivonka P, Komarova SV (2010) Mathematical modeling in bone biology: from intracellular signaling to tissue mechanics. Bone 47(2):181–189
27. Pivonka P, Zimak J, Smith DW, Gardiner BS, Dunstan CR, Sims NA, Martin TJ, Mundy GR (2008) Model structure and control of bone remodeling: a theoretical study. Bone 43(2):249–63

28. Pivonka P, Zimak J, Smith DW, Gardiner BS, Dunstan CR, Sims NA, Martin TJ, Mundy GR (2010) Theoretical investigation of the role of the RANK-RANKL-OPG system in bone remodeling. J Theor Biol 262(2):306–16

29. Podlubny I (2000) Matrix approach to discrete fractional calculus. Fractional Calc Appl Anal 3(4):359–386

30. Podlubny I (2012) Matrix approach to distributed-order ODEs and PDEs. http://www.mathworks.com/matlabcentral/fileexchange/36570-matrix-approach-to-distributed-order-odes-and-pdes

31. Podlubny I, Chechkin A, Skovranek T, Chen YQ, Jara BMV (2009) Matrix approach to discrete fractional calculus II: partial fractional differential equations. J Comput Phys 228:3137–3153

32. Podlubny I, Skovranek T, Jara BMV, Petras I, Verbitsky V, Chen YQ (1990) Matrix approach to discrete fractional calculus III: non-equidistant grids, variable step length and distributed orders. Philos Trans R Soc A 371:2013

33. Raggatt LJ, Partridge NC (2010) Cellular and molecular mechanisms of bone remodeling. J Biol Chem 285(33):25103–8

34. Rahimy M (2010) Applications of fractional differential equations. Appl Math Sci 4(50):2453–2461

35. Roodman GD (2004) Mechanisms of bone metastasis. New England J Med 360(16):1655–1664

36. Ryser MD, Komarova SV, Nigam N (2010) The cellular dynamics of bone remodeling: a mathematical model. SIAM J Appl Math 70(6):1899–1921

37. Ryser MD, Nigam N, Komarova SV (2009) Mathematical modeling of spatio-temporal dynamics of a single bone multicellular unit. J Bone Mineral Res 24(5):860–70

38. Ryser MD, Qu Y, Komarova SV (2012) Osteoprotegerin in bone metastases: mathematical solution to the puzzle. PLoS Comput Biol 8(10):e1002703

39. Samko SG, Kilbas AA, Marichev OI (1993) Fractional integrals and derivatives. Gordon and Breach, Yverdon

40. Savageau MA (1988) Introduction to S-systems and the underlying power-law formalism. Math Comput Modell 11:546–551

41. Scheiner S, Pivonka P, Smith DW, Dunstan CR, Hellmich C (2014) Mathematical modeling of postmenopausal osteoporosis and its treatment by the anti-catabolic drug denosumab. Int J Numer Methods Biomed Eng 30(1):1–27

42. Scheiner S, Pivonka P, Hellmich C (2013) Coupling systems biology with multiscale mechanics, for computer simulations of bone remodeling. Comput Methods Appl Mech Eng 254:181–196

43. Sierociuk D, Malesza W, Macias M (2016) Numerical schemes for initialized constant and variable fractional-order derivatives: matrix approach and its analog verification. J Vib Control 22(8):2032–2044

44. Suva LJ, Washam C, Nicholas RW, Griffin RJ (2011) Bone metastasis: mechanisms and therapeutic opportunities. Nat Rev Endocrinol 7(4):208–18

45. Valério D, Coelho R, Vinga S (2016) Fractional dynamic modelling of bone metastasis, microenvironment and therapy. In: International conference on fractional differentiation and its applications

46. Valério D, da Costa JS (2011) Introduction to single-input, single-output Fractional Control. IET Control Theor Appl 5(8):1033–1057

47. Valério D, da Costa JS (2011) Variable-order fractional derivatives and their numerical approximations. Sign Process 91(3):470–483

48. Valério D, da Costa JS (2013) An introduction to fractional control. IET, Stevenage. ISBN 978-1-84919-545-4

49. Wang Y, Pivonka P, Buenzli PR, Smith DW, Dunstan CR (2011) Computational modeling of interactions between multiple myeloma and the bone microenvironment. PloS one 6(11):e27494

50. Zumsande M, Stiefs D, Siegmund S, Gross T (2011) General analysis of mathematical models for bone remodeling. Bone 48(4):910–7

Mathematical Modelling of Spatio-temporal Cell Dynamics Observed During Bone Remodelling

Madalena M. A. Peyroteo, Jorge Belinha, Susana Vinga, R. Natal Jorge and Lúcia Dinis

Abstract Bone is an active tissue capable to adapt its activity according to external stimuli. The consequent changes in mass, composition and shape are a result of this remodelling process, in which old bone is continuously replaced by new tissue. In order to mathematically describe this remodelling process, several mathematical models have been proposed. In this chapter, a novel model simulating the biological events that occur during bone remodelling is presented, based on Komarova's (Bone 33:206–215, 2003 [1]) and Ayati's (Biol Direct 5:28, 2010 [2]) models. Also, a thorough temporal and spatial analysis is performed using numerical methods, namely the finite element method (FEM), the radial point interpolation method (RPIM) and the natural neighbour radial point interpolation method (NNRPIM), being the last two meshless approaches. Results show that the combination of this model with distinct numerical approaches allows an accurate reproduction of the biological event, as described in the literature. Besides this, although all numerical techniques have been successfully applied, better quality results have been obtained with the NNRPIM.

M. M. A. Peyroteo
Institute of Mechanical Engineering and Industrial Management (INEGI), Rua Dr. Roberto Frias, s/n, 4600-465 Porto, Portugal
e-mail: mmgomes@inegi.up.pt

J. Belinha (✉)
Department of Mechanical Engineering, School of Engineering, Polytechnic of Porto (ISEP), Rua Dr. António Bernardino de Almeida, 431, 4200-072 Porto, Portugal
e-mail: job@isep.ipp.pt

S. Vinga
Institute of Mechanical Engineering (IDMEC), IST, Avenida Rovisco Pais 1, 1049-001 Lisbon, Portugal
e-mail: susanavinga@tecnico.ulisboa.pt

R. Natal Jorge · L. Dinis
Faculty of Engineering, Mechanical Engineering Department, University of Porto (FEUP), Rua Dr. Roberto Frias, s/n, 4600-465 Porto, Portugal
e-mail: rnatal@fe.up.pt

L. Dinis
e-mail: ldinis@fe.up.pt

J. Belinha et al. (eds.), *The Computational Mechanics of Bone Tissue*,
Lecture Notes in Computational Vision and Biomechanics 35,
https://doi.org/10.1007/978-3-030-37541-6_5

1 Introduction

The skeletal system is constituted by bone and cartilage and has two main functions in the organism—mechanical and metabolic. Regarding the metabolic action, the skeleton is considered a mineral repository, especially of calcium and phosphate, responsible for the maintenance of serum homoeostasis. Besides this mineralized structure of hydroxyapatite, bone is also a porous material composed by cells, vessels and collagen [3].

Moreover, bone tissue is considered an active and adaptive system, since its activity generates changes in mass, composition and shape, in response to its environmental vicinity [4]. This phenomenon called remodelling is a complex process by which old bone is continuously replaced by new tissue. The process is sequential, starting by an activation phase followed by resorption, reversal, formation, mineralization and quiescence phases. The bone cells active in the process are the osteoclasts and the osteoblasts, which are temporally and spatially coupled, closely collaborating within bone multi-cellular units (BMUs). Thus, in response to biochemical messages due to external physical stimuli, a BMU is activated and the remodelling process is initialized. This way, the resorption phase starts with the activation of a quiescent bone surface through a cascade of signals to osteoclastic precursors, causing their migration to the bone's surface, where they form multinucleated osteoclasts [5]. Simultaneously, bone lining cells disappear from the bone's surface, allowing osteoclasts to adhere to the bone matrix and resorb bone [3]. After the completion of resorption, mononuclear cells appear on the bone surface, releasing signals for osteoblasts' differentiation and migration. This phase is known as the reversal phase, since it prepares the bone's surface for the arrival of new osteoblasts and consequently the formation of new bone [3]. During the formation phase, osteoblasts lay down bone until the resorbed bone is completely replaced by new one. In this phase, osteoblasts start by synthesizing and depositing collagen. Then, the production of collagen decreases, and a secondary full mineralization of the matrix takes place. In this step, the collagen matrix, built previously, acts as the scaffold in which minerals, such as phosphate and calcium, begin to crystalize and form bone [3, 5]. Following the mineralization phase, the mature osteoblasts gradually flatten and can either become quiescent lining cells or differentiate into osteocytes when embedded in the bone. Thus, bone's resorption and formation are the results of a series of events transforming a physical information, such as mechanical stimuli, into a biological response [4].

The cellular events responsible for remodelling have to be coordinated, being this task accomplished by a variety of autocrine and paracrine factors, such as hormones, cytokines and growth factors [6]. For instance, receptor activator of NF-kB ligand (RANKL), receptor activator of NF-Kb (RANK) and osteoprotegerin (OPG) that form the OPG/RANKL/RANK system are essential for the resorption phase. The major biological action of RANKL is inducing osteoclasts' activation, promoting bone resorption [7]. But, to counteract the differentiation and activation of osteoclasts, osteoblasts produce and secrete OPG, a decoy receptor that can block

RANKL/RANK interactions and, in this way, inhibit osteoclasts' function and accelerate osteoclasts' apoptosis [5]. Also, the transforming growth factor-β (TGF-β) possesses an important role during bone remodelling, having a two-directional effect on osteoblasts, depending on the state of maturation of these bone cells [8]. Thus, TGF-β stimulates the recruitment, migration and proliferation of osteoblasts' precursors, while it inhibits terminal osteoblastic differentiation [9]. The actions of TGF-β on bone resorption are also different depending on its levels of concentration. At low concentrations, it enhances osteoclasts' formation, whereas at high concentrations it induces osteoclasts' apoptosis [10]. Additionally, systemic hormones, such as oestrogen or parathyroid hormone (PTH), are known to affect bone turnover. Being an important regulator of calcium homoeostasis, PTH influences calcium concentrations by increasing renal calcium reabsorption and consequently stimulating bone resorption [5]. But, besides these examples, there are many other paracrine and autocrine effectors that also represent an important role in bone remodelling, acting as either stimulators or inhibitors of bone's resorption or formation.

Since the understanding of bone remodelling is continuously developing, several researchers have been working on creating semi-empirical mathematical descriptions of this process to gain a better insight into the nature of bone remodelling [11]. This way, mathematical modelling becomes a powerful tool to simulate in silico and predict experimental results. By including the current knowledge about bone cell activity, these mathematical models aim to study the cell dynamics and its effects [12].

The osteoclast/osteoblast interactions in BMUs and the involving microenvironment during bone remodelling have been thoroughly analysed with distinct models. Lemaire [6] has been the first to propose a mathematical model incorporating the RANKL/RANK/OPG pathway. Besides considering the reaction kinetics of RANK, RANKL and OPG, the model also included the action of TGF-β on osteoblastic and osteoclastic cells and the catabolic effect induced by continuous administration of PTH. Later, the model proposed by Pivonka [13] extended Lemaire's model, including the production of RANKL and OPG on both osteoblastic cell lines. Additionally, other modifications were incorporated, such as the inclusion of a rate equation (describing changes in bone volume), a rate equation (describing TGF-β concentration as function of resorbed bone volume) and new activator/repressor functions based on enzyme kinetics [13]. In a subsequent study [14], Pivonka has applied this model to identify successful treatment strategies for different bone diseases.

A different approach for modelling bone cell population dynamics was used by Komarova [1]. Instead of including the activities of specific factors, the authors use the power-law formalism developed by Savageau [15], in which the exponents represent the combined action of different possible autocrine and paracrine regulators. Thus, in this approach, all factors leading to a cell response, such as RANKL, RANK or OPG, are lumped together in a single exponential parameter [1]. The model successfully simulates bone remodelling in Paget's disease, a metabolic bone disorder characterized by accelerated rates of local bone resorption and formation that result in production of structurally deficient bone [16]. Based on this model, the first spatio-temporal model recreating the dynamics of a single BMU has been proposed by Ryser [17]. The model predicts new roles for RANKL and OPG in forming spatially specific gradients, which strongly regulates the direction and speed of osteoclasts' movement [18]. Ayati et al. [2] also reinterpreted Komarova's model, creating a new diffusion model for normal bone remodelling and for the dysregulated bone remodelling that occurs in myeloma bone disease. In this spatio-temporal model, therapeutic effects targeting both myeloma cells and cells of the bone marrow microenvironment are examined, illustrating how treatment approaches may be investigated using computational approaches.

In this chapter, a novel model simulating the biological events during bone remodelling is proposed. Based on both Komarova's [1] and Ayati's [2] models, but incorporating an additional spatial variable, the aim of this study is to spatially and temporally describe this process, using for the first time discretization techniques to analyse the results, namely the finite element method (FEM), the radial point interpolation method (RPIM) and the natural neighbour radial point interpolation method (NNRPIM), being the last two meshless approaches.

Therefore, after this brief introduction, this study includes a detailed description of each one of the numerical methods used, followed by an explanation of the new biological model proposed. Then, this new algorithm is applied in a 2D bone patch analysis, in which the main results and conclusions are presented.

2 Numerical Methods

The task of computationally simulated natural phenomena is very complex, since it often implies several dimensions and nonlinear effects. The bone remodelling process is no exception, requiring the use of partial differential equations and nonlinear differential equations. However, in general, none of these equations can be solved symbolically or analytically, so researchers need to use numerical methods [19].

Thus, due to the complexity of the problems being studied, in the 1940s, the first approximation method was proposed, known as the FEM. The works of Hrennikoff [20] and McHenry [21] were the starting point, having then Courant [22] introduced the concept of piecewise interpolation, or shape functions, over triangular sub-regions. Many subsequent studies have improved and extended this method of obtaining approximate numerical solutions, and since then, the finite elements became the most popular approach to solve continuum problems in many areas, such as structural engineering, fluid dynamics, heat or mass transfer [23]. But, even though the FEM has been used with great success in many fields of engineering, this method is not free of limitations [24]. The main one is related to the mesh-based interpolation, since low-quality meshes lead to high values of error. Also, the FEM presents many difficulties when analysing numerical problems with highly irregular complex geometries.

Therefore, since the FEM's performance relies greatly on the quality of the mesh, other numerical methods were created and offered as solid options, such as the meshless methods. Meshless methods are characterized by the discretization of the problem physical domain with an unstructured nodal distribution. Meshless methods are classified into two categories. Regarding the strong formulation category, the partial differential equations describing the phenomenon are used directly to obtain the solution [25]. On the other hand, the weak formulation aims to minimize the residual weight of each differential equation, given by an approximated function affected by a test function and not by the exact solution of the differential equation [25]. In this category, the first meshless methods that emerged used approximation functions, such as the diffuse element method (DEM) proposed by Nayroles [26] and the element-free Galerkin method (EFGM) proposed by Belytschko [27]. However, these methods did not possess the Kronecker delta property, making it difficult to impose the essential and natural boundary conditions. Due to this fact, interpolant meshless methods were developed, such as the point interpolation method (PIM) [28]. The PIM [28, 29] is a meshfree method developed using Galerkin weak form, in which its interpolant shape functions can have two different forms of basis functions: polynomial basis functions [28, 30] and radial basis functions (RBFs) [31–33]. When RBFs are used to construct shape functions, the method is termed as RPIM. This method has proven to be very stable and robust, and has been successfully applied to many problems, such as 1D, 2D and 3D solid mechanics [34, 35] and plate and shell structures [36]. But this method requires the use of a global background cell structure to evaluate the integration in the Galerkin weak form, jeopardizing the meshfree concept. Thus, other meshless methods proposing integration schemes using the nodal distribution were presented, such as the NNRPIM and the natural radial element method (NREM) [37–39]. The NNRPIM has been extended to many fields of computational mechanics, such as the analysis of isotropic and orthotropic plates [40], the analysis of functionally graded material plate [41], the 3D shell-like approach [42–47], the dynamic analysis of several solid-mechanic problems [48], the analysis of large deformations [49] and crack mechanics [50, 51].

Regarding the focus of this research work, bone remodelling has been successfully analysed using the FEM [52–57] and the NNRPIM [58–60], but only considering the mechanical behaviour of bone tissue. In fact, when it comes to describing only the biological response that occurs in bone remodelling, there is a marked lack of studies analysing this phenomenon with numerical approaches.

Therefore, for the first time, three distinct numerical approaches (the FEM, the RPIM and the NNRPIM) are used to describe the cell dynamics during an event of bone remodelling. Thus, the following subsections focus on these numerical methods, addressing with more detail the meshless methods considered in this study.

2.1 Finite Element Method

The FEM involves modelling the domain using small interconnected elements called finite elements. As can be depicted in Fig. 1a, every interconnected element is linked, directly or indirectly, to every other element through common interfaces. The total number of elements used and their variation in size and type within a given body

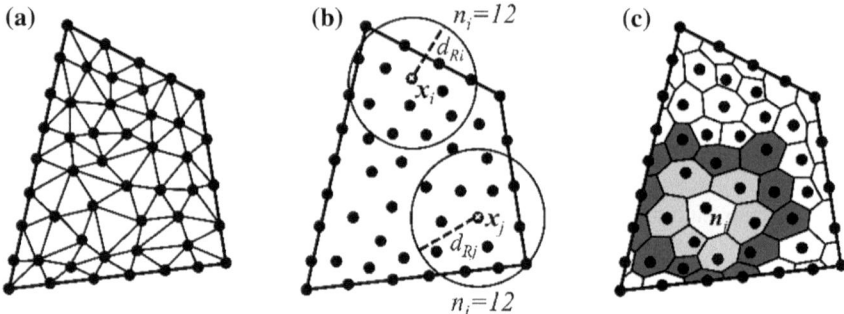

Fig. 1 Imposition of the nodal connectivity using **a** the FEM, **b** the RPIM and **c** the NNRPIM [25]

is a choice that weighs the computational cost and the accuracy of the results [60, 61]. These elements can also be irregular and possess different properties, enabling the discretization of structures with mixed properties [62]. In this chapter, triangular elements were used, due to the simplicity of the expressions associated with this element. Thus, by using a simple formulation of the FEM, it is possible to perform a fair comparison between these methods, the RPIM and the NNRPIM.

Using the FEM, a continuous quantity is approximated by a discrete model composed of a set of piecewise continuous functions defined within each finite domain or finite element [61]. These shape functions are defined using the nodal values of each element, being isoparametric since the same interpolation function defines the element's geometric shape and any unknown variable, such as the displacement within an element. Thus, it is possible to integrate the integro-differential equations governing the phenomenon under study, by simply using a background mesh. The Gauss–Legendre numerical scheme has proven to be a very useful tool to evaluate this integral. Since this quadrature is also used in meshless methods, it will be analysed with more detail in the next section.

2.2 Meshless Methods

As stated above, in meshless methods, the only information required is the spatial location of each node discretizing the problem domain. This way, there is no need of any previous information about the relation between each node, ensuring that the nodal distribution does not form a mesh. So, after discretization, nodal connectivity is imposed by using either influence-domains or Voronoï diagrams. This step is followed by the construction of a background integration mesh that allows the establishment of the interpolation functions of the unknown variable field functions.

Thus, it is possible to notice that every meshless method requires the presence and combination of three basic components: nodal connectivity, numerical integration scheme and shape functions. Therefore, these three concepts will be separately analysed for the RPIM and the NNRPIM in order to fully understand these methods.

2.2.1 Nodal Connectivity

In the RPIM, instead of the no-overlap rule between elements imposed in the FEM, the nodal connectivity is obtained by the overlap of the influence-domains of each node. Influence-domains are found by searching enough nodes inside a certain area or volume and can have a fixed or a variable size. Many meshless methods [33, 63, 64] use fixed-size influence-domains, but the RPIM uses a fixed number of neighbour nodes instead. By using variable size influence-domains, with a constant number of nodes inside the domain, even irregular nodal spatial distributions can be easily dealt with. Thus, performing a radial search and using the interest point x_I as centre, the n closest nodes are found. In Fig. 1b, this process is illustrated, culminating in a constant nodal connectivity, where $d_{Ri} \neq d_{Rj}$.

Regarding the NNRPIM, the nodal connectivity is obtained considering the natural neighbour concept [60, 65] by partitioning the discretized domain into a set of Voronoï cells. Thus, considering a set $N = \{n_1, n_2, \ldots, n_N\}$ of N distinct nodes, the Voronoï diagram of N is the partition of the domain defined by N in sub-regions V_I, closed and convex. Each Voronoï cell V_I is the geometric place where all points in the interior of V_I are closer to the node n_I than to any other node n_J, being $n_J \in N(J \neq I)$. The assemblage of the complete set of Voronoï cells $V = \{V_1, V_2, \ldots, V_N\}$ defines the Voronoï diagram.

In the NNRPIM [66], the Voronoï diagram is used to build the "influence-cells". This way, the nodal connectivity can be imposed by the overlap of these organic influence-domains. There are two distinct types of influence-cells: the "first-degree influence-cell" and the "second-degree influence-cell". In this work, it was only considered the "second-degree influence-cell". To establish it, a point of interest, x_I, starts by searching for its neighbour nodes following the natural neighbour Voronoï construction, considering only its first natural neighbours. Then, based again on the Voronoï diagram, the natural neighbours of the first natural neighbours of x_I are added to the influence-cell, as represented in grey in Fig. 1c.

2.2.2 Numerical Integration

The field functions are approximated within the influence-domain (or the influence-cell), which is obtained with the establishment of the nodal connectivity. But, this approximation is accomplished using a numerical integration scheme.

Thus, regarding the RPIM, the Gauss–Legendre quadrature scheme is used. So, initially, the solid domain is divided into an auxiliary grid as Fig. 2a indicates. Then, it

Fig. 2 **a** Fitted Gaussian integration mesh and **b** transformation of each triangular element into isoparametric triangles and application of the quadrature point rule [25, 60]

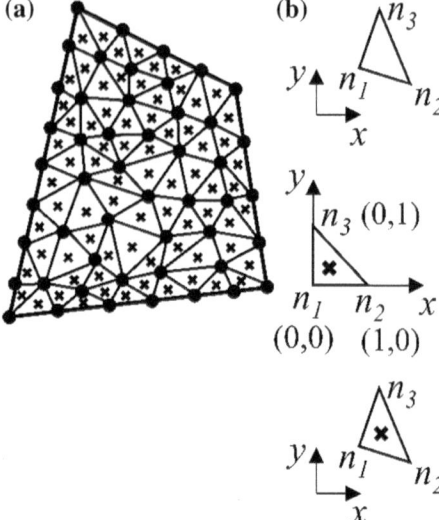

is possible to fill each grid cell with integration points, respecting the Gauss–Legendre quadrature rule, as illustrated in Fig. 2b.

To obtain the Cartesian coordinates of the quadrature points, isoparametric interpolation functions, N_i, are used. These functions are presented for quadrilaterals and triangles in Eq. (1) and Eq. (2), respectively.

$$N_1(\xi, \eta) = \frac{1}{4}(1 - \xi)(1 - \eta)$$

$$N_2(\xi, \eta) = \frac{1}{4}(1 - \xi)(1 + \eta)$$

$$N_3(\xi, \eta) = \frac{1}{4}(1 + \xi)(1 + \eta)$$

$$N_4(\xi, \eta) = \frac{1}{4}(1 + \xi)(1 - \eta) \tag{1}$$

$$N_1(\xi, \eta) = 1 - \xi - \eta$$
$$N_2(\xi, \eta) = \eta$$
$$N_3(\xi, \eta) = \xi \tag{2}$$

In the end, the Cartesian coordinates are given by

$$x = \sum_{i=1}^{m} N_i(\xi, \eta) \cdot x_i$$

$$y = \sum_{i=1}^{m} N_i(\xi, \eta) \cdot y_i \tag{3}$$

in which m is the number of nodes inside the grid cell and x_i and y_i are the Cartesian coordinates of the cells' nodes.

The integration weight of the quadrature point is obtained by multiplying the isoparametric weight of the quadrature point with the inverse of the determinant of the Jacobian matrix (Eq. 4) of the respective grid cell.

$$[J] = \begin{pmatrix} \frac{\partial x}{\partial \xi} & \frac{\partial x}{\partial \eta} \\ \frac{\partial y}{\partial \xi} & \frac{\partial y}{\partial \eta} \end{pmatrix} \tag{4}$$

Thus, to approximate the integral, the function is evaluated at several sampling points. Then, each value, multiplied by the appropriate weight, ω_i, is added, as depicted by Eq. (5). Gauss's method chooses the sampling points so that for a given number of points, the best possible accuracy is obtained.

$$\int_{-1}^{1} \int_{-1}^{1} f(\mathbf{x}) \mathrm{d}x \mathrm{d}y = \sum_{i=1}^{m} \sum_{j=1}^{n} \omega_i \omega_j f(\mathbf{x}) \tag{5}$$

For the numerical integration of the NNRPIM interpolation functions, the Delaunay triangles are applied to create a node depending on background mesh. The Delaunay triangulation is the geometrical dual of the Voronoï diagram, and it is constructed by connecting the nodes whose Voronoï cells have common boundaries with, since it is implied that a Delaunay edge exists between two nodes in the plane if and only if their Voronoï cells share a common edge.

Using the Delaunay triangulation, the area of each Voronoï cell is subdivided into several sub-areas. Firstly, considering the Voronoï cell, V_I, the intersection points, P_{Ii}, of the neighbour edges of V_I are settled. Then, the middle points, M_{Ii}, between n_I and its neighbour nodes are obtained, as Fig. 3 indicates. This way, the Voronoï cells are divided into n quadrilateral sub-cells, S_{Ii}, being n the number of natural neighbours of node n_I.

Therefore, the size of any Voronoï cell, V_I, can be determined using the size of the n sub-cells, S_{Ii},

$$A_{V_I} = \sum_{i=1}^{n} A_{S_{Ii}}, \forall A_{S_{Ii}} \geq 0 \tag{6}$$

being A_{V_I} the Voronoï cell area and $A_{S_{Ii}}$ the sub-cell area. So, it is important to notice that if the set of Voronoï cells is a partition, without gaps, of the global domain, then the set of sub-cells is also a partition, without gaps, of the global domain. Thus, these

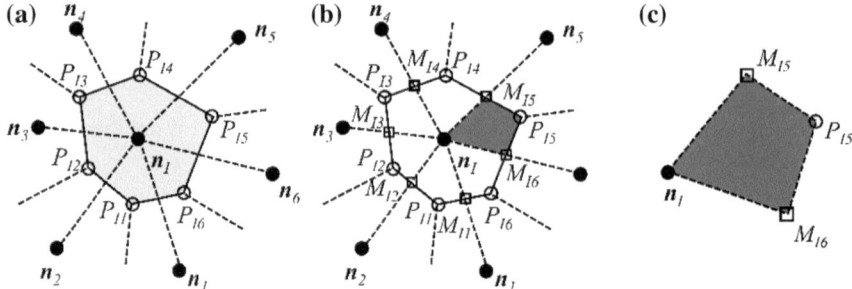

Fig. 3 a Initial Voronoï cell. **b** Voronoï cell and the corresponding P_{Ii} intersection points and middle points M_{Ii}. **c** Respective generated quadrilaterals [25]

sub-cells can be used in several integration schemes such as the Gauss–Legendre numerical scheme.

Studies found that the use of only one quadrature point in each sub-cell led to a more efficient integration scheme and that the second-degree influence-cell is the type of "influence-cell" that produces more accurate and stable results [60, 66]. Therefore, those parameters were the ones used in this research work.

2.2.3 Interpolation Functions

Considering the RPIM and the NNRPIM, the interpolation functions for both methods possess the Kronecker delta property, satisfying the following condition

$$\varphi_i\left(x_j\right) = \delta_{ij} \tag{7}$$

where δ_{ij} is the Kronecker delta, $\delta_{ij} = 1$ if $i = j$ and $\delta_{ij} = 0$ if $i \neq j$. This property is considered an asset, because it simplifies greatly the process of imposition of the boundary conditions.

The interpolation functions for both methods are determined using the RPI technique [33], which requires the combination of a polynomial basis with a RBF. So, considering the function $u(x_I)$ defined in the domain $\Omega \subset \mathbb{R}^2$, the value of function $u(x_I)$ at the point of interest x_I is defined by

$$u(x_I) = \sum_{i=1}^{n} R_i(x_I) \cdot a_i(x_I) + \sum_{j=1}^{m} p_j(x_I) \cdot b_j(x_I)$$
$$= R^T(x_I) \cdot a(x_I) + p^T \cdot b(x_I) \tag{8}$$

where $R_i(x_I)$ is the RBF, $p_j(x_I)$ is the polynomial basis function and $a_i(x_I)$ and $b_j(x_I)$ are non-constant coefficients of $R_i(x_I)$ and $p_j(x_I)$, respectively [59]. The variable defined on the RBF is the distance r_{Ii} between the relevant node x_I and the neighbour node x_i, given by $r_{Ii} = |x_i - x_I|$. The RBF used in this work is the multiquadric RBF [67], $R_i(x_I) = R(r_{Ii}) = (r_{Ii}^2 + c^2)^p$, in which shape parameter c depends on the volume occupied by the interest point and takes a value close to zero, $c \cong 0$, and p close to one, $p \cong 1$ [40, 66]. Regarding Eq. (8), it is still needed to obtain the non-constant coefficients a and b. The polynomial basis functions used have the following monomial term

$$p^T(x_I) = [1, x, y, x^2, xy, y^2, \ldots]$$
(9)

Considering Eq. (9) for each node inside the influence-cell domain and including an extra equation, $\sum_{i=1}^{n} p_j(x_I)a_i = 0$, in order to guarantee a unique solution, a system of equations as defined in Eq. (10) is obtained.

$$\begin{bmatrix} R & p \\ p^T & 0 \end{bmatrix} \begin{Bmatrix} a \\ b \end{Bmatrix} = \begin{Bmatrix} u_S \\ 0 \end{Bmatrix}$$
(10)

Through this system of equations, and being the vector of the nodal function values for the nodes on the influence-cell defined by: $u_S = \{u_1, u_2 \ldots u_n\}^T$, these coefficients are determined (Eq. 11)

$$\begin{Bmatrix} a \\ b \end{Bmatrix} = \begin{bmatrix} R & p \\ p^T & 0 \end{bmatrix}^{-1} \begin{Bmatrix} u_S \\ 0 \end{Bmatrix} \Rightarrow \begin{Bmatrix} a \\ b \end{Bmatrix} = M^{-1} \begin{Bmatrix} u_S \\ 0 \end{Bmatrix}$$
(11)

Recalling that a certain field variable value for an interest point x_I is interpolated using the shape function values obtained at the nodes inside the support domain of x_I, it is now possible to define the interpolation function, by substituting in Eq. (8) the result from Eq. (11). The interpolation function $\Phi(x_I) = \{\varphi_1(x_I), \varphi_2(x_I), \ldots, \varphi_n(x_I)\}$ for an interest point x_I is then defined by

$$u(x_I) = \{R^T(x_I), p^T(x_I)\} M^{-1} \begin{Bmatrix} u_S \\ 0 \end{Bmatrix} = \Phi(x_I) \begin{Bmatrix} u_S \\ 0 \end{Bmatrix}$$
(12)

In order to compute the partial derivatives of the interpolated field function, it is necessary to obtain the respective RPI shape functions partial derivatives. So, for the 2D problems, the partial derivative of $\Phi(x_I)$ is defined as

$$\Phi_x(x_I) = \{R^T(x_I) p^T(x_I)\}_x M^{-1}$$
(13)

$$\Phi_y(x_I) = \{R^T(x_I) p^T(x_I)\}_y M^{-1}$$
(14)

The first-order partial derivative of the RBF vector with respect to the same 2D problem is defined as

$$R(x_I)_x = \left\{ R_1(x_I)_x \ R_2(x_I)_x \ \ldots \ R_n(x_I)_x \right\}^T = \left\{ \frac{\partial R_1(x_I)}{\partial x} \ \frac{\partial R_2(x_I)}{\partial x} \ \ldots \ \frac{\partial R_n(x_I)}{\partial x} \right\}^T \tag{15}$$

$$R(x_I)_y = \left\{ R_1(x_I)_y \ R_2(x_I)_y \ \ldots \ R_n(x_I)_y \right\}^T = \left\{ \frac{\partial R_1(x_I)}{\partial y} \ \frac{\partial R_2(x_I)}{\partial y} \ \ldots \ \frac{\partial R_n(x_I)}{\partial y} \right\}^T \tag{16}$$

being the partial derivatives of the MQ-RBF obtained with

$$\frac{\partial R_i(x_I)}{\partial x} = -2p\left(r_{iI}^2 + c^2\right)^{p-1}(x_i - x_I) \tag{17}$$

$$\frac{\partial R_i(x_I)}{\partial y} = -2p\left(r_{iI}^2 + c^2\right)^{p-1}(y_i - y_I) \tag{18}$$

3 New Spatio-temporal Model

As stated previously, many mathematical models have been proposed in the field of bone biology. But, among these various models, the model proposed by Komarova [1] stands out, due to its unique way of expressing the actions of paracrine and autocrine regulatory factors. Few years later, Ayati et al. [2] have reinterpreted Komarova's model and created a new spatio-temporal model of bone remodelling. Thus, in this chapter, the model proposed is a spatial extension of the model presented by Ayati, analysed using numerical approaches.

3.1 Komarova's Model

Komarova's approach is to model bone cell population dynamics, examining the osteoclast/osteoblast interactions at a single site within a BMU [1]. Thus, this model is considered a local and purely temporal model [17]. Also, Komarova proposes that all factors leading to a bone cell response can be lumped together in a single exponential parameter, using the power-law formalism developed by M. A. Savageau [15]. This way, in order to describe the dynamics of bone cell populations, the following system of differential equations has been presented

$$\frac{\partial C}{dt} = \alpha_C C^{g_{CC}} B^{g_{BC}} - \beta_C C \tag{19}$$

$$\frac{\partial B}{dt} = \alpha_B C^{g_{CB}} B^{g_{BB}} - \beta_B B \tag{20}$$

where C and B are the number of osteoclasts and osteoblasts, respectively. Parameter α_i is the rate of overall production of each cell population, i, reflecting the net effect of recruitment of precursors and the formation of mature cells. Parameter β_i defines the rates of cell removal, reflecting cell death, as well as differentiation of osteoblasts into osteocytes and bone lining cells. Finally, parameters g_{ij} represent the net effectiveness of osteoclast- or osteoblast-derived autocrine or paracrine factors. This way, the ability of bone cells to interact with each other via effectors, released or activated by these cells, is expressed. This can be done in an autocrine manner, when it locally affects its cell type of origin, or in a paracrine manner, when it affects the other cell type [1]. Therefore, parameter g_{CC} describes the combined effects of all factors produced by osteoclasts that regulate osteoclast formation. Parameter g_{CB} includes the combined effects of all factors produced by osteoclasts that regulate osteoblasts' formation. Parameter g_{BB} defines the combined effects of all factors produced by osteoblasts that regulate osteoblasts' formation, while parameter g_{BC} includes the combined effects of all factors produced by osteoblasts that regulate osteoclasts' formation. The regulatory factors considered in Komarova's model are listed in Table 1 as well as their effect on the dynamics of bone cell populations.

Equations (19) and (20) have well-defined steady-state solutions, denoted by $\bar{C}$ and $\bar{B}$, respectively. Thus, regarding the variation of bone mass, Komarova assumed that cell numbers above steady-state levels indicated proliferation and differentiation of precursors into mature cells which are able to remove or build bone. On the other hand, populations of osteoclasts and osteoblasts under steady-state conditions are assumed to consist of less differentiated cells not actively involved in the processes of resorption and production of bone matrix, but able to participate in autocrine and paracrine signalling [1]. Thus, the number of cells exceeding steady-state levels is given by $\max[0, C - \bar{C}]$ for osteoclasts and $\max[0, B - \bar{B}]$ for osteoblasts. Considering that the rates of bone resorption and formation are proportional to the number

Table 1 Description of the regulatory factors considered in Komarova's model and their respective effect on bone cell dynamics

	Types of signalling	Factor	Effect	Parameter
Osteoclasts	Autocrine	TGF-β	Stimulation of osteoclast formation	g_{CC}
	Paracrine	RANKL	Stimulation of osteoclast formation	g_{BC}
		OPG	Inhibition of osteoclast formation	g_{BC}
Osteoblasts	Autocrine	IGF	Stimulation of osteoblast formation	g_{BB}
	Paracrine	TGF-β	Stimulation of osteoblast formation	g_{CB}
		IGF	Stimulation of osteoblast formation	g_{CB}

Identification of the parameter in which each factor is considered in the model's equations of osteoclasts and osteoblasts, Eq. (21) describes the changes in bone mass,

$$\frac{\partial \rho}{\partial t} = -k_C \max[0, C - \bar{C}] + k_B \max[0, B - \bar{B}] \tag{21}$$

in which ρ is total bone mass and k_i is normalized activity of bone resorption (k_C) and formation (k_B). The number of cells exceeding steady state is the number of cells actively resorbing or forming bone.

3.2 Ayati's Model

Following Komarova's work, Ayati created a new diffusion model of bone remodelling [2]. Since Komarova's model is a dynamical system with zero explicit space dimensions, Ayati has presented an updated version of that model and proposed a one-dimensional spatial model. Thus, it added a new term by determining the second-order derivative. Using this term, it is possible to describe in space the diffusion of any property as time progresses. So, assuming that both osteoclasts and osteoblasts are diffusing in the spatial domain Ω, the following equations have been proposed:

$$\frac{\partial}{\partial t} C(t, x) = \sigma_C \frac{\partial^2}{\partial x^2} C(t, x) + \alpha_C C(t, x)^{g_{CC}} B(t, x)^{g_{BC}} - \beta_C C(t, x) \tag{22}$$

$$\frac{\partial}{\partial t} B(t, x) = \sigma_B \frac{\partial^2}{\partial x^2} B(t, x) + \alpha_B C(t, x)^{g_{CB}} B(t, x)^{g_{BB}} - \beta_B B(t, x) \tag{23}$$

$$\frac{\partial}{\partial t} \rho(t, x) = \sigma_\rho \frac{\partial^2}{\partial x^2} \rho(t, x) - k_C \max[0, C(t, x) - \bar{C}(x)]$$
$$+ k_B \max[0, B(t, x) - \bar{B}(x)] \tag{24}$$

The variables of the model are $C(t, x)$ and $B(t, x)$ that define the density of osteoclasts and osteoblasts, respectively, at a time t with respect to $x \in \Omega$. Additionally, the variable $\rho(t, x)$ reflects the changes in bone mass, also as a function of x and t. Parameters σ_C and σ_B represent the migration of bone stromal cells, and σ_ρ represents stochasticity in bone dynamics.

3.3 Proposed Model

3.3.1 Model Description

The main goal of this study is to simulate, spatially and temporarily, the bone remodelling process, in regard to the biological response. The proposed new model is based on both Komarova's and Ayati's models, but incorporates an additional spatial variable, and it is combined with discrete numerical methods. Therefore, the apparent density is dependent on the dynamic behaviour of osteoclasts and osteoblasts, but its variation is described by a nonlinear equation that has been adapted to work with the three numerical methods under study, namely the FEM, the RPIM and the NNRPIM. Thus, the bone mass is a temporal–spatial function, $\rho(\pmb{x}, t) : \mathbb{R}^{d+1} \mapsto \mathbb{R}$, discretized along a one-dimensional temporal line and a d-dimensional space. Presenting it as a differential equation and minimizing it with respect to time, this equation's expression is given by

$$\frac{\partial \rho(\pmb{x}, t)}{\partial t} \cong \frac{\Delta \rho(\pmb{x}, t)}{\Delta t} = \frac{(\rho_{\text{model}})_{t_{j+1}} - (\rho_{\text{model}})_{t_j}}{t_{j+1} - t_j} = 0 \qquad (25)$$

In this model, any problem being analysed is discretized in space and time. So, the d-dimensional spatial domain is assumed to be discretized in N nodes: $X = \{\pmb{x}_1, \pmb{x}_2, \dots, \pmb{x}_n\} \in \pmb{\Omega}$, leading to Q interest points: $\pmb{Q} = \{\pmb{x}_1, \pmb{x}_2, \dots, \pmb{x}_Q\} \in \pmb{\Omega}$, being $\pmb{x}_i \in \mathbb{R}^d$ and $\pmb{Q} \cap X = \emptyset$. The temporal domain is discretized in iterative fictitious time steps $t_j \in \mathbb{R}$, with $j \in \mathbb{N}$.

The medium apparent density for the complete model domain is defined by $(\rho_{\text{model}})_{t_j}$ at a fictitious time t_j. Consequently, considering the same iterative step, the apparent density for the complete model domain, ρ_{model}, can be determined as

$$\rho_{\text{model}} = Q^{-1} \sum_{i=1}^{Q} \rho_i \qquad (26)$$

being ρ_i the infinitesimal apparent density on interest point $\pmb{x}_I$. The determination of the apparent density on interest point $\pmb{x}_I$, ρ_I, is performed using the new diffusion model proposed in this chapter. By adding mixed derivatives to describe the diffusion process, the equations proposed in this novel approach are the following:

$$
\begin{aligned}
C(t_k, x_I, y_I) = {}& C(t_{k-1}, x_I, y_I) \\
& + \Delta t \left[\sigma_{Cx} \frac{\partial^2}{\partial x^2} C(t_{k-1}, x_I, y_I) + \sigma_{Cy} \frac{\partial^2}{\partial y^2} C(t_{k-1}, x_I, y_I) \right] \\
& + \sigma_{Cxy} \frac{\partial^2}{\partial x \partial y} C(t_{k-1}, x_I, y_I) \\
& + \alpha_C C(t_{k-1}, x_I, y_I)^{g_{CC}} B(t_{k-1}, x_I, y_I)^{g_{BC}} \\
& - \beta_C C(t_{k-1}, x_I, y_I)
\end{aligned} \qquad (27)
$$

$$
\begin{aligned}
B(t_k, x_I, y_I) = {}& B(t_{k-1}, x_I, y_I) \\
& + \Delta t \left[\sigma_{Bx} \frac{\partial^2}{\partial x^2} B(t_{k-1}, x_I, y_I) + \sigma_{By} \frac{\partial^2}{\partial y^2} B(t_{k-1}, x_I, y_I) \right]
\end{aligned}
$$

$$+ \sigma_{Bxy} \frac{\partial^2}{\partial x \partial y} B(t_{k-1}, x_I, y_I)$$
$$+ \alpha_B C(t_{k-1}, x_I, y_I)^{g^{CB}} B(t_{k-1}, x_I, y_I)^{g^{BB}}$$
$$- \beta_B B(t_{k-1}, x_I, y_I) \tag{28}$$

$$\rho(t_k, x_I, y_I) = \rho(t_{k-1}, x_I, y_I)$$
$$+ \Delta t \left[\sigma_{\rho x} \frac{\partial^2}{\partial x^2} \rho(t_{k-1}, x_I, y_I) + \sigma_{\rho y} \frac{\partial^2}{\partial y^2} \rho(t_{k-1}, x_I, y_I) \right]$$
$$+ \sigma_{\rho xy} \frac{\partial^2}{\partial x \partial y} \rho(t_{k-1}, x_I, y_I)$$
$$- k_C \max\left[0, C(t_{k-1}, x_I, y_I) - \bar{C}(x_I, y_I)\right]$$
$$+ k_B \max\left[0, B(t_{k-1}, x_I, y_I) - \bar{B}(x_I, y_I)\right] \tag{29}$$

The variables of the model are the density of osteoclasts, $C(t_k, x_I, y_I)$, the density of osteoblasts, $B(t_k, x_I, y_I)$, and the bone mass, $\rho(t_k, x_I, y_I)$. All variables are calculated in respect of a time t, at each instant k, and for each integration point I, with spatial coordinates x_I and y_I. In Eqs. (27) and (28), the diffusion coefficients σ_{ix} and σ_{iy} represent the migration of the bone cell type i in the direction x and y, respectively. In Eq. (29), $\sigma_{\rho x}$ and $\sigma_{\rho y}$ represent the stochasticity in the bone dynamics, as proposed by Ayati [2]. The diffusion coefficient σ_{ixy} is given by $\sigma_{ixy} = \frac{\sigma_{ix} + \sigma_{iy}}{2}$, for each i variable of the model.

However, since in this chapter, discrete numerical methods are being considered, the second-order spatial partial derivates of $C(t_k, x_I, y_I)$, $B(t_k, x_I, y_I)$ and $\rho(t_k, x_I, y_I)$ can be obtained using the data. Thus, consider a nodal set $= \{n_1, n_2, \ldots, n_N\}$, with the following spatial coordinated $X = \{x_1, x_2, \ldots, x_N\}$, discretizing the two-dimensional problem domain $\Omega \in \mathbb{R}^2$. Then, consider an interest point $x_I \in \Omega$ and $x_I \notin X$. For the interest point x_I, the influence-domain (or the element) is defined. Then, the shape function is obtained, $\varphi(x_I)$. If the nodal data is known, $U = \{u_1, u_2, \ldots, u_N\}$, it is possible to interpolate the data to the interest point x_I with

$$u(x_I) = \varphi(x_I)U \tag{30}$$

Thus, if the first and second partial derivatives are known, it is possible to directly approximate the first and second partial derivatives of the data field,

$$\frac{\partial u(x_I)}{\partial x} = \frac{\partial \varphi(x_I)}{\partial x} U \tag{31}$$

$$\frac{\partial u(x_I)}{\partial y} = \frac{\partial \varphi(x_I)}{\partial y} U \tag{32}$$

$$\frac{\partial^2 u(\pmb{x}_I)}{\partial x^2} = \frac{\partial^2 \varphi(\pmb{x}_I)}{\partial x^2} U \tag{33}$$

$$\frac{\partial^2 u(\pmb{x}_I)}{\partial y^2} = \frac{\partial^2 \varphi(\pmb{x}_I)}{\partial y^2} U \tag{34}$$

$$\frac{\partial^2 u(\pmb{x}_I)}{\partial x \partial y} = \frac{\partial}{\partial x} \left(\frac{\partial \varphi(\pmb{x}_I)}{\partial y} U \right) \tag{35}$$

However, the three-node 2D finite element method possesses a linear shape function, for which only one partial derivative is possible (such shape functions have C^1 continuity). Thus, for this kind of shape function, another approach is required. So, in order to obtain the second partial derivative of the data it is necessary first to obtain the interpolated partial derivative of the data field. So, for each node $\pmb{x}_J \in X$, it will define its influence-domain (or element) and then, for each node, the shape functions and the first partial derivatives of the shape functions are constructed. Then, it is possible to obtain the partial derivative field of the data field using the following relations

$$\frac{\partial u(\pmb{x}_J)}{\partial x} = \frac{\partial \varphi(\pmb{x}_J)}{\partial x} U \tag{36}$$

$$\frac{\partial u(\pmb{x}_J)}{\partial y} = \frac{\partial \varphi(\pmb{x}_J)}{\partial y} U \tag{37}$$

In the end, two new vectors are obtained: $\pmb{U}_x = \{\partial u(\pmb{x}_1)/\partial x, \ldots, \partial u(\pmb{x}_N)/\partial x\}$ and $\pmb{U}_y = \{\partial u(\pmb{x}_1)/\partial y, \ldots, \partial u(\pmb{x}_N)/\partial y\}$, which are the partial derivative fields of the scattered data.

Now, using the same procedure presented first, it is possible to obtain the second-order partial derivatives of the data,

$$\frac{\partial^2 u(\pmb{x}_I)}{\partial x^2} = \frac{\partial \varphi(\pmb{x}_I)}{\partial x} \pmb{U}_x \tag{38}$$

$$\frac{\partial^2 u(\pmb{x}_I)}{\partial y^2} = \frac{\partial \varphi(\pmb{x}_I)}{\partial y} \pmb{U}_y \tag{39}$$

$$\frac{\partial^2 u(\pmb{x}_I)}{\partial x \partial y} = \frac{\partial \varphi(\pmb{x}_I)}{\partial x} \pmb{U}_y \quad \text{or} \quad \frac{\partial^2 u(\pmb{x}_I)}{\partial x \partial y} = \frac{\partial \varphi(\pmb{x}_I)}{\partial y} \pmb{U}_x \tag{40}$$

This procedure induces accumulative numerical errors; however, it is sufficiently accurate to be used in such application [68].

3.3.2 Remodelling Procedure

The algorithm implemented can be divided into three different phases, as presented in Fig. 4. This algorithm is initialized by a pre-processing phase. The problem domain is discretized with an unstructured nodal mesh $X \in \Omega$, and a background integration mesh is constructed, according to the numerical method chosen by the user (the FEM, the RPIM or the NNRPIM). The shape functions for each integration point x_I are constructed, $\varphi(x_I)$, and the material properties are applied. Also, to initialize the remodelling process, it is necessary to impose first the initial conditions.

Then, the second phase begins, consisting of the determination of the spatial partial derivatives of the density of osteoclasts, density of osteoblasts and bone mass.

Afterwards, it is possible to move on to the bone remodelling phase, in which a new cell density of osteoclasts, osteoblasts and bone mass is determined using Eq. (27), Eq. (28) and Eq. (29), respectively. Consequently, the medium apparent density of the model, given by Eq. (26), is updated and the remodelling procedure finishes when the following condition is achieved

$$\frac{\Delta \rho}{\Delta t} = 0 \vee (\rho_{\text{model}})_{t_j} = \rho_{\text{control}} \tag{41}$$

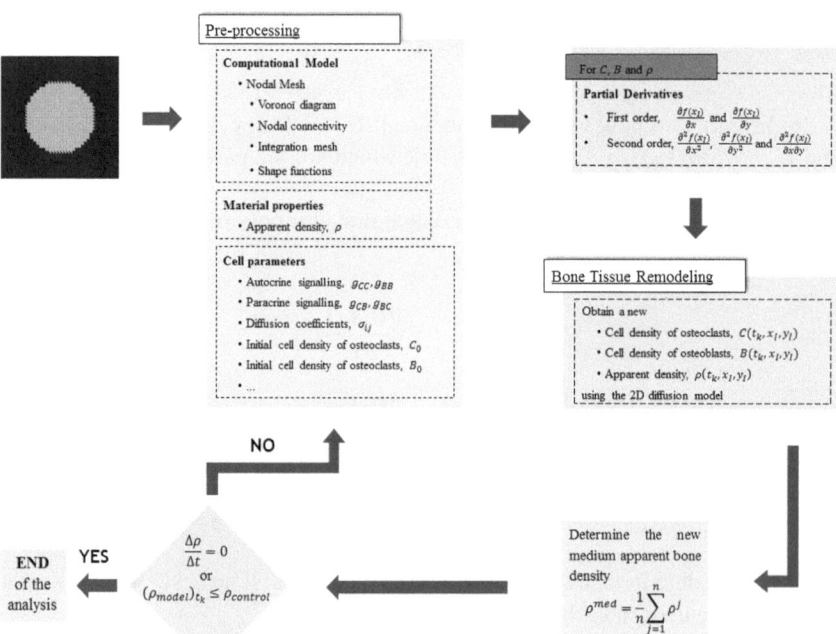

Fig. 4 Proposed bone remodelling algorithm

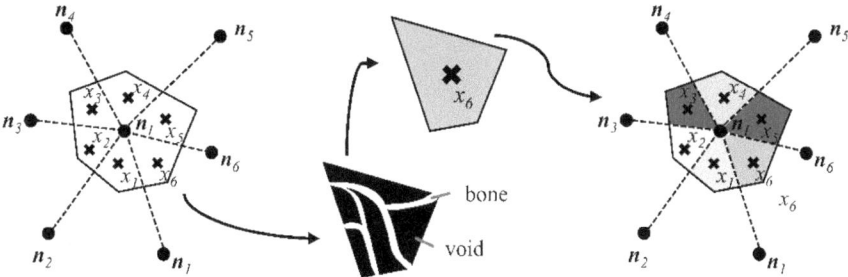

Fig. 5 Representation of the local apparent density of each sub-cell and the obtained medium apparent density for that Voronoï cell [25]

In the end, the results are represented with isomaps. In the case of the NNRPIM, these maps are obtained using the sub-cells, S_{Ii}, already discussed previously. So, for each of these cells, the volume porosity, $p(\boldsymbol{x}_J)$, is obtained. Then, the local apparent density, $\rho(\boldsymbol{x}_J) = \rho_0(1 - p(\boldsymbol{x}_J))$, being $\rho_0 = 2.1\,\mathrm{g/cm^3}$ the compact bone density, is determined. Thus, as represented in Fig. 5, each integration point can present a distinct local apparent density, allowing to obtain a local medium apparent density for each Voronoï cell V_i.

So, knowing the local apparent density of each field node $\boldsymbol{x}_i$, the following expression can be applied,

$$\rho(\boldsymbol{x}_i) = \frac{\sum_{J=1}^{k} \hat{w}_J \cdot \rho(\boldsymbol{x}_J)}{\sum_{J=1}^{k} \hat{w}_J} \tag{42}$$

where $\hat{w}_J$ is the integration weight of an integration point $\boldsymbol{x}_J$ belonging to the Voronoï cell V_i of the field node $\boldsymbol{x}_i$.

An analogous procedure occurs in the FEM and in the RPIM. The difference is that both use the Gauss–Legendre quadrature scheme, instead of the Voronoï diagram. However, in the end, all methods allow the definition of a local apparent density field, with Eq. (42), and its representation in isomaps.

Therefore, the results presented in this chapter, in all numerical examples, are presented as grey tone isomaps. In these grey tone isomaps, the white colour represents the considered maximum apparent density $\rho_0 = 2.1\,\mathrm{g/cm^3}$ and the dark-grey colour represents the minimum apparent density $\rho_0 = 0.1\,\mathrm{g/cm^3}$ admitted in the analysis. All the other grey tones in the middle represent transitional apparent densities, following a linear colour greyscale gradient.

4 2D Bone Patch Analysis

4.1 Initial Conditions

The bone remodelling algorithm developed has been applied to 2D bone micro-patches. A single cycle of remodelling is simulated using unit square patches, as illustrated in Fig. 6a. In this figure, it is possible to detect a central portion represented in light-grey, considered as bone tissue, while the involving area is considered as surrounding tissue. So, distinct initial conditions for each type of tissue have been applied, having been stated as follows (Table 2).

Regarding the bone tissue, as Komarova has proposed, the initiation of bone remodelling is manually induced by choosing initial values of osteoclasts' density higher than their steady state. On the other hand, it is assumed that, initially, there are no osteoblasts active. Thus, their density equals their steady-state value. Additionally, the initial value of bone mass is set to 100%. Regarding the materials' properties, this tissue presents a uniform apparent density distribution characteristic of cortical bone [69]. In the surrounding tissue, there are no bone cells present so bone mass is assumed as 0%. Consequently, the apparent density of this tissue is very low.

Concerning the spatial distribution of bone cells, since osteoblasts can differentiate into osteocytes or bone lining cells, it is assumed that osteoblasts can be found anywhere in bone. However, since osteoclasts only attach to the bone's surface during the resorption phase and then abandon this local, this model assumes that, initially,

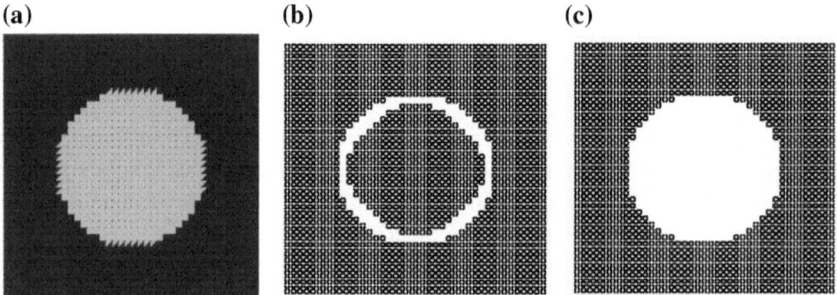

Fig. 6 a Mesh (1681 nodes) used for the analysis, in which nodes inside the light-grey area are considered bone. **b** Initial distribution of osteoclasts indicated in white. **(c)** Initial distribution of osteoblasts indicated in white

Table 2 Initial conditions

Bone tissue	Surrounding tissue
$C(0, x, y) = \bar{C} + 10$	$C(0, x, y) = 0$
$B(0, x, y) = \bar{B}$	$B(0, x, y) = 0$
$\rho(0, x, y) = 100\%$	$\rho(0, x, y) = 0\%$

Table 3 Parameters of the biological model

$\bar{C}$	$1.06\,\text{cell m}^{-2}$	k_C	$0.24\%\,\text{cell}^{-1}\,\text{day}^{-1}$
$\bar{B}$	$212.13\,\text{cell m}^{-2}$	k_B	$0.0017\%\,\text{cell}^{-1}\,\text{day}^{-1}$
α_C	$3.0\,\text{cell day}^{-1}$	σ_{Cx}	$0.000001\,\text{cell m}^{-1}$
α_B	$4.0\,\text{cell day}^{-1}$	σ_{Cy}	$0.000001\,\text{cell m}^{-1}$
β_C	$0.2\,cell\,day^{-1}$	σ_{Bx}	$0.000001\,\text{cell m}^{-1}$
β_B	$0.02\,\text{cell day}^{-1}$	σ_{By}	$0.000001\,\text{cell m}^{-1}$
g_{CC}	0.5	$\sigma_{\rho x}$	$0.000001\,\text{bone mass m}^{-1}$
g_{CB}	1.0	$\sigma_{\rho y}$	$0.000001\,\text{bone mass m}^{-1}$
g_{BC}	-0.5	Δt	$0.1\,\text{day}$
g_{BB}	0.0		

osteoclasts can only be found at this site. The initial spatial distribution described is illustrated in Fig. 6b, c.

For this analysis, the bone cells' parameters have the values proposed by Komarova and Ayati for a single cycle of bone remodelling. This data is based on histomorphometric studies, and it is listed in Table 3 [70].

Analysing with detail the effectiveness of the regulatory factors, the analysis of the temporal model [1] has showed that the parameter g_{CC} plays a critical role in controlling the dynamic behaviour of remodelling when acting as a positive feedback [1]. Moreover, it has been shown that positive autocrine regulation of osteoclasts is necessary to describe the complex behaviour observed in osteoclast cultures in vitro [17, 71]. Therefore, g_{CC} is assumed to be positive and equal to 0.5. Komarova has also proved that the factor g_{CB} has to be strictly positive [1] and is crucial for the coupling of osteoclasts and osteoblasts [72]. The factor g_{BB} has not influenced the dynamical behaviour of the temporal model [1], and therefore, it is assumed as null. Finally, for the parameter g_{BC}, it assumed a negative feedback on osteoclasts' production.

Lastly, the parameters used in the RPIM and the NNRPIM analyses are listed in Table 4.

Using the algorithm proposed, several tests have been performed. Firstly, it has studied the convergence of the three numerical methods tested. Thus, the example illustrated in Fig. 6 is analysed, using a solid domain discretized in increasingly denser nodal distributions. Then, the spatio-temporal evolution of a single bone remodelling cycle is analysed, comparing the results obtained with each numerical approach tested.

Table 4 Parameters of the RPIM and the NNRPIM

Order of the influence-cell	16
c	0.0001
p	0.9999
Polynomial basis	Constant
Gauss quadrature	1

4.2 Results and Discussion

4.2.1 Convergence Test

In the convergence study, the mesh sizes tested are 11×11, 21×21, 31×31 and 41×41 nodes. So, initially, for each mesh size tested, the medium apparent density for the complete model domain, defined by $\rho_{\text{model}}(t_j)$ at a fictitious time t_j, is determined using Eq. (26). Then, the medium apparent density is normalized using the following equation:

$$\gamma(t_j) = \left(\frac{\rho_{\text{model}}(t_j) - \rho_{\text{min}}}{\rho_{\text{max}} - \rho_{\text{min}}} \right) \times 100 \tag{43}$$

being $\rho_{\text{min}} = 0\text{g/cm}^3$ and $\rho_{\text{max}} = \max(\rho_{\text{model}})$. Using as an example the results obtained with the FEM, in Fig. 7a, it is possible to observe the variation of bone mass along time, $\gamma(t_j)$, for each mesh size.

Then, it is possible to evaluate the convergence of each numerical approach, by determining the integral of function $\gamma(t_j)$ along time, as defined by λ in Eq. (44).

$$\lambda = \int_{t_0}^{t} \gamma(t_j)\mathrm{d}t \tag{44}$$

The results obtained in this convergence test are represented in Fig. 7b. In this graph, it is the similitude visible between the FEM, the RPIM and the NNRPIM

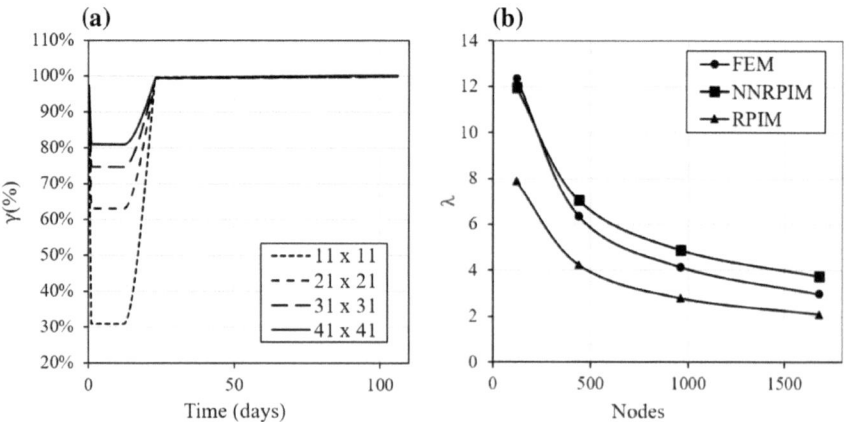

Fig. 7 **a** Bone mass variation during time for different types of mesh obtained with FEM. **b** Convergence of the three numerical methods varying the number of nodes discretizing the problem domain

convergence lines. Also, all convergence paths lead to a converged solution, as is already suggested in Fig. 7a. When comparing the three numerical methods, the denser mesh of 41 × 41 nodes was the one that produced a closer solution. With this study, it is possible to conclude that the model depends on the refinement of the mesh. Thus, the following tests were performed using only a 41 × 41 nodal mesh.

4.2.2 Bone Remodelling

The remodelling model proposed in this chapter is applied in the simulation of a single event of bone remodelling. Since this is a spatio-temporal model, results are expected to describe accurately both the temporal and the spatial behaviours of bone cells.

Thus, starting with a temporal analysis, Figs. 8, 9 and 10 show the changes with time in the cell density of osteoclasts and osteoblasts obtained using the FEM, the RPIM and the NNRPIM, respectively. Each figure represents the variation profile of each one of the integration points used in the respective numerical approach. Graphics representing the variation of osteoblasts' density present two distinct profiles, one for the osteoblasts located at the bone's surface and the other for the internal bone cells, as can be depicted in Fig. 8b. Only the osteoblasts at the surface are indeed actively involved in the remodelling process. For this reason, in this temporal analysis, it is only considered the variational profile of these cells, indicated with the symbol "*".

So, as was already mentioned, the bone remodelling cycle is initiated by a momentary increase in the number of osteoclasts at time $t = 0$, which can be depicted in the graphs by the initial peak of osteoclasts. Then, while the density of osteoclasts is elevated, osteoclasts stimulate the slower process of osteoblasts' formation, through paracrine signalling, leading to an increase in the number of osteoblasts. Consequently, osteoblasts reach their peak later. Osteoblasts' paracrine signalling can also

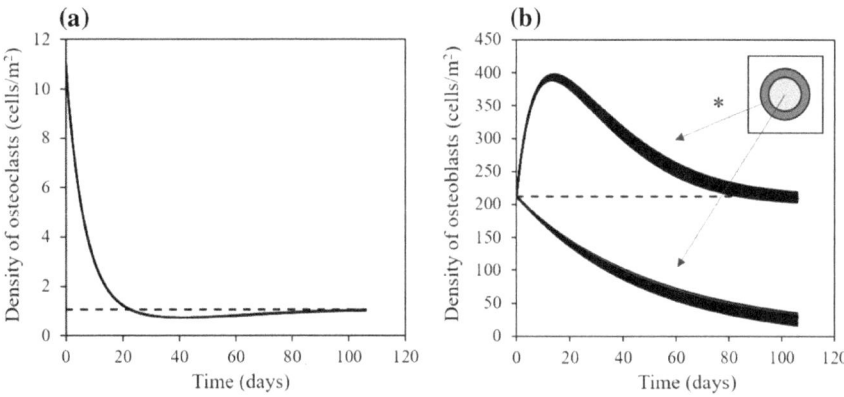

Fig. 8 Simulation of a bone remodelling cycle using the FEM: **a** changes with time in the cell density of osteoclasts and **b** osteoblasts. The dashed line represents the steady-state solution

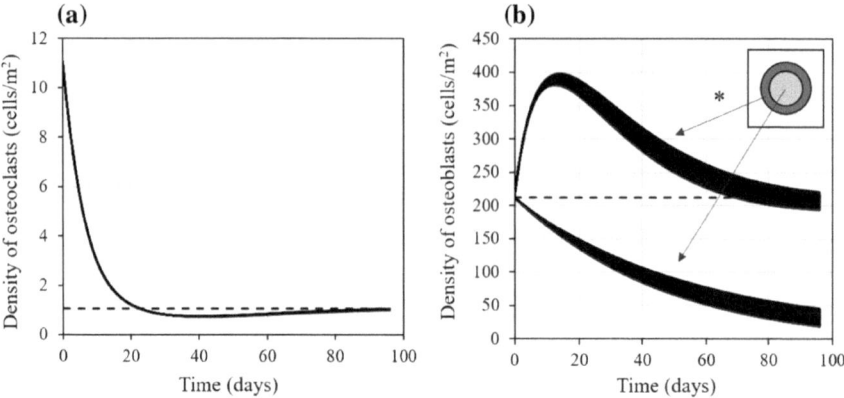

Fig. 9 Simulation of a bone remodelling cycle using the RPIM: **a** changes with time in the cell density of osteoclasts and **b** osteoblasts. The dashed line represents the steady-state solution

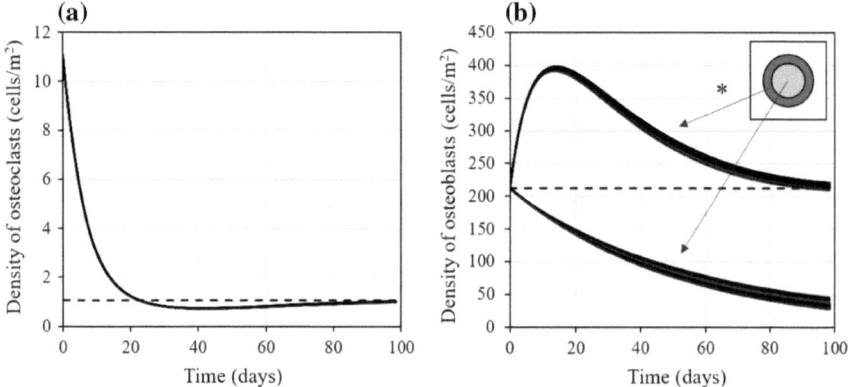

Fig. 10 Simulation of a bone remodelling cycle using the NNRPIM: **a** changes with time in the cell density of osteoclasts and **b** osteoblasts. The dashed line represents the steady-state solution

be detected in these graphics, since it causes a slight decrease in osteoclasts' density. Due to osteoclasts' autocrine signalling, approximately 100 days after perturbation, osteoclasts' density returns to a basal level. Osteoblasts are also capable to slowly return to their steady state. The described behaviour is in very good agreement with the temporal pattern of changes observed in Komarova's and Ayati's models. Additionally, results obtained with different numerical approaches do not present any significant differences.

The consequent changes in bone mass, as illustrated in Fig. 11, can accurately reproduce all phases of the bone remodelling process. In this case, bone remodelling is induced manually, by the increase of osteoclasts' density. This stimulus causes the activation of BMUs, and the resorption phase is initialized. These initial phases are indicated in Fig. 11, by the drastic decrease of medium bone apparent density. Then,

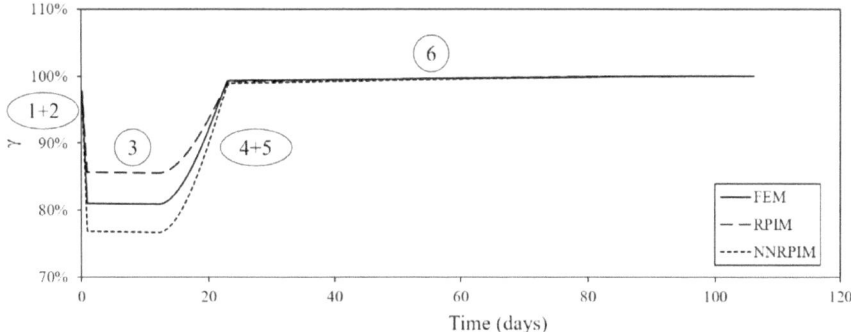

Fig. 11 Simulation of a bone remodelling cycle: Changes with time in bone mass obtained using the FEM, the RPIM and the NNRPIM. Numbers indicate the several phases of bone remodelling: activation (1), resorption (2), reversal (3), formation (4), mineralization (5) and quiescence (6)

bone mass stabilizes as the reversal phase occurs. This is followed by a slower phase of bone formation and mineralization, which is visible by the recovery of the initial bone apparent density. The final phase, the quiescent phase, does not imply any bone mass variation, since it consists of the differentiation of osteoblasts. Therefore, in the end, the stabilization of the medium bone apparent density suggests the occurrence of such quiescent phase.

When comparing the performance of each numerical method, it is possible to observe that bone mass undergoes a minor reduction in the FEM. This suggests that the surface layer of osteoclasts is thinner in the FEM and thicker in the NNRPIM. In fact, results show precisely this. Since the number of integration points varies with respect to the numerical approach used, this layer of osteoclasts also varies according to the method. However, in all methods, the lost bone mass is fully recovered at the end, indicating that the observed difference is not significant.

Additionally, for the first time, the biological response during a bone remodelling event is simulated spatially using numerical methods. Thus, in Fig. 12, it is the spatial variation of bone apparent density visible at different time frames. Again, observing the variation profile of bone mass, as expected, it seems to be occurring removal of bone only in the area where osteoclasts are present. Since osteoblasts are also present, bone mass is able to recover the portion of resorbed bone.

Moreover, the spatial evolution of bone mass varies slightly according to the numerical technique used and, again, the use of denser integration meshes is the reason for these differences. The meshless methods tested in this chapter imply a higher computational cost, since they use a higher number of integration points. However, this also conveys an advantage, since it leads to more stable and accurate results. This can be detected in the images presented in Fig. 12. Especially in the NNRPIM, it is possible to detect a more defined bone's surface, both before and after bone's resorption.

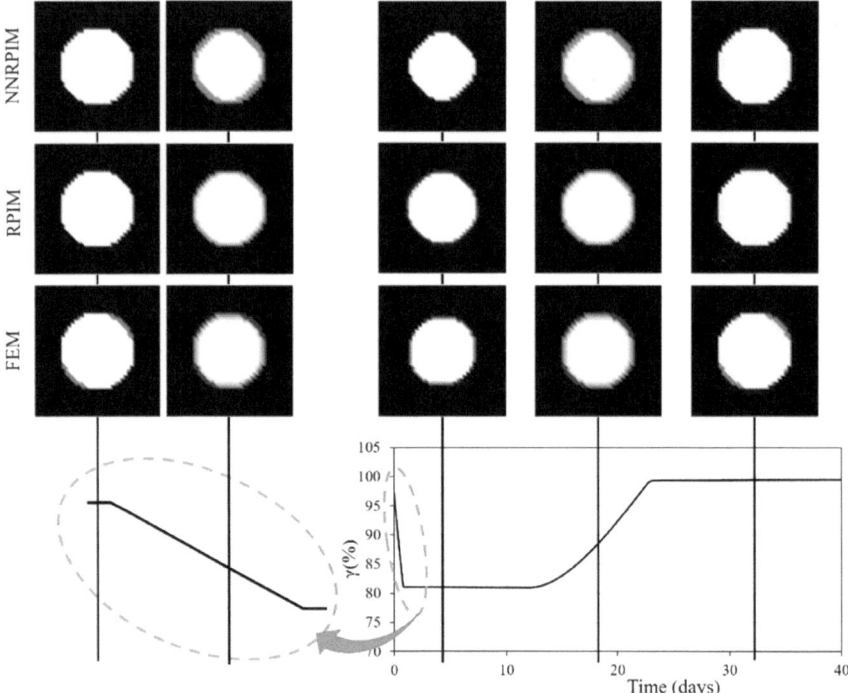

Fig. 12 Simulation of a bone remodelling cycle: spatial variation of bone mass along time for each numerical method under study

5 Conclusion

In this chapter, a novel model is presented, capable to accurately recreate the biological response observed during an event of bone remodelling. The thorough analysis performed, both in time and in space, is in very good agreement with the literature. Regarding the temporal analysis, the variation profile of bone mass, as well as the density of osteoclasts and osteoblasts, allows the identification of all phases occurring during a bone remodelling cycle. The combination of this model with distinct numerical methods, accomplished for the first time in this chapter, enables also a very relevant spatial analysis. All numerical approaches are able to accurately reproduce the expected spatial variation of bone mass. However, it was observed that better results are obtained using the NNRPIM.

To conclude, with this biological model, it is possible to temporally and spatially reproduce accurately the bone remodelling process, using any of the numerical methods tested. In future, other types of dynamic behaviour could be tested, such as cyclic bone remodelling or even unstable behaviour similar to pathological bone remodelling disorders. Also, a mechanical analysis could be added to this model, allowing a mechanobiological description of bone's behaviour.

References

1. Komarova SV, Smith RJ, Dixon SJ et al (2003) Mathematical model predicts a critical role for osteoclast autocrine regulation in the control of bone remodeling. Bone 33:206–215. https://doi.org/10.1016/S8756-3282(03)00157-1

2. Ayati BP, Edwards CM, Webb GF, Wikswo JP (2010) A mathematical model of bone remodeling dynamics for normal bone cell populations and myeloma bone disease. Biol Direct 5:28. https://doi.org/10.1186/1745-6150-5-28

3. Hadjidakis DJ, Androulakis II (2006) Bone remodeling. Ann NY Acad Sci 1092:385–396. https://doi.org/10.1196/annals.1365.035

4. Lemaire T, Naili S, Lemaire T (2013) Multiscale approach to understand the multiphysics phenomena in bone adaptation. Stud Mechanobiol Tissue Eng Biomater 14:31–72. https://doi.org/10.1007/8415_2012_149

5. Post TM, Cremers SCLM, Kerbusch T, Danhof M (2010) Bone physiology, disease and treatment. Clin Pharmacokinet 49:89–118. https://doi.org/10.2165/11318150-000000000-00000

6. Lemaire V, Tobin FL, Greller LD et al (2004) Modeling the interactions between osteoblast and osteoclast activities in bone remodeling. J Theor Biol 229:293–309. https://doi.org/10.1016/j.jtbi.2004.03.023

7. Tsukii K, Shima N, Mochizuki S et al (1998) Osteoclast differentiation factor mediates an essential signal for bone resorption induced by 1α,25-Dihydroxyvitamin D3, Prostaglandin E2, or parathyroid hormone in the microenvironment of bone. Biochem Biophys Res Commun 246:337–341. https://doi.org/10.1006/bbrc.1998.8610

8. Simmons DJ, Grynpas M (1990) Mechanisms of bone formation in vivo. In: Hall BK (ed) Bone, vol I, A treatrise. CRC Press, Boca Raton

9. Eriksen EF (2010) Cellular mechanisms of bone remodeling. Rev Endocr Metab Disord 11:219–227. https://doi.org/10.1007/s11154-010-9153-1

10. Greenfield EM, Bi Y, Miyauchi A (1999) Regulation of osteoclast activity. Life Sci 65:1087–1102. https://doi.org/10.1016/S0024-3205(99)00156-3

11. Pettermann HE, Reiter TJ, Rammerstorfer FG (1997) Computational simulation of internal bone remodeling. Arch Comput Methods Eng 4:295–323. https://doi.org/10.1007/BF02737117

12. García-Aznar JM, Rueberg T, Doblare M (2005) A bone remodelling model coupling microdamage growth and repair by 3D BMU-activity. Biomech Model Mechanobiol 4:147–167. https://doi.org/10.1007/s10237-005-0067-x

13. Pivonka P, Zimak J, Smith DW et al (2008) Model structure and control of bone remodeling: a theoretical study. Bone 43:249–263. https://doi.org/10.1016/j.bone.2008.03.025

14. Pivonka P, Zimak J, Smith DW et al (2010) Theoretical investigation of the role of the RANK-RANKL-OPG system in bone remodeling. J Theor Biol 262:306–316. https://doi.org/10.1016/j.jtbi.2009.09.021

15. Savageau MA (1976) Biochemical systems analysis—a study of function and design in molecular biology. Addison-Wesley Pub. Co., Massachusetts

16. Noor M, Shoback D (2000) Paget's disease of bone: diagnosis and treatment update. Curr Rheumatol Rep 2:67–73. https://doi.org/10.1007/s11926-996-0071-x

17. Ryser MD, Nigam N, Komarova SV (2009) Mathematical modeling of spatio-temporal dynamics of a single bone multicellular unit. J Bone Miner Res 24:860–870. https://doi.org/10.1359/jbmr.081229

18. Pivonka P, Komarova SV (2010) Mathematical modeling in bone biology: from intracellular signaling to tissue mechanics. Bone 47:181–189. https://doi.org/10.1016/j.bone.2010.04.601

19. Pepper D, Kassab A, Divo E (2014) Introduction to finite element, boundary element, and meshless methods: with applications to heat transfer and fluid flow. ASME Press, USA

20. Hrennikoff A (1941) Solution of problems in elasticity by the frame work method. J Appl Mech 8:169–175

21. McHenry D (1943) A lattice analogy for the solution of plane stress problems. J Inst Civ Eng 21:59–82. https://doi.org/10.1680/ijoti.1943.13967

22. Courant R (1943) Variational methods for the solution of problems of equilibrium and vibrations. Bull Am Math Soc 49:1–24. https://doi.org/10.1090/s0002-9904-1943-07818-4
23. Zienkiewicz OC, Taylor RL, Zhu JZ (2013) The finite element method: its basis and fundamentals. Elsevier, Amsterdam
24. Nguyen VP, Rabczuk T, Bordas S, Duflot M (2008) Meshless methods: a review and computer implementation aspects. Math Comput Simul 79:763–813. https://doi.org/10.1016/j.matcom. 2008.01.003
25. Belinha J (2014) Meshless methods in biomechanics—bone tissue remodelling analysis, 1st edn. Springer International Publishing Switzerland, Cham
26. Nayroles B, Touzot G, Villon P (1992) Generalizing the finite element method: diffuse approximation and diffuse elements. Comput Mech 10:307–318. https://doi.org/10.1007/BF00364252
27. Belytschko T, Gu L, Lu YY (1994) Fracture and crack growth by element free Galerkin methods. Model Simul Mater Sci Eng 2:519–534. https://doi.org/10.1088/0965-0393/2/3A/007
28. Liu GR, Gu YT (2001) A point interpolation method for two-dimensional solids. Int J Numer Methods Eng 50:937–951. https://doi.org/10.1002/1097-0207(20010210)50:4%3c937:AID-NME62%3e3.0.CO;2-X
29. Liu GR, Gu YT (2001) A local point interpolation method for stress analysis of two-dimensional solids. Struct Eng Mech 11:221–236. https://doi.org/10.12989/sem.2001.11.2.221
30. Liu GR, Gui-R (2010) Meshfree methods : moving beyond the finite element method. CRC Press, Boca Raton
31. Liu GR, Zhang GY, Gu YT, Wang YY (2005) A meshfree radial point interpolation method (RPIM) for three-dimensional solids. Comput Mech 36:421–430. https://doi.org/10.1007/s00466-005-0657-6
32. Wang JG, Liu GR (2002) On the optimal shape parameters of radial basis functions used for 2-D meshless methods. Comput Methods Appl Mech Eng 191:2611–2630. https://doi.org/10.1016/S0045-7825(01)00419-4
33. Wang JG, Liu GR (2002) A point interpolation meshless method based on radial basis functions. Int J Numer Methods Eng 54:1623–1648. https://doi.org/10.1002/nme.489
34. Liu GR, Gu YT (2001) A local radial point interpolation method (LRPIM) for free vibration analyses of 2-D solids. J Sound Vib 246:29–46. https://doi.org/10.1006/jsvi.2000.3626
35. Gu YT, Liu GR (2001) A local point interpolation method for static and dynamic analysis of thin beams. Comput Methods Appl Mech Eng 190:5515–5528. https://doi.org/10.1016/S0045-7825(01)00180-3
36. Liu GR, Yan L, Wang JG, Gu YT (2002) Point interpolation method based on local residual formulation using radial basis functions. Struct Eng Mech 14:713–732. https://doi.org/10.12989/sem.2002.14.6.713
37. Belinha J, Dinis LMJS, Jorge RMN (2013) Composite laminated plate analysis using the natural radial element method. Compos Struct 103:50–67. https://doi.org/10.1016/j.compstruct.2013.03.018
38. Belinha J, Dinis LMJS, Jorge RMN (2013) Analysis of thick plates by the natural radial element method. Int J Mech Sci 76:33–48. https://doi.org/10.1016/j.ijmecsci.2013.08.011
39. Belinha J, Dinis LMJS, Jorge RMN (2013) The natural radial element method. Int J Numer Methods Eng 93:1286–1313. https://doi.org/10.1002/nme.4427
40. Dinis LMJS, Jorge RMN, Belinha J (2008) Analysis of plates and laminates using the natural neighbour radial point interpolation method. Eng Anal Bound Elem 32:267–279. https://doi.org/10.1016/j.enganabound.2007.08.006
41. Dinis LMJS, Jorge RMN, Belinha J (2010) An unconstrained third-order plate theory applied to functionally graded plates using a meshless method. Mech Adv Mater Struct 17:108–133. https://doi.org/10.1080/15376490903249925
42. Moreira S, Belinha J, Dinis LMJS, Jorge RMN (2014) Análise de vigas laminadas utilizando o natural neighbour radial point interpolation method. Rev Int Métodos Numéricos para Cálculo y Diseño en Ing 30:108–120. https://doi.org/10.1016/j.rimni.2013.02.002

43. Belinha J, Araújo AL, Ferreira AJM et al (2016) The analysis of laminated plates using distinct advanced discretization meshless techniques. Compos Struct 143:165–179. https://doi.org/10.1016/j.compstruct.2016.02.021

44. Dinis LMJS, Jorge RMN, Belinha J (2011) A natural neighbour meshless method with a 3D shell-like approach in the dynamic analysis of thin 3D structures. Thin-Walled Struct 49:185–196. https://doi.org/10.1016/j.tws.2010.09.023

45. Dinis LMJS, Jorge RMN, Belinha J (2011) Static and dynamic analysis of laminated plates based on an unconstrained third order theory and using a radial point interpolator meshless method. Comput Struct 89:1771–1784. https://doi.org/10.1016/j.compstruc.2010.10.015

46. Dinis LMJS, Jorge RMN, Belinha J (2010) Composite laminated plates: a 3D natural neighbor radial point interpolation method approach. J Sandw Struct Mater 12:119–138. https://doi.org/10.1177/1099636209104735

47. Dinis LMJS, Jorge RMN, Belinha J (2010) A 3D shell-like approach using a natural neighbour meshless method: isotropic and orthotropic thin structures. Compos Struct 92:1132–1142. https://doi.org/10.1016/j.compstruct.2009.10.014

48. Dinis LMJS, Jorge RMN, Belinha J (2009) The natural neighbour radial point interpolation method: dynamic applications. Eng Comput 26:911–949. https://doi.org/10.1108/02644400910996835

49. Dinis LMJS, Natal Jorge R, Belinha J (2009) Large deformation applications with the radial natural neighbours interpolators. Comput Model Eng Sci 44:1–34

50. Azevedo JMC, Belinha J, Dinis LMJS, Natal Jorge RM (2015) Crack path prediction using the natural neighbour radial point interpolation method. Eng Anal Bound Elem 59:144–158. https://doi.org/10.1016/j.enganabound.2015.06.001

51. Belinha J, Azevedo JMC, Dinis LMJS, Jorge RMN (2016) The natural neighbor radial point interpolation method extended to the crack growth simulation. Int J Appl Mech 08:1650006. https://doi.org/10.1142/S175882511650006X

52. Rossi J-M, Wendling-Mansuy S (2007) A topology optimization based model of bone adaptation. Comput Methods Biomech Biomed Engin 10:419–427. https://doi.org/10.1080/10255840701550303

53. Jacobs CR, Levenston ME, Beaupré GS et al (1995) Numerical instabilities in bone remodeling simulations: the advantages of a node-based finite element approach. J Biomech 28:449–459. https://doi.org/10.1016/0021-9290(94)00087-k

54. Weinans H, Huiskes R, Grootenboer HJ (1992) The behavior of adaptive bone-remodeling simulation models. J Biomech 25:1425–1441. https://doi.org/10.1016/0021-9290(92)90056-7

55. Beaupré GS, Orr TE, Carter DR (1990) An approach for time-dependent bone modeling and remodeling-application: a preliminary remodeling simulation. J Orthop Res 8:662–670. https://doi.org/10.1002/jor.1100080507

56. Huiskes R, Weinans H, Grootenboer HJ et al (1987) Adaptive bone-remodeling theory applied to prosthetic-design analysis. J Biomech 20:1135–1150

57. Carter DR, Fyhrie DP, Whalen RT (1987) Trabecular bone density and loading history: regulation of connective tissue biology by mechanical energy. J Biomech 20:785–794. https://doi.org/10.1016/0021-9290(87)90058-3

58. Belinha J, Jorge RMN, Dinis LMJS (2012) Bone tissue remodelling analysis considering a radial point interpolator meshless method. Eng Anal Bound Elem 36:1660–1670. https://doi.org/10.1016/j.enganabound.2012.05.009

59. Belinha J, Dinis LMJS, Natal Jorge RM (2016) The analysis of the bone remodelling around femoral stems: A meshless approach. Math Comput Simul 121:64–94. https://doi.org/10.1016/j.matcom.2015.09.002

60. Peyroteo M, Belinha J, Dinis L, Natal Jorge R (2018) The mechanologic bone tissue remodeling analysis: a comparison between mesh-depending and meshless methods. In: Numerical methods and advanced simulation in biomechanics and biological processes. Academic Press, Cambridge, pp 303–323

61. Logan DL (2011) A first course in the finite element method, 5th edn. Cengage Learning, Boston

62. Chao TY, Chow WK, Kong H (2002) A review on the applications of finite element method to heat transfer and fluid flow. Int J Archit Sci 3:1–19

63. Liu WK, Jun S, Zhang YF (1995) Reproducing kernel particle methods. Int J Numer Methods Fluids 20:1081–1106. https://doi.org/10.1002/fld.1650200824

64. Atluri SN, Zhu T (1998) A new Meshless Local Petrov-Galerkin (MLPG) approach in computational mechanics. Comput Mech 22:117–127. https://doi.org/10.1007/s004660050346

65. Sibson R (1981) A brief description of natural neighbour interpolation. In: Barnett V (ed) Interpreting multivariate data. Wiley, Chichester, pp 21–36

66. Dinis LMJS, Jorge RMN, Belinha J (2007) Analysis of 3D solids using the natural neighbour radial point interpolation method. Comput Methods Appl Mech Eng 196:2009–2028. https://doi.org/10.1016/j.cma.2006.11.002

67. Hardy RL (1990) Theory and applications of the multiquadric-biharmonic method 20 years of discovery 1968–1988. Comput Math with Appl 19:163–208. https://doi.org/10.1016/0898-1221(90)90272-L

68. Seibel MJ, Robins SP BJ (2006) Dynamics of bone and cartilage metabolism: principles and clinical applications, 2nd edn. Academic Press, Cambridge

69. Eilertsen K, Vestad A, Geier O, Skretting A (2008) A simulation of MRI based dose calculations on the basis of radiotherapy planning CT images. Acta Oncol (Madr) 47:1294–1302. https://doi.org/10.1080/02841860802256426

70. Parfitt AM (1994) Osteonal and hemi-osteonal remodeling: the spatial and temporal framework for signal traffic in adult human bone. J Cell Biochem 55:273–286. https://doi.org/10.1002/jcb.240550303

71. Akchurin T, Aissiou T, Kemeny N et al (2008) Complex dynamics of osteoclast formation and death in long-term cultures. PLoS ONE 3:e2104. https://doi.org/10.1371/journal.pone.0002104

72. Komarova SV (2005) Mathematical model of paracrine interactions between osteoclasts and osteoblasts predicts anabolic action of parathyroid hormone on bone. Endocrinology 146:3589–3595. https://doi.org/10.1210/en.2004-1642

A Mechanostatistical Approach to Multiscale Computational Bone Remodelling

X. Wang and J. Fernandez

Abstract Computational models in biomechanics are generally unable to incorporate mechanical and anatomical data over the entire range of relevant spatial scales. This chapter proposes the construction of a framework, which unites several methodologies that operate on traditionally different aspects of bone remodelling, bridging the gap between previously incompatible data. The presented framework is used to solve the load adaptation response of the femoral neck as an application and consists of passing data from different sources across a multitude of spatial scales to solve for both organ-level and Haversian-level biomechanical states. The solutions are then stored in a database, to be utilised by a statistical method which can quickly estimate new load adaptation responses for which solutions were not previously generated, cutting down computation time.

1 Introduction

Bone damage and fracture from osteoporosis remain a costly medical condition with significant implications for the quality of life among those who have suffered injuries. For this reason, much research has been devoted to the understanding, treatment, and prevention of bone-related diseases, especially among the elderly. Currently, the widely known mechanobiological model of Wolff's law of bone adaptation and its successor, the *mechanostat* of the Utah Paradigm [1], still holds great explanatory power for its conceptual simplicity and remains a core component to the many existing computational bone adaptation models primarily informed by biomechanics.

X. Wang (✉)
Auckland Bioengineering Institute, 70 Symonds St, Auckland, New Zealand
e-mail: xm.wang@auckland.ac.nz

J. Fernandez
Department of Engineering Science, Auckland Bioengineering Institute, 70 Symonds St, Auckland, New Zealand
e-mail: j.fernandez@auckland.ac.nz

© Springer Nature Switzerland AG 2020
J. Belinha et al. (eds.), *The Computational Mechanics of Bone Tissue*,
Lecture Notes in Computational Vision and Biomechanics 35,
https://doi.org/10.1007/978-3-030-37541-6_6

159

Although there is a wealth of literature on the topic, most attempts at constructing an in silico model for bone remodelling are immediately confronted with three competing factors, which limit the available computational resources for a feasible evaluation. These are: (i) the geometric detail; (ii) the spatiotemporal scale; and (iii) the applied constitutive mechanical laws. The existing studies have attempted to balance the amount of detail considered for each of these factors. Fernandez [2] and McNamara [3] have demonstrated bone remodelling behaviour in two dimensions at the microscale in cortical and trabecular bone, respectively, complete with changes in microgeometry; the model given by Fernandez has further demonstrated the merging of pores as an emulation of osteoporosis, while the study by McNamara has explored various modes of damage and recovery in the calculation of bone adaptation. Beaupre et al. [4] constructed a model which simulates the bone density changes of a long-term load response of a 2D macrolevel model to external loads, providing a foundation for models investigating the response to sustained exercise regimes. Coelho et al. [5] introduced a 3D multiscale hierarchical approach, characterising bone spatial variation with repeating microstructures in several discrete regions. Other important aspects of computational bone remodelling have produced studies focussing on modelling osteoclast biochemistry [6], bone resorption and stress shielding from orthopaedic implants [7], and reduction of solution time with neural network approaches [8].

Apart from the difficulty of capturing complex geometries and scales, most existing literature disproportionately focus on trabecular bone remodelling when the cortical to trabecular loading ratio is estimated to be as high as 65:35 [9].

This chapter proposes the development of a multiscale modelling framework within the context of hip fracture for the swift prediction of bone strain and the estimation of its adaptation response for a given exercise regime. The development of the framework is divided into three parts: Part I utilises finite elements (FE) to solve for the mechanical state of bones at the macroscale; Part II incorporates a collection of algorithms based on a previous study [10], which addresses microscale Haversian-level bone adaptation in response to loading based on ideas from the mechanostat; and Part III describes a statistical surrogate model using partial least squares regression (PLSR), which addresses the problem of computation time. The integration of multiscale information allows the framework to remain anatomically and physiologically relevant at all spatial scales and features high compatibility with clinically important measurements and biomechanical data.

In Part I, we describe the construction of a biomechanical model from the visible human (VH) data set of muscles and bones in the hip area and subject it to loading obtained from gait analyses, from which we obtain stresses at the femoral neck cortical bone. This femoral neck cortex henceforth will be referred to as the framework's region of interest (ROI). In Part II, we summarise the construction of Haversian-scale FE models and link the ROI with the set of Haversian models through the propagation of stresses from the macroscale down to the microscale. Furthermore, we describe two load transduction algorithms, which affect cortical bone at the Haversian scale;

the first alters localised bone strength via a change in bone mineral density, and the second alters Haversian microstructure. When combined, these algorithms form our approach to bone remodelling and load adaptation. In Part III, we detail the construction of a database of stress scenarios and resulting homogenised material properties and utilise PLSR to predict the evolution of material properties given non-simulated load cases.

All FE simulations utilised the software package SIMULIA Abaqus (www.3ds. com). A schematic of the framework and its information sources is shown in Fig. 1. The framework is categorised into two phases: the construction phase details the process behind building each part of the framework, and the application phase is where the framework is used for solving the load adaptation problem. The large savings in computation time occur due to the replacement of Part II in the framework construction phase, the *adaptation response*, with Part II in the framework application phase, the *response database*.

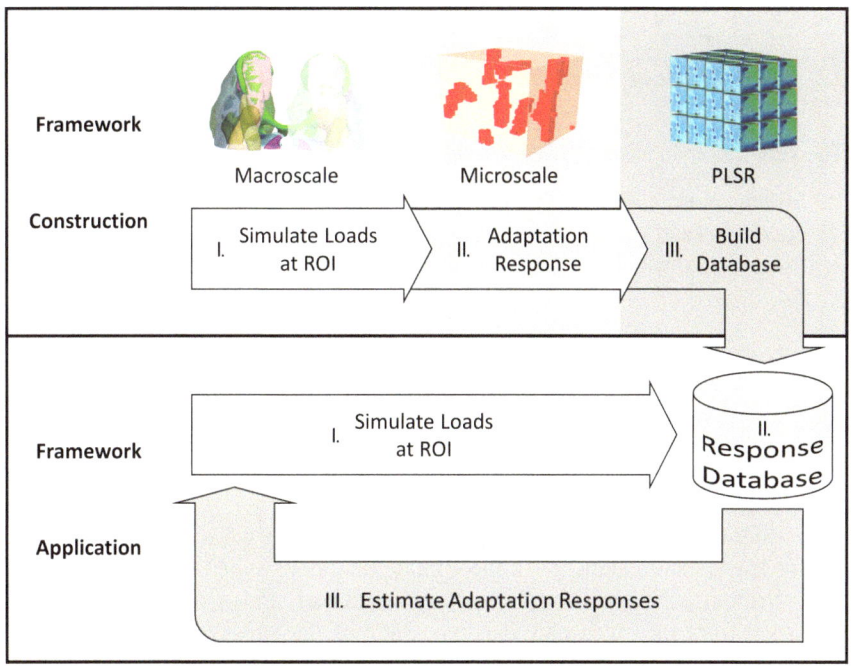

Fig. 1 Framework diagram. Shades refer to the different parts of the framework. (Light grey) Part I, collection of load information from macroscale; (grey) Part II, receiving load information from Part I and formulating the adaptation response which takes the form of homogenised material properties; (dark grey) Part III, statistical model of responses

2   Part I: Macroscale FE Model

2.1   Model Construction

Bone and muscle shapes were extracted and meshed with hexahedral 2 mm reduced integration elements from the VH data set [11] from 269 colour photograph slices in the original resolution of 0.33 mm × 0.33 mm per pixel and 1 mm of spacing between slices.

Anatomically realistic porosity variations $P^e = \{p_i^e\}$, $1 \leq i \leq n^e$, with n^e as the total number of ROI elements, were statistically generated from Gaussian distributions based on experimentally determined values and uncertainties for circumferentially varying femoral cortex porosities [12]. The generated porosities for the femoral cortex elements provide a realistic femoral neck mesh on which FE mechanics simulations are run.

2.2   Mechanical Simulations

The simulation consists of two sections:

1. Initialisation, which estimates key parameters under the prescribed initial conditions and
2. Progression, which emulates the evolution of the bone state.

2.2.1   Macroscale Initialisation

Bones were subjected to muscle forces recorded from walking gait analysis [13] with simulation under linear elasticity and isotropic material properties shown in Table 1.

Maximum absolute principal stresses $S^e = \{s_i^e\}$, $1 \leq i \leq n^e$ for each element were recorded at the ROI element integration points as the load transduction signal for Haversian modelling. To incorporate the observation that bone fibres align in a direction, which maximally resists stresses and strains, fibre directions were estimated from the eigenvalue decomposition of the stress tensors from FE load simulations, as shown in Fig. 2.

2.2.2   Macroscale Progression

The evolution of the mechanical state of the femoral cortex was simulated for a total of $T = 90$ days of simulation time, with iteration steps of one day. In each iteration, S^e is passed to the PLSR model in Part III, which outputs a set of evolved homogenised

Table 1 Macroscale anatomical structures and their respective material properties as applied in FE simulations

Structure	Category	E (GPa)	ν	Notes
Adductor brevis, longus, magnus Biceps femoris Gluteus maximus, medius, minimus Gracilis Iliacus Rectus femoris Sartorius Semitendinosus	Muscle	0.27	0.4999	E from [15]; ν from [16]
Hip bone	Cortical	14.65	0.332	[17]
Femur[a]	Cortical Cartilage Trabecular	14.65 0.580 3.386	0.332 0.39 0.12	Not ROI for $t > 0$[b]; [18] Femoral head[c]; [19] Cancellous bone[d]

All are modelled as isotropic elastic materials

[a]Femur is composed of three different types of materials

[b]ROI material properties evolve after the initialisation step ($t = 0$). Rest of the cortex does not change

[c]Femoral head is modelled with a layer of cartilage surrounding the cortex

[d]Femur is modelled with a thick layer of cortex elements and filled with trabecular elements

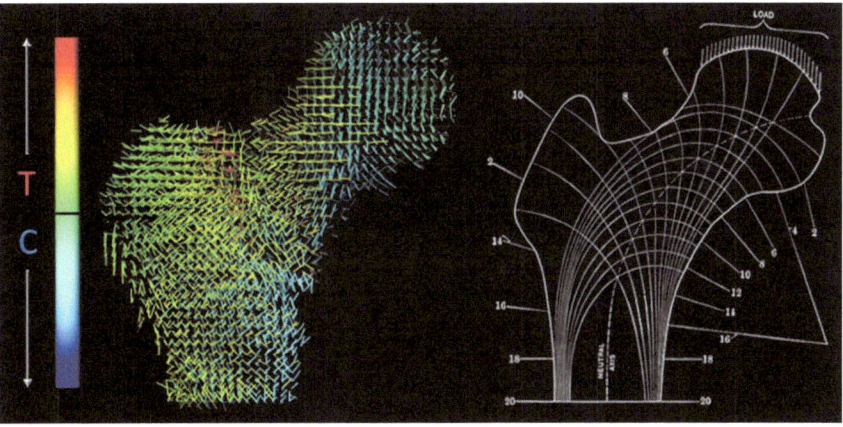

Fig. 2 Left: stress tensor eigendecomposition of internal simulated stresses of the proximal femur at the trabecular bone. Colour and line direction indicates magnitude and direction of maximum absolute eigenvalue and eigenvector, respectively; with reference to the colour bar, green to blue colouration indicates a compressive minimum principal stress (C), and green to red colouration indicates a tensile maximum principal stress (T). Eigenvectors match closely with stress lines found in other studies of bone stress, e.g. Koch's mathematical analysis. Right: adapted from Gray [14]; note the compressive stresses running in the proximal-distal direction along the medial shaft through the femoral head, and the tensile stresses running up the lateral shaft then in the direction of the femoral neck axis. The criss-cross pattern through the core indicates the co-dominance of both tensile and compressive stress directions and closely resembles trabecular anatomy (see Fig. 7)

material properties as a result of the construction of a database of material property evolutions built in Part II.

3   Part II: Microscale Haversian Model

3.1   *Model Construction*

Haversian anatomy was constructed and voxel meshed with 5 μm elements from microcomputed tomography (μCT) images (Table 2) at 5 μm resolution of an equine cortical bone biopsy of dimensions 4 mm × 3.5 mm × 2 mm (Fig. 3, left). The use of equine data was justified in our framework as (i) equine models are highly translational to human contexts due to similar Haversian anatomy and have been

Table 2 Micro-CT imaging parameters

X-ray energy	60 kV, 10 W
Exposure time for each projection	60 s
Total number of projections	721
Objective magnification	4×
Source to sample distance	120 mm
Detector to sample distance	40 mm
Pixel numbers	1024 × 1024 × 1024
Effective voxel size	5 μm

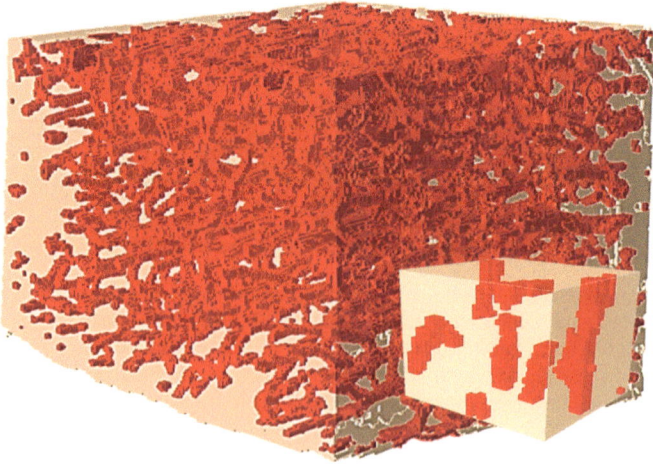

Fig. 3 Left: equine biopsy mesh showing Haversian canals. Right: example Haversian representative mesh cut from the biopsy mesh (not to scale)

recommended by the US Federal Drugs Administration (FDA) for comparative joint research [20], and (ii) there exists in vivo bone strain and biomarker data from equine studies, unlike human data which only reports post-mortem information. A database of Haversian models was formed from cutting $n^m = 4$ samples of on average 0.003 mm^3 from the voxel mesh, chosen as a size which contains a representative number of Haversian canals for microscale anatomy and material strength changes to be observed (Fig. 3, right). These samples were chosen based on their volume fraction of canal elements versus dense cortical bone elements, and thus contained porosity information to be matched to the porosity variation generated in Part I, obtained as

$$P^m = \left\{ p_j^m \right\}, \ 1 \le j \le n^m \tag{1}$$

3.2 Macro-microscale Link

3.2.1 Haversian Link to ROI Elements

For each ROI element i and associated porosity p_i^e, lower and upper bounding porosities $p_i^{m\alpha}$ and $p_i^{m\beta}$ were extracted from P^m. Subsequently, the link from the ROI elements to the Haversian models is defined by weights w_i^α and w_i^β:

$$
w_i^\alpha = \begin{cases} \frac{p_i^{m\beta} - p_i^e}{p_i^{m\beta} - p_i^{m\alpha}}, & p_i^{m\alpha} < p_i^e < p_i^{m\beta} \\ \frac{1}{2}, & p_i^{m\alpha} = p_i^e = p_i^{m\beta} \end{cases}
$$
$$
w_i^\beta = \begin{cases} \frac{p_i^e - p_i^{m\alpha}}{p_i^{m\beta} - p_i^{m\alpha}}, & p_i^{m\alpha} < p_i^e < p_i^{m\beta} \\ \frac{1}{2}, & p_i^{m\alpha} = p_i^e = p_i^{m\beta} \end{cases} \tag{2}
$$

This allows any parameter of the ith ROI element to be represented by a weighted sum of its corresponding parameters of the linked Haversian models.

3.2.2 Stress Propagation from ROI to Haversian Models

The local positive z directions for each Haversian model were defined to be parallel to the fibre directions estimated from the elements in the ROI. To determine the appropriate load range for each Haversian model j from the propagation, we formulate sets of reverse links by constructing a set of ROI element indices, which is associated with these elements:

$$I_j^m = \left\{ i : p_j^m \in \left\{ p_i^{m\alpha}, p_i^{m\beta} \right\} \right\}, \quad 1 \le i \le n^e \tag{3}$$

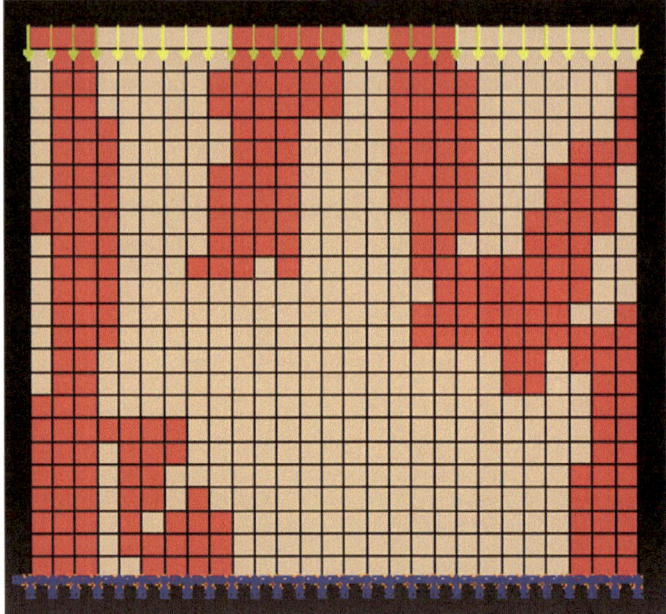

Fig. 4 Side view of application of load on a Haversian specimen. Blue symbols (bottom) indicate spatial and rotational fixed boundary conditions. Yellow arrows (top) indicate load direction

Taking the magnitude s_i of the maximum absolute principal stress on the ROI element centroid as the most significant component of the load, we calculate a set of loads L_j which appear in the macroscale initialisation step (Sect. 2.2.1) as

$$L_j = \frac{s_i}{a_j}, \quad i \in I_j^{\mathrm{m}} \tag{4}$$

where a_j is the surface area of the load application. An example of a typical loading scenario of a Haversian specimen is given in Fig. 4.

3.3 Haversian Simulation

3.3.1 Microscale Initialisation

For each Haversian model j, the models were simulated under five different evenly spaced loading regimes between $\min(L_j)$ to $\max(L_j)$ under linear elasticity with isotropic elements. An isotropic Young's modulus was assigned to the cortical elements based on the homogenised cortical and canal values and the fraction of cortical and canal elements in the model (see Table 3).

Table 3 Microscale model parameters

Name/description	Parameter values	Notes
Cortical ν	0.3	Cortical E varies with P^m
Canal E	2 MPa	
Canal ν	0.167	
Canal ρ	1 kg mm^{-3}	Same as water
c_1	4621.36 MPa	From Eq. (7) parameters at $t = 0$
c_2	1.075 kg m^{-3} s^{-1}	Remodelling constant; from [2]
b	1.54	Power law exponent; from [21]
Canal closing rate	1.9 nm s^{-1}	Adapted from canine data in [22]
Canal opening rate	2.083 nm s^{-1}	From [6]
L^{I}	$-500\ \mu\varepsilon$	Adapted from [2]
L^{II}	$+500\ \mu\varepsilon$	
L^{III}	$+2000\ \mu\varepsilon$	

Formulation of the Mechanostat

The initial simulations allowed the determination of the model's elementwise mechanostat across Haversian models j and their corresponding cortical elements k, defined by the piecewise density evolution parameter

$$\varepsilon_{jkt} = \begin{cases} \varepsilon_{jkt}^{\mathrm{VMS}} - L_{jkt}^{\mathrm{I}}, & \varepsilon_{jkt}^{\mathrm{VMS}} < L_{jkt}^{\mathrm{I}} \\ 0, & L_{jkt}^{\mathrm{I}} \leq \varepsilon_{jkt}^{\mathrm{VMS}} < L_{jkt}^{\mathrm{II}} \\ \varepsilon_{jkt}^{\mathrm{VMS}} - L_{jkt}^{\mathrm{II}}, & L_{jkt}^{\mathrm{II}} \leq \varepsilon_{jkt}^{\mathrm{VMS}} < L_{jkt}^{\mathrm{III}} \\ L_{jkt}^{\mathrm{III}} - \varepsilon_{jkt}^{\mathrm{VMS}}, & \varepsilon_{jkt}^{\mathrm{VMS}} \geq L_{jkt}^{\mathrm{III}} \end{cases} \tag{5}$$

where $\varepsilon_{jkt}^{\mathrm{VMS}}$ is the von Mises stimulus strain at time step t.

The four conditions governing the piecewise function determine the four *zones* of the mechanostat (Fig. 5), where the region below L^{I} is the resorption zone which encourages bone resorption due to lack of loading, the region between L^{I} and L^{II} is the adapted state (homeostatic) zone where no changes occur, the region between L^{II} and L^{III} is the growth zone which encourages an increase in bone mineral density, and the region above L^{III} is the failure zone which also causes bone resorption. In the initialisation step ($t = 0$), where the strain stimulus is defined to be in the centre of the adapted state zone, the piecewise function's boundaries are given by

$$L_{jk0}^{\mathrm{I}} = \varepsilon_{jk0}^{\mathrm{VMS}} + L^{\mathrm{I}}$$
$$L_{jk0}^{\mathrm{II}} = \varepsilon_{jk0}^{\mathrm{VMS}} + L^{\mathrm{II}}$$

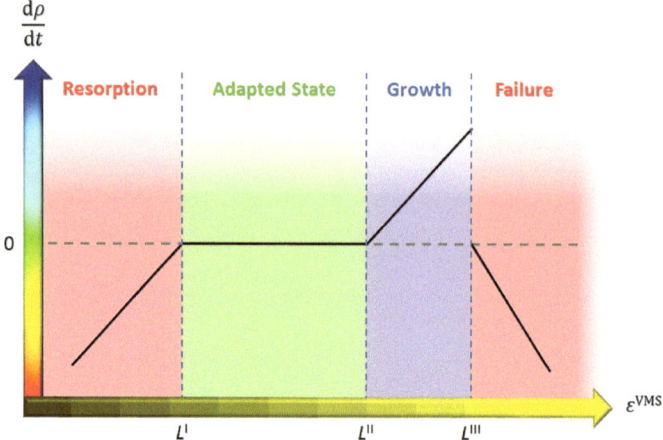

Fig. 5 Illustration of the zones of the mechanostat, indicating the rate of change of bone density as a function of strain

$$L_{jk0}^{\mathrm{III}} = \varepsilon_{jk0}^{\mathrm{VMS}} + L^{\mathrm{III}} \tag{6}$$

Initialisation of Element-wise Density

The elementwise material properties also allowed the determination of an initial elementwise density ρ_{jk0} through a power law relation [21], given by

$$E_{jkt} = c_1 \rho_{jkt}^b \tag{7}$$

where E_{jk0} at $t = 0$ is the initial Young's modulus.

3.3.2 Microscale Progression

The Haversian models were simulated for a total of $T = 90$ days, with iteration steps of one day. Each Haversian model j across the five initial loads was subjected to three different evenly spaced excitation loads between values from $0.375 \times \min(L_j)$ to $1.5625 \times \max(L_j)$ across its loading surface. Each excitation load was further applied at seven evenly spaced angles between $0°$ and $90°$. The strain state of the model was obtained after each iteration, where for each Haversian cortical element the von Mises stimulus $\varepsilon_{jkt}^{\mathrm{VMS}}$ was calculated from the strain tensor.

Calculation of the Strain Stimulus

The data given in Table 4 shows that bone growth continues in the rest period after the exercise regime has ceased, indicating that the strain stimulus does not immediately revert back to pre-exercise levels. The von Mises strain stimulus $\varepsilon_{jkt}^{\mathrm{VMS}}$ in Eq. (5) is thus formulated as a weighted moving average on the actual von Mises equivalent

Table 4 Cortical bone growth rates in horse specimens determined from oxytetracycline/fluorescein complexone staining in a histological sample taken from the dorsal cortex in equine third metacarpal bone, from Davies [23]

Days	Growth rate per day, μm $\pm$ SD		Significance (p-value), control versus exercised
	Control	Exercised	
40	1.8 $\pm$ 1.1	2.6 $\pm$ 2.6 (Trot)	0.510
40	3.2 $\pm$ 4.3	2.2 $\pm$ 3.3 (Canter)	0.680
40	2.1 $\pm$ 2.9	12.7 $\pm$ 8.6 (Gallop)	0.017
40	1.2 $\pm$ 0.7	4.3 $\pm$ 2.5 (Rest)	0.014

strain ε_{jkt}^{VM}. The von Mises equivalent strain is used to determine ε_{jkt}^{VMS} by

$$\varepsilon_{jkt}^{VMS} = \sum_{\tau=1}^{\tau_{max}} U(\tau) H_{jk}(\tau) \tag{8}$$

where $\tau_{max} = 30$ and $U(\tau)$ and $H_{jk}(\tau)$ are sets of weights and historical von Mises strains, respectively, given by

$$T^{W} = \{1, 2, \ldots, \tau_{max}\}$$
$$M^{W} = \{t + \tau - \tau_{max}\}, \quad \tau \in T^{W}$$
$$U' = \{\exp(\tau - \tau_{max})\}, \quad \tau \in T^{W}$$
$$U = \frac{U'}{\sum U'}$$
$$H_{jk} = \{\varepsilon_{jk\mu}^{VM}\}, \quad \mu \in M^{W} \tag{9}$$

Here, exp is the exponential function and $\sum U = 1$. The mechanostat boundaries from Eq. (6) shift by an amount equal to the difference between the current and previous stimulus strains:

$$L_{jkt}^{I} = L_{jk,t-1}^{I} + \varepsilon_{jkt}^{VMS} - \varepsilon_{jk,t-1}^{VMS}$$
$$L_{jkt}^{II} = L_{jk,t-1}^{II} + \varepsilon_{jkt}^{VMS} - \varepsilon_{jk,t-1}^{VMS}$$
$$L_{jkt}^{III} = L_{jk,t-1}^{III} + \varepsilon_{jkt}^{VMS} - \varepsilon_{jk,t-1}^{VMS} \tag{10}$$

The von Mises equivalent strain is used as the stimulus criteria as it considers both normal and shear deformations. This is formulated as

$$\varepsilon^{VM} = \frac{\sqrt{(\varepsilon_{11} - \varepsilon_{22})^2 + (\varepsilon_{22} - \varepsilon_{33})^2 + (\varepsilon_{33} - \varepsilon_{11})^2 + 6(\varepsilon_{12}^2 + \varepsilon_{23}^2 + \varepsilon_{31}^2)}}{\sqrt{2}(1 + v')} \tag{11}$$

with the strain components ε_{qr} as that found from the strain tensor ε_{jkt}, and v' is the equivalent Poisson's ratio, which is equal to the material Poisson's ratio v under elastic constitutive laws.

Density and Young's Modulus Evolution
The density evolution parameter ϵ_{jkt} is calculated from Eq. (5). This causes a change in the density according to the forward Euler formulation

$$\rho_{jkt} = \rho_{jk,t-1} + c_2 \epsilon_{jkt} \tag{12}$$

The updated density is subsequently converted to a Young's modulus via Eq. (7).

Haversian Microstructure Evolution
Haversian microstructure changes according to osteoclast and osteoblast activity at the head and tail, respectively, of Haversian canals; bone resorption and deposition occur at the head and tail, and the regions where these activities occur are known as the *cutting* and *closing* regions.

TIFF image stacks were generated from the element states of the Haversian model. With each element represented as a pixel, black and white binary colour values were assigned to the cortical and canal elements, respectively. The image stacks were subsequently passed to Fiji (fiji.sc/Fiji) for morphological thinning via medial axis skeletonisation (Skeletonize3D plugin, fiji.sc/Skeletonize3D) and a shape analysis on the generated skeleton (fiji.sc/AnalyzeSkeleton). This analysis allowed the determination of the cutting and closing regions of the canal and their present evolution directions.

Figure 6 shows the guidelines behind the determination of cutting and closing regions, with respective rates of 2.083 nm s^{-1} and 1.9 nm s^{-1} as given in Table 3. For all canal ends which are classified as cutting, the cutting direction $\boldsymbol{g}$ was determined by

$$\boldsymbol{g} = \sum_{\gamma=1}^{n^l} \boldsymbol{V}_\gamma^{\mathrm{T}} \boldsymbol{\lambda}_\gamma \tag{13}$$

where T is the matrix transpose operation, n^l is the number of cortical elements at the boundary between the cutting region and the canal elements, and $\boldsymbol{V}_\gamma$, $\boldsymbol{\lambda}_\gamma$ are a horizontally concatenated matrix of eigenvectors and a vector of eigenvalues, respectively, found through the eigendecomposition of the strain tensor $\boldsymbol{\varepsilon}_{jyt}$. The direction $\boldsymbol{g}$ typically aligns with the longitudinal direction in cortical bone towards the zones of highest strain and is heavily influenced by the angle of load application, as shown in Fig. 7.

Elements undergoing resorption receive canal material properties as specified in Table 3, and elements which have been mineralised by the closing cone receive material properties equal to the average of the surrounding cortical elements in the closing regions.

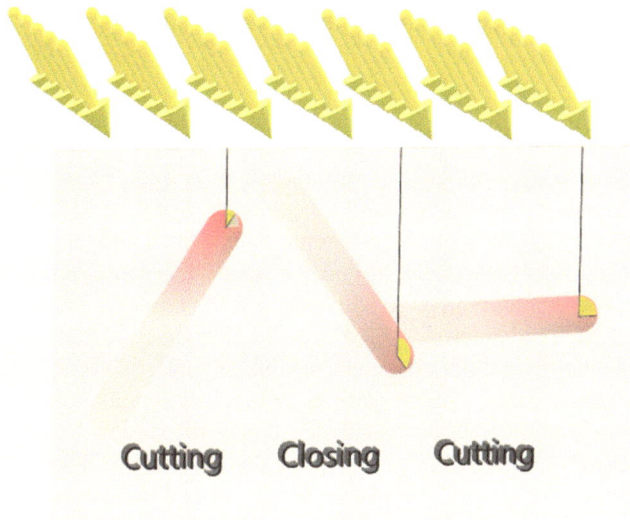

Fig. 6 Determination of Haversian canal activity type based on canal geometry and load application. The type of activity that the Haversian canal undergoes is determined through the angle between the current canal direction and the closest point of load application. An acute or right angle categorises the canal end as a cutting cone, while an obtuse angle categorises the canal end as a closing cone

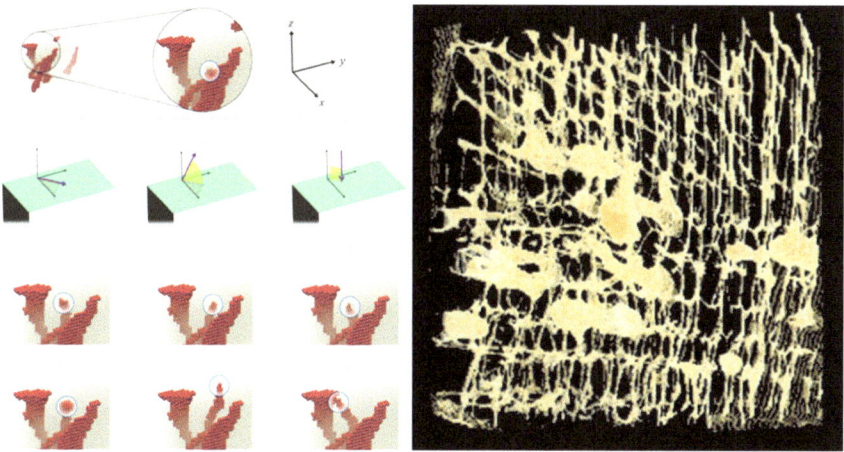

Fig. 7 Comparison between simulated canal evolution and anatomical morphology. Left: example of the change in canal evolution due to different load angle applications (purple) along the surface (green) in the first row. Red pie sectors specify the angular deviation from the x-axis, and yellow pie sectors indicate the angular deviation from the xy plane. A single Haversian model was subjected to three different loading angles of the same magnitude; the region in the solid blue circle shows a cutting cone. Left column: pure shear load 45° from the x-axis; middle column: tensile load 45° from the x-axis and 67.5° from the xy plane; right column: pure normal compressive load to the xy plane. Compare tunnelling behaviour with example (right) of equine Haversian canals reconstructed from μCT data and estimated trabecular fibre directions in Fig. 2. Adapted from Wang et al. [10] with permission from Springer Nature

4 Part III: Statistical Modelling Using PLSR

4.1 Database Construction

Partial least squares regression [24] is used to relate two sets of matrices, the sample predictors X and corresponding sample responses Y, of data via a linear multivariate model, with the capability of analysing non-independent and incomplete variables in either matrix. Once the model has been trained on X and Y, it can be utilised to compute new responses from predictors which were not in the training sample.

The database is constructed on 420 Haversian simulations, consisting of all parameter combinations of the four porosities, five initial loads, three excitation loads, and seven application angles, with the addition of the time iteration step of each simulation. The values of these parameters in each combination form X for a total of 3780 parameter combinations.

The response Y is chosen as the homogenised Young's modulus E_{jt}^{H} as a result of the Haversian simulations. This is given by

$$E_{jt}^{H} = \frac{1}{n^{e}} \sum_{k} E_{jkt} \tag{14}$$

The error in the PLSR response predictions was evaluated using a leave-one-out analysis, where each of the 3780 predictor/response sets was left out of the training data set and predicted using the rest of the training data. This is repeated for each set. Table 5 shows results from two samples of Haversian models for a time iteration at 30 days which indicate that homogenised Young's modulus was predicted with less than 0.4 and 0.2% error, respectively.

Table 5 Error of PLSR predictions for homogenised Young's modulus in two different specimens, varying load angle and magnitude

Excitation magnitude[a]	Angle from xy plane (degrees)						
	0	15	30	45	60	75	90
Specimen 1							
0.375	0.02	0.06	<0.01	<0.01	0.03	0.06	0.25
1	0.50	0.12	0.13	0.21	0.15	0.12	0.30
1.5625	0.15	0.37	0.24	0.09	0.05	0.42	0.16
Specimen 2							
0.375	0.08	0.57	0.19	0.05	0.04	0.38	0.21
1	0.13	0.22	0.08	0.16	0.25	0.11	0.06
1.5625	0.16	0.04	0.36	0.02	0.16	0.03	0.07

[a]Excitation load magnitude is given as a factor of the initial magnitude

4.2 Macroscale-PLSR Link

The index link from the ROI elements to the Haversian models is retrieved as the weights from Eq. (2). For each iteration of the femoral cortex mechanical state, the homogenised material properties E_{it}^H for the ith ROI element are updated as

$$E_{it}^H = w_i^\alpha E_i^{P\alpha} + w_i^\beta E_i^{P\beta} \tag{15}$$

where E^P is the PLSR response function for the homogenised Young's modulus and

$$E_i^{P\alpha} = E^P\left(p_i^{m\alpha}, l^0, l^x, \vartheta\right)$$
$$E_i^{P\beta} = E^P\left(p_i^{m\beta}, l^0, l^x, \vartheta\right) \tag{16}$$

with l^0, l^x, and ϑ as the initial load, excitation load, and angle of the excitation load, respectively.

4.3 Example Prediction

Figure 8 shows the strains simulated from the right femur ROI to material property changes predicted from PLSR, stimulated by walking and running exercises. The femur was originally conditioned to a walking exercise regime and subsequently subjected to the same walking exercise or changed to a running exercise regime for 90 simulated days, and PLSR was used to predict the Young's modulus at days 50 and 90. The effect of the adaptation is shown by the change in von Mises strains since day 1 when subjected to the same exercises since day 1.

The von Mises strains from the running exercises effected higher remodelling stimuli than the walking stimulus, as seen in the top row at $t = 1$. This is reflected in the strain response, where the walking exercise shows smaller strain changes than the running exercise. The former corresponds to much of the ROI falling in the adapted state zone in Fig. 5, while growth and failure regions are more abundant in the running series and appear earlier as seen in the strain changes at $t = 50$ and $t = 90$.

Large regions of high strains are found at the posterior lateral region of the femoral neck cortex, while more moderately high strains are found at the anterior lateral region, agreeing with calculations of femoral neck cortical strains from hip muscle contractions [25]. The higher strains at the posterior lateral region correspond to bone weakening and the mechanostat failure zone, with large zones of bone weakening appearing in the running series and smaller, and moderate amounts of weakening appearing in the walking series. In contrast, the large regions of moderately high strains appearing at the anterior lateral region corresponded to a growth response in the running series and little to no response in the walking series.

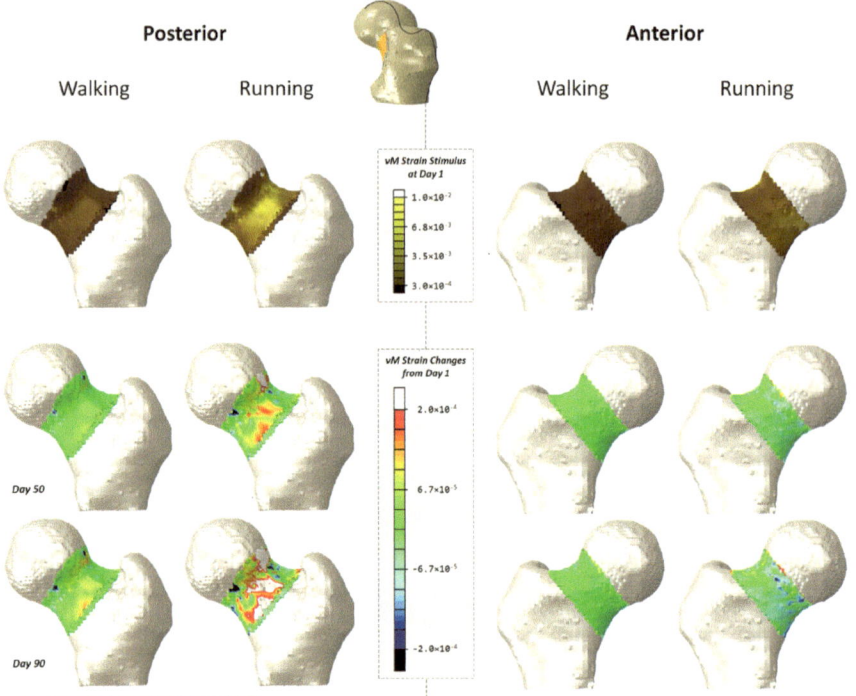

Fig. 8 Von Mises strain changes due to PLSR-predicted homogenised Young's moduli from different exercise regimes over 90 simulated days. With reference to Fig. 5, strengthening is shown through negative strain changes, corresponding to the growth zone of the mechanostat, while weakening is shown through positive strain changes, corresponding to the failure zone. Stable areas with little to no strain changes fall in the adapted state zone

From these results, the PLSR is shown to be able to capture spatially varying patterns of bone growth and resorption as governed by the mechanostat. Adapted state regions are generally more widespread in the walking series, and the running series exhibited bone strengthening in moderately high strain regions and bone weakening in extremely high strain regions.

5   Conclusions

The presented work in this chapter proposes a framework incorporating detailed anatomical and biomechanical data across all relevant spatial scales to solve for the strain state and estimate an adaptation response for given stresses obtained from bone exercise regimes. In particular, the adaptation response is modelled at the microscale and the effects emerge at the macroscale, reflecting in vivo bone density changes at the Haversian level. We demonstrate that it is possible to circumvent lengthy computation times and mitigate the loss of important detail through the application of statistical techniques, namely a partial least squares regression in this case, resulting in rapid prediction of the state of strains and bone strength from given exercise regimes. The presented techniques may be adapted to other bone joints to make use of the rich information available at different spatial scales. For example, in this study, we used information from high-resolution μCT of Haversian canals and biomarker data showing rates of bone turnover from equine models to inform macroscale responses. This modelling pipeline exhibits strong relations between changes observed at Haversian levels and the homogenised whole bone response, linking microscale adaptation with functional behaviour such as walking and running measured at the whole organ level. Outside of standard functional behaviour, the pipeline has the potential to explore disease states at the Haversian microscale and how these states manifest as changes at larger scales.

References

1. Frost HM (2000) The Utah paradigm of skeletal physiology: an overview of its insights for bone, cartilage and collagenous tissue organs. J Bone Miner Metab 18(6):305–316
2. Fernandez JW, Das R, Cleary PW, Hunter PJ, Thomas CD, Clement JG (2013) Using smooth particle hydrodynamics to investigate femoral cortical bone remodelling at the Haversian level. Int J Numer Methods Biomed Eng 29(1):129–143. https://doi.org/10.1002/cnm.2503
3. McNamara LM, Prendergast PJ (2007) Bone remodelling algorithms incorporating both strain and microdamage stimuli. J Biomech 40(6):1381–1391. https://doi.org/10.1016/j.jbiomech. 2006.05.007
4. Beaupre GS, Orr TE, Carter DR (1990) An approach for time-dependent bone modeling and remodeling-application: a preliminary remodeling simulation. J Orthop Res: Off Publ Orthop Res Soc 8(5):662–670. https://doi.org/10.1002/jor.1100080507
5. Coelho PG, Fernandes PR, Rodrigues HC, Cardoso JB, Guedes JM (2009) Numerical modeling of bone tissue adaptation—a hierarchical approach for bone apparent density and trabecular structure. J Biomech 42(7):830–837. https://doi.org/10.1016/j.jbiomech.2009.01.020
6. Pivonka P, Buenzli PR, Scheiner S, Hellmich C, Dunstan CR (2013) The influence of bone surface availability in bone remodelling—a mathematical model including coupled geometrical and biomechanical regulations of bone cells. Eng Struct 47:134–147. https://doi.org/10.1016/ j.engstruct.2012.09.006
7. Turner AW, Gillies RM, Sekel R, Morris P, Bruce W, Walsh WR (2005) Computational bone remodelling simulations and comparisons with DEXA results. J Orthopaed Res : Off Publ Orthop Res Soc 23(4):705–712. https://doi.org/10.1016/j.orthres.2005.02.002
8. Hambli R, Katerchi H, Benhamou CL (2011) Multiscale methodology for bone remodelling simulation using coupled finite element and neural network computation. Biomech Model Mechanobiol 10(1):133–145. https://doi.org/10.1007/s10237-010-0222-x
9. White AA, Panjabi MM (1990) Clinical biomechanics of the spine, 2nd edn. Lippincott, Philadelphia
10. Wang X, Thomas CD, Clement JG, Das R, Davies H, Fernandez JW (2015) A mechanostatistical approach to cortical bone remodelling: an equine model. Biomech Model Mechanobiol 15(1):29–42. https://doi.org/10.1007/s10237-015-0669-x
11. Ackerman MJ, Spitzer VM, Scherzinger AL, Whitlock DG (1995) The Visible Human data set: an image resource for anatomical visualization. Medinfo. MEDINFO 8(Pt 2):1195–1198
12. Bell KL, Loveridge N, Power J, Garrahan N, Meggitt BF, Reeve J (1999) Regional differences in cortical porosity in the fractured femoral neck. Bone 24(1):57–64

13. Sartori M, Reggiani M, Lloyd DG, Pagello E (2011) A neuromusculoskeletal model of the human lower limb: towards EMG-driven actuation of multiple joints in powered orthoses. In: IEEE international conference on rehabilitation robotics [proceedings] 2011, 5975441 (2011). https://doi.org/10.1109/icorr.2011.5975441

14. Gray H (1918) Anatomy of the human body, 20th edn. Lea & Febiger, Philadelphia

15. Shinohara M, Sabra K, Gennisson JL, Fink M, Tanter M (2010) Real-time visualization of muscle stiffness distribution with ultrasound shear wave imaging during muscle contraction. Muscle Nerve 42(3):438–441. https://doi.org/10.1002/mus.21723

16. Herzog W (2000) Skeletal muscle mechanics: from mechanisms to function. Wiley, Chichester

17. Smit TH, Huyghe JM, Cowin SC (2002) Estimation of the poroelastic parameters of cortical bone. J Biomech 35(6):829–835

18. Hayes WC, Keer LM, Herrmann G, Mockros LF (1972) A mathematical analysis for indentation tests of articular cartilage. J Biomech 5(5):541–551

19. Brown TD, Ferguson AB Jr (1980) Mechanical property distributions in the cancellous bone of the human proximal femur. Acta Orthop Scand 51(3):429–437

20. U.S. FDA Cellular Tissue and Gene Therapies Advisory Committee: Meeting #38 (2005) Cellular products for joint surface repair briefing document. In: U.S. Food and Drug Administration (ed) Rockville, MD

21. Keller TS, Mao Z, Spengler DM (1990) Young's modulus, bending strength, and tissue physical properties of human compact bone. J Orthopaed Res Off Publ Orthopaed Res Soc 8(4):592–603. https://doi.org/10.1002/jor.1100080416

22. Lee WR (1964) Appositional bone formation in canine bone: a quantitative microscopic study using tetracycline markers. J Anat 98:665–677

23. Davies HMS (1995) The adaptive response of the equine metacarpus to locomotory stress. PhD Thesis, University of Melbourne

24. Wold S, Sjostrom M, Eriksson L (2001) PLS-regression: a basic tool of chemometrics. Chemometr Intell Lab 58(2):109–130. https://doi.org/10.1016/S0169-7439(01)00155-1

25. Martelli S (2017) Femoral neck strain during maximal contraction of isolated hip-spanning muscle groups. Comput Math Method Med 2017:2873789. https://doi.org/10.1155/2017/2873789

Numerical Simulation of Bone Tissue and Adjacent Structures

Finite Element Analysis of Bone and Experimental Validation

Francisco M. P. Almeida and António M. G. Completo

Abstract This chapter describes the application of the finite element (FE) method to bone tissues. The aspects that differ the most between bone and other materials' FE analysis are the type of elements used, constitutive models and experimental validation. These aspects are observed from a historical evolution point of view. Several types of elements can be used to simulate similar bone structures, and within the same analysis, many types of elements may be needed to realistically simulate an anatomical part. Special attention is made to constitutive models, including the use of density–elasticity relationships enabled through CT scanned images. Other more complex models are also described, such as viscoelastic and anisotropic models. The importance of experimental validation is discussed, describing several methods used by different authors in this challenging field. The use of cadaveric human bones is not always possible or desirable and other options are described, as the use of animal or artificial bones. Strain and strain rate measuring methods are also discussed, such as rosette strain gauges and optical devices.

1 Introduction

Probably, the first-ever published work on finite element (FE) analysis in the field of biomechanics was the article of Brekelmans et al. [1]. They developed a two-dimensional FE model of a human femur. The bone material was considered as homogeneous, isotropic and linear, possessing Young's modulus of 20 GPa and a Poisson's ratio of 0.37. Brekelmans et al. divided the model into 936 triangular elements and applied to its simple forces and boundary conditions. Today, such simulation would be considered extremely simple and basic.

F. M. P. Almeida · A. M. G. Completo (✉)
Department of Mechanical Engineering, University of Aveiro, Aveiro, Portugal
e-mail: completo@ua.pt

F. M. P. Almeida
e-mail: f.almeida@ua.pt

© Springer Nature Switzerland AG 2020
J. Belinha et al. (eds.), *The Computational Mechanics of Bone Tissue*,
Lecture Notes in Computational Vision and Biomechanics 35,
https://doi.org/10.1007/978-3-030-37541-6_7

However, the biomechanical study of bone did not start there. Before that, many researchers dedicated their time to studying mechanical properties of anatomical parts, especially bone. It is reported that Galileo published work on bone mechanics as early as 1638 [2], and it is known that this has been the subject matter of many studies since then.

By 1983, Huiskes and Chao [3] described the first ten years of FE analysis in biomechanics. They were optimistic about the evolution of the field, especially in view of developments in mechanical engineering and the rapid evolution of computers. Their main concerns were to understand the complexity of clinical problems and the behaviour of biological structures. They considered that after the acquisition of this knowledge, its correct implementation into an FE analysis was also a major hurdle to overcome in the field of biomechanics.

From then on, the FE analysis of bone has been evolving almost exponentially. If original studies were in two dimensions, whether in plain stress or axisymmetric, and constitutive models assumed only simple isotropic linear elastic laws, present studies are easily performed in three dimensions and constitutive models can be very complex, as described later in this chapter. The number of elements in an FE model is also representative of this evolution; while, in the 1980 s, these were counted by the hundreds, today FE models can certainly have more than 100,000 elements [4]. Recently, two noteworthy articles were published reporting on the first four decades of FE analysis applied to biomechanics, the first one on lumbar intervertebral discs [5]and the other on orthopaedic devices [4]. Both articles' tone is one that conveys the advanced state of a field that has proved its value, but also has some new challenges like accounting for patient variability and other uncertainties.

The implementation of an FE analysis requires several inputs. including geometry, element type, constitutive models of materials, meshing considerations, boundary conditions and loading, verification and validations. Bone FE analyses, as any other FE analysis, require all these data. However, the parts that raise more questions and differ the most from other types of FE analysis are the types of elements, constitutive models and validation. For these reasons, these will be the topics discussed in this chapter.

2 Element Types

During the process of creating an FE model, choices have to be made regarding the type of element representing diverse parts of the structure being studied. This choice is related to various factors including but not limited to:

- Type of analysis, e.g. 2D, axisymmetric, 3D;
- Geometry, whether it is simple or complex, or if it can be simplified;
- Expected behaviour, e.g. small deformations or large deformations;
- Constitutive model, some element types may not be compatible with some constitutive models.

There are innumerable types of elements, each with its specific mathematical model; nonetheless, there are some types of elements that fall within the same set of basic properties. Biomechanics can be a quite demanding field in this regard; within the same analysis, one can easily find several types of elements. What follows does not aim to be a detailed description of the elements used by the referred authors, which is absent in the published articles most of the time, but rather a general view of different approaches to similar problems regarding bone tissue discretization.

When modelled in three-dimensional studies, both cortical and cancellous bones are frequently represented through eight-node hexahedral elements, e.g.: [6–16]. Although with some disadvantages related to accuracy and excessive stiffness [17, 18], four-node tetrahedral elements are sometimes used for modelling cortical and cancellous bone because they are very easy to use in complex geometries, e.g.: [19–22]. As a means to overcome first-order tetrahedral elements' disadvantages, in some studies cortical and cancellous bones are modelled through second-order (ten-node) tetrahedral elements, e.g.: [23–25].

In some FE analyses, cancellous bone and cortical bone are modelled with distinct elements, as is the case in Denozière and Ku [26] and Ezquerro et al. [27], where cortical bone was represented through hexahedral elements, and cancellous bone was represented through tetrahedral elements. In other words, 3D solid elements (hexahedral or tetrahedral) are used for cancellous bone, while shell elements are used for cortical bone, e.g.: [28–32].

Stiffness of bones is several orders of magnitude higher than soft tissues' stiffness. For this reason, in some types of analyses, the error committed by considering bones as rigid structures is small. Several authors have opted for this solution, therefore saving computer processing time, e.g.: [33–40].

3 Constitutive Models of Bone

The choice of constitutive models to represent different bony tissues depends on many factors. Along the history of FE analysis in biomechanics, there has been a noticeable evolution in the complexity of these models, resulting from the increasing understanding of tissues' behaviour and simultaneously the rapid progress in computing power. Although the simplest models can be described by a restrict set of parameters, a comprehensive analysis on the development of constitutive models would be incomplete without a short description of its implementation. Many models overlap in some aspects, most are not completely original and build on previously developed ones. The selection of constitutive models presented here seeks to display original studies and their evolution through subsequent works.

Bone has been a deeply studied topic within the subject of biomechanics. Besides the consequences of malformations and diseases, such as osteoporosis, its constitutive model is very important in the field of orthopaedic surgery. In many types of surgeries, bone is the primary material to which implants and screws are attached. For these reasons, there are numerous quantities of FE studies considering appropriate constitutive model of bone.

Most bones are composed of two forms of bone tissue. Cortical bone constitutes the structural shell of nearly all bones. Cancellous bone is contained within cortical bone and forms a continuous mass, made of a three-dimensional lattice comprising rod-like and plate-like portions, the trabeculae. The compactness of the trabeculae defines the level of porosity and density of cancellous bone. By definition, cancellous bone exhibits a relative density varying from 0.05 to 0.7, while cortical bone shows a relative density between 0.7 and 0.95, where relative density is the ratio of specimen density to that of fully dense cortical bone. Cortical bone shows a porosity of approximately 5–30% and the porosity of cancellous bone varies between approximately 30 and 90%. Apparent density is another important parameter defined by the mineralized mass divided by total tissue volume. It can be used for measuring mechanical properties of bone and it shows an almost linear relation with porosity [41–43].

Bone tissue is composed of cells surrounded by a matrix. This matrix consists mainly of collagen (mostly type I) and a mineral phase (mostly calcium phosphate and calcium carbonate). Most of this mineral phase is arranged in hydroxyapatite crystals, which are the main source of bones rigidity [42, 43]. This composition and variability do not result in a simple constitutive model, as can be seen by the work of many researchers.

The understanding of this complex behaviour has increased during the course of history. Testing several types of human bones, Dempster and Liddicoat [44] found that cortical bone was non-isotropic and showed inelastic behaviour before the breaking point. For the longitudinal direction, their measured values of Young's moduli were on average 2.5 and 3.0 million psi (17.2 and 20.7 GPa) for dry bone, in tension and compression, respectively; for wet bones, the values obtained were lower by about 0.5 million psi (3.4 GPa) in compression and 1 million psi (6.9 GPa) in tension. Adding to this, for radial and tangential directions, the values obtained were both 52% of the modulus for the longitudinal direction. In the same study, Dempster and Liddicoat also measured the ultimate compressive strength, obtaining values for the longitudinal direction in the order of 25,000 psi (172 MPa) for dry bone, and 15,000 psi (103 MPa) for wet bone and slightly lower values for transverse directions.

Throughout their study, Dempster and Liddicoat compared the mechanical properties of bone with that of other materials such as wood, concrete and steel. They went further and reported that bone exhibits a similar behaviour to wood, in part because both materials present mainly orthotropic mechanical properties.

In most FE studies, bone is considered a linear elastic isotropic time-independent material. This constitutes a pronounced simplification, especially in the case of cancellous bone where several studies have been trying to implement more complex models with varying site-specific properties. In the case of cortical bone, its inherent anisotropy is the major issue. Considering simpler models represents an advantage in terms of model construction and computational resources. However, for some case studies, it is necessary, or advantageous, to use more complex representations than can replicate more accurately the reality.

Brekelmans et al. [1] completely ignored the distinction between cancellous and cortical bone, and for their 2D FE analysis of a human femur, a homogenous, isotropic, linear elastic model was used. They applied Young's modulus of 20 GPa and a Poisson's ratio of 0.37.

In another early study of FE analysis applied to the field of biomechanics, Huiskes et al [45] studied the behaviour of a human femur, considering the cortical material linear elastic, homogeneous and transversely isotropic. Again, cancellous material was not considered. Even so, they found excellent agreement with experimental results. Huiskes et al. considered that FE analysis was able to accurately represent at least the diaphysis of the femur. In the same study, an axisymmetric FE analysis of the femur was performed. The values for the Young's moduli considered were 20 and 13.6 GPa for the longitudinal and transverse directions, respectively, and Poisson's ratio was taken as 0.37.

In one timely attempt to capture cancellous bone variability and complexity, Brown and Ferguson [46] tested large numbers of 5 mm edge length cubic specimens of human proximal femur cancellous bone in three orthogonal directions. Their results showed the markedly anisotropic behaviour of this structure but also a clear proportionality between Young's modulus and yield strength, independent of direction. The values for the elasticity modulus obtained were rather high, ranging from 1 to 9.7 GPa, and the yield strengths measured were 120 and 310 MPa.

Adapting the available information about cancellous bone to FE analysis, Taylor et al. [47] used a very simple FE model of the femoral part of a hip prosthesis with all materials considered homogenous, isotropic and perfectly elastic for the exception of cancellous bone which was considered elastic perfectly plastic. This was done as an attempt to gain a better comprehension of the interface between bone and implant.

Kopperdahl and Keaveny [48] proposed that tensile yield strain of cancellous bone is independent of apparent density for human bone, while compressive yield strains have a linear relation with apparent density. For vertebral cancellous bone, Kopperdahl and Keaveny arrived at the following average values: Young's modulus in compression and tension are 291 MPa and 301 MPa, respectively; yield strain in compression and tension are 0.84% and 0.78%, respectively; ultimate strain in compression and tension are 1.45% and 1.59%, respectively; yield stress in compression and tension are 1.92 MPa and 1.75 MPa, respectively; ultimate stress in compression and tension are equal and show the magnitude of 2.23 MPa (Fig. 1).

Morgan and Keaveny [49], through the analysis of several specimens of cancellous bone from four anatomical sites, found that both yield strain and stress could be better predicted when a site-specific model was adopted. It was hypothesized that this had to do with the particular architecture and hard tissue properties of cancellous bone at those sites. On the same subject, Chang et al. [50] found yield strains of cancellous bone to be isotropic. Studying on bovine cancellous bone, characterized by high density and strong plate-like anisotropic architecture, Chang et al. found that this bone showed similar yield strains between on-axis and $90°$ off-axis. From the particular characteristics of these bones, they made extrapolations to other bones, including humans.

Fig. 1 Typical stress–strain
behaviour of cancellous bone
of different densities

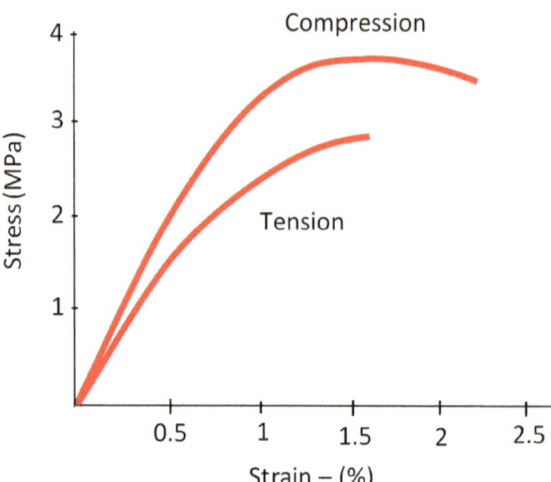

In reality, bone can be said to have an anisotropic viscoelastic behaviour, which is adequately simplified for most studies. However, Iyo et al. [51], considering that the viscoelastic anisotropy was important for implant fixations, proposed a model from which the Young's modulus of cortical bone could be derived as a function of time. This study shows how complex the description of bone's constitutive model can be.

3.1 Use of CT Scans and Density–Elasticity Relationship

Because mechanical properties of bone show high variability, there has been a stimulus to create FE models that reflect these changing values from element to element. The relation between bone density and its mechanicals properties has been known for some time [41, 52–57]. In the last fifteen years, it has been possible to apply this relationship to FE models, mostly by making use of values acquired through quantitative computed tomography (QCT) scanning. These models have been increasing in complexity with the improvement of imaging technologies and computer power.

QCT is a technique that allows the measurement of bone density using computed tomography (CT) scanners calibrated through the use of phantoms, such as the European Spine Phantom (ESP) [58]. The ESP was created in 1995 and its original purpose was to calibrate CT scanners and DXA devices in order to obtain correctly evaluated diagnoses of osteoporosis regardless of where the exam was performed [55]. However, as for QCT in general, its use proved to be extremely useful in FE bone analysis, as it allowed increasing confidence in the collected data.

Most studies try to match the relationship between a mechanical property and bone density through an empirical equation of the form [41]:

$$\gamma = A\rho^B \tag{1}$$

where γ is the material property (mostly Young's modulus or strength), ρ is the apparent density and A and B are experimentally derived constants.

Galante et al. [52] defined two different densities regarding cancellous bone: apparent density that equals the weight divided by the total volume of the sample and real density that is equal to the weight divided by the volume of the matrix excluding the marrow vascular spaces. They tested samples from human lumbar vertebras under compression. Their results showed a very good relation between apparent density and compressive strength, obtaining the following equation:

$$Y = -6.9 + 128.02X \tag{2}$$

where Y is the strength in kp/cm^2, and X is the apparent density in g/cm^3.

Galante et al [52] proposed that apparent density was more important than real density in the evaluation of strength of cancellous bone. In their study, time-dependent and anisotropic results were also noticed, leading them to suggest that cancellous bone has a complex rheological behaviour.

Using samples of human and bovine cancellous bones, Carter and Hayes [41] arrived at the following relationship:

$$E = 3790\varepsilon_r^{0.06}\rho^3 \tag{3}$$

where E is the Young's modulus in MPa, ε_r is the strain rate in s^{-1} and ρ is apparent density in g/cm^3.

Lotz et al. [53] studied the relationship between human proximal femur cancellous bone apparent density and mechanical properties. The samples were initially scanned using QCT, then mechanical properties were measured and finally, the density was experimentally measured. This process permitted to establish a direct relation between QCT data and mechanical properties. They obtained highly positive correlations with compressive Young's modulus and also with compressive strength through the following equations:

$$E = 1310\rho^{1.40} \tag{4}$$

$$\sigma = 25\rho^{1.8} \tag{5}$$

where E is the Young's modulus in MPa, ρ is the apparent density in g/cm^3 and σ is the compressive strength in MPa.

Hodgskinson and Currey [54] studied the relationship between density and Young's modulus for a wide variety of cancellous bone types, corresponding to a wide range of densities. They found a very strong correlation for the entire interval even when considering bones from different species such as human, equine and bovine.

Having FE analysis in mind and trying to overcome the fact that density–elasticity ratios do not provide any information regarding the anisotropy of bones, Rho et al.

[56] provided a series of orthotropic ratios between measured density and Young's modulus for several human bones. In their equations, axial elasticity is a function of density, while radial and circumferential elasticity are functions of axial elasticity. These relationships were found to be stronger for cancellous bone than for cortical bone.

By means of a meta-analysis of literature, Wirtz et al. [57] provided a very comprehensive study of this type of relationship. At an early stage of this method, Wirtz et al. described that each finite element can be characterized by its QCT derived apparent density. For both cortical and cancellous bone, relations between density and Young's modulus, strength, shear modulus, Poisson's ratio and viscoelastic behaviour were provided as follows.

Thus, the Young's modulus (E) in MPa—apparent density (ρ) in g/cm^3 relationships for cortical femoral bone in the axial and transverse direction are obtained, respectively, with:

$$E_{\text{axial}} = 2065\rho^{3.09} \tag{6}$$

$$E_{\text{tranv}} = 2314\rho^{1.57} \tag{7}$$

Regarding the cancellous femoral bone, the Young's modulus—apparent density relationships in the axial and transverse direction can be defined with the following expressions:

$$E_{\text{axial}} = 1904\rho^{1.64} \tag{8}$$

$$E_{\text{tranv}} = 1157\rho^{1.78} \tag{9}$$

Concerning the compressive strength (σ_b)—apparent density relationships for cortical femoral bone in the axial and transverse direction, the following relations are proposed:

$$\sigma_b^{\text{axial}} = 72.4\rho^{1.88} \tag{10}$$

$$\sigma_b^{\text{tranv}} = 37\rho^{1.51} \tag{11}$$

Alternatively, for cancellous femoral bone, the compressive strength—apparent density relationships in the axial and transverse direction are obtained with:

$$\sigma_b^{\text{axial}} = 40.8\rho^{1.89} \tag{12}$$

$$\sigma_b^{\text{tranv}} = 21.4\rho^{1.37} \tag{13}$$

For Poisson's ratio, Wirtz et al. proposed average values of 0.3 for cortical bone and 0.12 for cancellous bone.

Regarding viscoelasticity, Wirtz et al. referred to the expression of Carter and Hayes [59] where ε_r equals the strain rate in s^{-1}:

$$\sigma_b = 68\varepsilon_r^{0.06}\rho^2 \tag{14}$$

Taylor et al. [60] derived the orthotropic elastic constants of a human femur by comparing experimentally measured natural frequencies with values obtained through an FE modal analysis. The model used for the FE construct resulted from CT scan data where an orthotropic density–elasticity relation was used and applied to each element as a function of its position along 16 different radial orientations. The three different Young's modulus and shear modulus equations were a function of density and of a maximum value for each modulus. Through the use of an FE analysis, Taylor et al. were able to validate the entire bone model instead of site-specific values acquired through strain gauges.

As an example of application of density–elasticity relation to a specific problem, Pancanti et al. [61] used the equations derived by Wirtz et al. [57] in order to obtain a more precise FE model of a cementless total hip replacement.

When there was already some accumulated experience regarding the application of CT scanned data to FE models, Taddei et al. [62] presented a very clear set of possible ways to implement this technique:

- From the simpler 'voxel mesh' where cubic elements were generated from the information contained in a pre-set number of voxels. The constitutive model of each element being derived from the average density of the voxels that fall within it.
- Through the use of a structured mesh, where facets of elements could be made to coincide with tissue boundaries, similarly to the above-described method, each element properties was derived from an average of the measured densities within it. The major disadvantage of this method being that some manual input was needed when defining the mesh geometry.
- Using an unstructured mesh, in which case there is no alignment of element facets with tissue boundaries. This procedure relies on a more automated approach since it becomes impossible to manually define element properties, but ultimately also depends on averaging densities within each element in order to derive its mechanical properties.

Additionally, Taddei et al. presented software that improves the automatization of this process, mapping CT scanned data into FE models. With increasingly more expedite processes, a trend starts to unravel towards the use of patient-specific models, allowing the study of interventions that account for specific bone characteristics of each individual.

Morgan et al. [63] studied different density–elasticity ratios obtained for cancellous bone from different anatomical sites. By investigating cancellous bone from vertebra, proximal tibia, femoral greater trochanter and femoral neck, they concluded

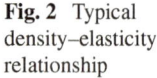

Fig. 2 Typical
density–elasticity
relationship

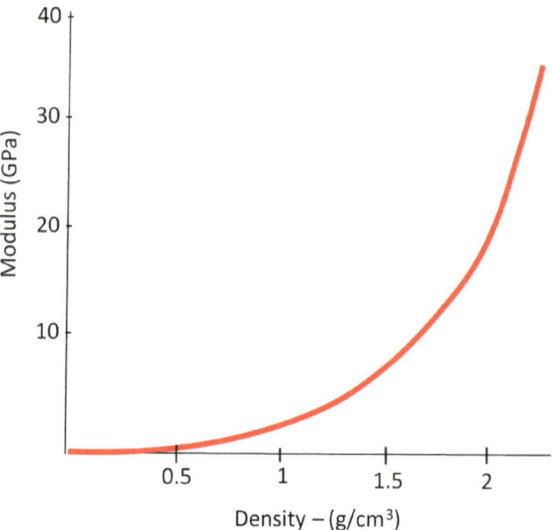

that the axial elasticity–density ratio for cancellous bone varied depending on the
anatomical site under study. Using the same type of mathematical relationship presented above [53, 57, 59], they derived several equations to account for this variation.
Morgan et al. attributed the varying relationships to differences in local cancellous
bone architecture (Fig. 2).

Taddei et al. [23], using data from a CT scanned femur, constructed two FE
models. In the first case, a different Young's modulus was attributed to each element,
based on the average density, numerically integrated through the element's volume
and using the following expression:

$$E = 10.5\rho_{\mathrm{ash}}^{2.29} \tag{15}$$

where ρ_{ash} is ash density. The application of this method provided a maximum
Young's modulus of 19.8 GPa and a 12.9 GPa average for cortical bone.

For the second model, only two discrete material models were used. Applying the
above equation, a calibrated homogeneous Young's modulus value of 19.3 GPa was
achieved for cortical bone and 590 MPa for cancellous bone. Cortical and cancellous
bones were distinguished at a predefined threshold density value from the CT data.

While both models correlated well with experimental values obtained for the
same bone, the first method proved more accurate. Taddei et al. considered that it
is possible to use automatic tools to generate FE models from CT data, and that
accuracy is influenced by the material models mapping strategy implemented.

In order to overcome some of the shortcomings resulting from considering
isotropy for the whole bone model, Marangalou et al. [64] used the data obtained
from micro-CT scans to attribute a site varying orthotropic model for cancellous
bone. Clinical CT scans do not have enough resolution to detail the micro-structure

of cancellous bone, and by simply applying a density–elasticity ratio, errors are made, whether by overestimation or underestimation of mechanical properties, depending on direction. Their micro-CT derived orthotropic model showed higher correlation with micro-CT measurements than other isotropic models. Marangalou et al. considered that this approach can lead to more precise estimations of strength and elasticity of osteoporotic bones, and around implants for surgery studies.

In order to study the evolution of osteoarthritis in subchondral proximal tibia, Nazemi et al. [25] evaluated several density–elasticity equations [54, 56, 63] comparing the results of FE analyses with macro-indentation tests. They concluded that, for this particular anatomical area, no single equation offered a good prediction of mechanical properties and underlined the importance of accounting for bone's heterogeneity when performing this type of FE study.

3.2 Micro-Finite Element Modelling

The first time a micro-FE model of a bone attempted was more than two decades ago, requiring the use of supercomputers [65]. Nowadays, with improvements in imaging technology, models with higher degree of definition have been made easier to create. In particular, high-resolution peripheral quantitative CT (HR-pQCT) is a technology that has been enabling the study of the micro-structure of peripheral bones in vivo.

This technology has been mostly used to predict the strength of bones as a diagnostic tool in cases of osteoporosis and other bone affecting diseases. This technique enables the acquisition of bone micro-structure, which can afterwards be studied using the effective Young's modulus, as opposed to the apparent Young's modulus, used in most analyses. In the future, this technology will probably be available to anatomical parts other than peripheral bones (in vivo), in which case other applications may arise, such as assisted surgery for complex orthopaedic interventions.

In most cases of micro-FE analysis derived from HR-pQCT, a voxel conversion approach is utilized. In such studies, bone tissue voxels are converted into brick elements of the same size while voxels representing soft tissues are ignored [65].

In order to evaluate which density–elasticity model better predicts the behaviour of a whole human ulna, Austman et al. [24] applied different equations [41, 53, 57, 63] to a micro-FE model and compared it with experimental results. They concluded that the equation that showed the best match varied with specimen, but overall, Morgan et al. [63] and Carter and Hayes's [41] equations were the ones that showed smaller errors.

In a similar study, Scholz et al. [66] compared different density–elasticity relationships [41, 63] in an FE modal analysis of pelvic bone specimens with results obtained experimentally. They used micro-CT scanned data to construct micro-FE models. All of the density–elasticity relationships used had been acquired from long bones' models, so there were doubts regarding their adequacy to the pelvic bone, as the cortical layer of this bone is thinner. Scholz et al. concluded that all the analyzed

FE models lacked stiffness, and the one that showed best results was the one using Morgan et al. [63] equations.

Liu et al. [67] compared the results of human distal tibial bones' models acquired through HR-pQCT with models acquired through micro-CT. Micro-CT is a technology considered the gold standard for this type of image acquisition, but cannot be practiced in vivo, much less in humans. They concluded that the results showed high correlation although HR-pQCT overestimated the mechanical properties of the bone. This may be the result of the combination of voxel size and the considered density threshold, because while micro-CT captures smaller details, HR-pQCT compensates for this by attributing a lower density value for each structural element; the result being thicker structures with lower Young's moduli, which at a global level ends up as a stronger and stiffer bone.

3.3   Complex Constitutive Models

As can be inferred by what has been described so far, the two main parts of bones, cortical and cancellous are very different in their structure. For this reason, in general, even the simplest FE studies take into account two different models describing these two different regions. From a structural stand view, it seems quite obvious that cortical bone is the most important part; however, the question may be posed in terms of how important cancellous bone is to the overall strength and stiffness. Different authors propose different methods that span from studying cancellous bone's resistance separately to considering that, for some purposes, bone can safely be studied ignoring cancellous bone entirely.

A good example of this duality is the study of Parr et al. [22] where six different models of a human talus bone were considered, each one with differing levels of complexity. Starting from images acquired through micro-CT, they were able to model the small intricacies of cancellous bone. Comparing the most realistic model, which included the micro-structure of cancellous bone and porosity in cortical bone, with other simpler models that deleted the cancellous bone or treated it as a homogeneous mass, the authors were able to measure the contribution of this interior network-shaped structure to the overall stiffness of bone. They concluded that the way cancellous bone was modelled had a great impact on the stiffness of the whole bone. Another secondary but interesting point shown by their study was the importance of the computational development in this field: contrary to studies conducted one or two decades prior, this one was performed on a commercial desktop computer, taking at most one hour to analyze the most complex model.

One of the more complex constitutive laws for cortical bone was developed by Carnelli et al. [68, 69]. In these studies, they developed a model that accounts for anisotropic elastic and post-yield behaviour, as well as tension–compression mismatch and direction-dependent yield stresses. They tested this model against nanoindentation experiments which they confirmed to show high correlation in most parameters. Carnelli et al. advocate that such complex model would help gaining a deeper knowledge of the microstructural behaviour of cortical bone tissue.

Contrary to the general trend, Koivumaki et al. [70], using models obtained through a multi-detector CT scanner, showed that the failure load of the proximal femur could be well predicted by ignoring cancellous bone. In their FE models, only the cortical part was considered, which allowed significant savings in computer resources, and the results showed high correlation with experimental data and only slightly worse results than a similar model in which cancellous bone had been considered. Cortical bone was modelled using a bi-linear elastoplastic constitutive law, where the post-yield modulus was 5% of the initial elastic modulus. Failure was considered to occur when the material reached a stress of 118.6 MPa in tension or 1.35% strain in compression.

In order to include plasticity and damage control with a relatively simple constitutive model, Kinzl et al. [14] identified a crushable foam model than could be adapted to different bones and densities using only three different parameters. In their study, they used QCT generated images where each voxel corresponded to an element. The major disadvantage of this material model was the absence of anisotropy, but when compared with other more complexes and harder to implement models, Kinzl et al. found that this readily available and easy to implement model showed similar results and it was a good predictor of bone strength and damage.

With the objective of studying the region of interest during a pull-out of a pedicle screw from a vertebra, Liu [15] used polyurethane foam as a substitute for cancellous bone and made an FE analysis where some of the material model parameters were changed in order to simulate different stages of osteoporosis. The experiments with polyurethane foam served for initial calibration of the FE model, which was subsequently used to gain a better understanding of the surrounding mechanics of a pedicle screw pull-out.

Concerning the study of cancellous bone at the interface with a press-fit proximal tibia implant, Nelly et al. [71] used a crushable foam with isotropic hardening model to improve the plastic deformation prediction. This type of model revealed itself to be superior to the traditional von Mises plasticity formulation, as it takes into account pressure-dependent yield. Besides proving to be a better fit for cancellous bone, this crushable foam model also showed a better representation of polyurethane foam used in experimental testing.

In order to estimate bone drilling forces during orthopaedic surgeries, Lughmani et al. [16] proposed a transversely isotropic elastic–plastic rate-dependent model of cortical bone. In addition to this, an element removal scheme was included in the FE analysis, simulating the advancement of the drilling bit through the cortical bone. Such a model is an example of the use of increasingly complex FE models that help improve the knowledge of very specific processes.

3.4 Observations Regarding Constitutive Models

Most material models described here, in particular the more recent ones, are probably too complex for the objectives of most FE studies. The choice of constitutive models depends ultimately on the type of study being performed. In a healthy human being, materials with very dissimilar mechanical behaviours are interconnected; these materials may exhibit properties, such as Young's modulus, differing by several orders of magnitude. For this reason, it is paramount to adapt constitutive models to each particular analysis. If, for instance, it is the researcher's aim to analyze the behaviour of a particular joint under normal physiological loads, then it is not expected for a bone adjacent to that joint to reach yield stresses or even to have any noticeable strains; in this case, bones, without perceptible errors, can be modelled as rigid elements. If, alternatively, it is a requirement of the study to understand a bone's load-carrying capacity, as would be the case for most orthopaedic surgeries, then the bone constitutive model should be as realistic as possible. Within this field of biomechanics where soft tissues are sometimes side by side with titanium alloys, a sensible choice of constitutive models seems imperative.

4 Experimental Validation

Experimental validation was not always seen as an important step in the process of assessing the quality of an FE analysis. When Brekelmans et al. [1] introduced the FE method to the field of biomechanics, it was meant to replace experimental analysis. Their intent was to affirm FE analysis as a superior tool to contemporary experimental techniques:

> A comparison of [finite element analysis] (…) possibilities with those afforded by the experimental techniques or classical analytical theories is clearly in favour of the analysis with the aid of the finite element method.

Throughout their historical article, the only time they did not disregard experimental techniques was when they enumerated reasons for choosing the femur as an object of analysis:

> The femur was chosen because (…) it is currently a focus of interest in the literature on theoretical and experimental investigations; this affords a possibility of comparing results.

The contempt for experimental analysis is very clear; nonetheless, this sentence touched a very important aspect in the development of FE analysis that would follow: the need to somehow validate results. The more obvious and direct method was to compare them with results obtained experimentally.

Providing a background, at the beginning of their article, Brekelmans et al. made a description of experimental methods existing at the time:

- Brittle coating technique: a technique that relied on the cracking of varnish previously applied to the surface of bone. The lines created by the cracked varnish when the specimens were tested would reveal deformations.
- Optical method (photo-stress technique): the bone to be analyzed was subjected to a treatment that would give its surface particular optical properties. When loaded and subjected to a polarized light, lines would appear on the bone surface and conclusions could be taken.
- Strain gauge measurements: a method that is still used today, in which strain gauges are glued to the bone surface, and can, therefore, take direct measurements of strains in relation with known loads.
- Photo-elastic technique: a plastic model of the bone was made; it was then illuminated with polarized light, resulting in the appearance of lines in the model. Conclusions were taken in accordance with those lines.

Almost a decade later, Huiskes et al. [45] claimed to be among the first investigators to perform a well-defined comparison between theoretical and experimental results. Using both femurs of the same cadaver, they used the left one to perform experimental studies, while the right one was cut into sections in order to acquire precise dimensions. Experimental measurements were made through the application of strain gauge rosettes, 100 in total. From the experimental results, they derived two constitutive models that were applied to FE analyses. Through this process, they obtained good agreement between results. Huiskes et al. accepted that some discrepancies between strain gauge and FE analysis values could result from inadequate mesh refinement, which was limited in view of computer costs at the time.

Much more recently, in an editorial paper regarding the quality of published works on FE analysis in biomechanics, Viceconti et al. [72] defined validation as:

> The process that ensures that the numerical modal accurately predicts the physical phenomenon it was designed to replicate.

However, the authors go on to state that validating a numerical model completely is generally impossible and compared this process to the more generalized process of science, where validation is reached through a slow procedure. Viceconti et al. also warned against the clinical use of FE models not correctly interpreted and thoroughly validated, while accepting that this is a very difficult task considering the nature of biological tissues.

4.1 Various Experimental Validation Techniques

While not exactly keeping pace with FE analysis' evolution, which has been exceptional, experimental methods have also shown some development. In some instances, these were very much dependent on the resources available to researchers, as is the case when cadavers are used. This method may raise some moral and availability issues in some societies, epochs or institutions. In order to bypass these difficulties and improve results in general, many researchers have made use of their ingenuity, creating different ways to experimentally validate FE analysis.

Taylor et al. [47] validated stress values obtained in their FE analysis of a hip prosthesis femoral component by comparing with existing values of clinically measured subsidence. Values of subsidence were acquired two years after the intervention and were all for the same type of surgical hardware. This was not a straightforward validation as they compared very different parameters: stress values obtained in an FE analysis versus subsidence measured two years after surgery. Nonetheless, they obtained good correlations when comparing magnitudes of those two parameters.

Kumaresan et al. [6] validated their FE model of a cervical spine through data derived from experiments on eight spinal cadaver units. They collected data through a built-in force gauge and a linear variable differential transformer connected in series to the electrohydraulic piston, an assembly that enabled recording applied force against deflection. Strain gauges were also used, glued to the anterior part of the vertebrae and the lateral masses of the middle vertebra. A correction factor was applied to account for the difference in age of the cadaver specimens used experimentally and the specimen used for the FE model.

With the objective of measuring anisotropic viscoelastic properties of cortical bone, Iyo et al. [51] collected two types of rectangular bovine femur samples: one with the long axis coinciding with the long axis of the bone and the other with the long axis coinciding with the transverse axis of the bone. These samples were subjected to three-point bending loads in saline solution at a constant temperature of 37 °C. Measurements were made using a set of devices that included a strain gauge transducer used as a force sensor and a position detector. After the initial load was applied, the reaction force produced by the bone sample was recorded as a function of time for up to 10^5 s. With the obtained values, it was possible to derive relaxation constitutive properties of cortical bone.

Regarding the validation of an FE analysis of a rat tibia, Evans et al. [20] used an original method which involved loading the specimen inside a micro-CT scanner. The 3D geometry of the bone was registered unloaded and loaded using a material testing stage inside a micro-CT scanner. Hundreds of landmarks were visually marked from the surface and interior of the bone, thus comprising cortical and cancellous bone. The movement of these landmarks upon loading was quantified and compared with values obtained for the same landmarks on the FE model. This process allowed calibration of the Young's modulus of the bone as well as a qualitative validation of the FE analysis. Overall, Evans et al. obtained an unrealistic Young's modulus of one to two orders of magnitude lower than expected, but the authors alleged that this might have been the result of rigid body motion and other related difficulties.

In order to estimate the fracture load of the proximal femur through a more expedite process, Koivumaki et al. [70] tested 61 human femur cadaver bones under loads simulating a side fall. Of these 61 femurs, 21 were used to define the threshold strain beyond which there was a fracture. The other 40 specimens were used to validate the FE analysis. Their study had the peculiarity of neglecting cancellous bone contribution in the overall resistance of bone.

Concerning the validation of an FE analysis of cortical bone drilling, Lughmani et al. [16] used the diaphysis of bovine femurs cut into approximately rectangular specimens. These specimens were mounted on a force transducer which measured the drilling force. The force transducer was attached to a rotating table that had its rotation restricted by a cantilever beam equipped with a strain gauge, thus providing the drilling torque. These forces were recorded at a rate of 1000 Hz by a data acquisition system. The experiments were repeated for a considerable number of specimens at varying rotations per minute applied by a DC servo motor, always using a 2.5-mm drill bit. The results showed good agreement with the FE analysis.

4.2   Use of Modal Frequencies

Taylor et al. [60] experimentally measured the modal frequencies of a femur using an apparatus comprising soft elastic straps, a unidirectional piezoelectric accelerometer fixed to the surface of the bone and an impact hammer containing a force transducer. Natural frequencies and corresponding mode shapes between 0 and 1000 Hz were recorded. An FE model of the same bone was created through CT scan imaging, and an orthotropic elastic constitutive model was calibrated in order to achieve the same modal frequencies and shapes. In a final step, the properties of the bone were measured using transmission ultrasound techniques, which revealed good correlation with FE analysis results.

Also using modal analysis to choose the best fit constitutive model, Scholz et al. [66] compared the experimentally measured frequencies on ten human pelvic bone specimens with those obtained in FE analysis using the specimens' geometry. Scholz et al. used the experimental results from Neugebauer et al. [73], who used a 3D laser vibrometer to obtain the resonance frequencies between 100 and 2000 Hz of the ten specimens. The setup consisted of three measuring laser heads, two aluminium rivets for suspending the bone, a force sensor connected through an aluminium plate, and eight markers attached to the bone for geometric referencing. Comparison between FE analysis and experimental observation showed that even the constitutive model providing the closest results, still produced lower resonance frequencies, indicating that FE models lacked stiffness.

4.3   Use of Synthetic Bone

The use of human cadaver bones poses several problems; an alternative in biomechanical studies is the use of artificial bones. These synthetic structures try to reproduce natural bones' mechanical properties with the added advantage of being more constant. Known advantages of artificial bones over natural human bones include: less geometric and mechanical variability, higher availability, easier handling and preservation [74]. For these reasons, several authors have opted for using artificial bones instead of natural bones in their studies.

In order to validate FE analyses of the distal femur part of total knee replacements, Completo et al. [75] used synthetic bones to replicate three different reconstruction techniques. Strains were measured on ten different locations of the bone using rosette strain gauges, which were connected to a computer. Two different load cases simulating physiological activities were applied by means of a pneumatic device and were repeated five times. Overall good agreement was reached between averaged experimentally obtained strain values and FE analysis values.

Also using synthetic bones, in this case tibiae, Nelly et al. [71] performed an experimental and FE analysis study to determine the influence of cancellous bone plasticity during press-fit implantation of a tibial component in total knee arthroplasty. They used seven specimens, in which the distal part was potted while the implant was being driven into the proximal part of the tibiae by means of a testing machine that simultaneously recorded loads applied at a prescribed displacement rate. Preceding implant insertion, a hole of 11 mm diameter was punched through the artificial cancellous bone. As the implant tapered from 12 to 10 mm diameter, it caused an interference fit of 1 mm in the proximal part. With the objective of accurately simulating the artificial cancellous bone, Nelly et al. also performed uniaxial compression tests on cubic samples of the same polyurethane material used in the artificial tibiae. These experiments were used to validate an FE analysis simulating the same press-fit implantation.

Regarding the validation of an FE analysis on a patellofemoral arthroplasty, Castro et al. [76] used the experimental results from Meireles et al. [77] for the same procedure. Meireles et al. used five synthetic femurs and a tibia. Five triaxial rosette strain gauges were glued to the femurs and were connected to a data acquisition system, which was itself connected to a computer. The joint was tested in a testing machine under various loading conditions simulating daily activities in the intact condition and in the post-surgery state. The readings obtained from the strain gauges showed good agreement with the FE analysis performed by Castro et al. Some differences were attributed to difficulties related to reproducing the exact location of load application and strain gauges on the FE model.

4.4 Use of Strain Gauges

The use of strain gauges is not a novelty [1], but with some improvements throughout the last decades, they have remained instruments of choice of many researchers. In many cases, they were used combined with other techniques, e.g. [6]. In other cases, they were the main means of information acquisition. Strain gauges have some known limitations such as only being capable of recording information from the location they are attached to, they can only be used on the surface, and they can disturb the specimens being tested. Nonetheless, the information collected by these instruments is sometimes sufficient to help validate FE analysis.

In order to estimate the accuracy of a simplified constitutive model, Taddei et al. [23] performed several experiments on a human cadaver femur. A total of thirteen rosette strain gauges were glued to the bone. The specimen was tested on a material testing machine in several configurations that covered a physiological range of loads corresponding to normal activities, including walking, single-leg stance, stair climbing and others. Strains were recorded during the different loading processes and after load removal for a period of 90 s. Each load configuration was repeated five times and the bone was kept moist by wrapping it in cloths soaked with a saline solution.

With the objective of determining which density–elasticity relationship best fitted the mechanical properties of the ulna, Austman et al. [24] compared experimentally obtained strain values with values acquired through the application of six different constitutive model equations found in the literature. They glued 12 uniaxial strain gauges to six different locations on eight ulna specimens. Loads were applied by means of materials testing machines and respective strains recorded. In the FE models, the elements located under each strain gauge were identified and their strain values averaged. This allowed a direct comparison between experiment and all six FE analyses that allowed the evaluation of each constitutive model, leading Austman et al. to identify the two best matches.

4.5 Use of Optical Recording Devices

A more recent method of measuring strains and strain rates has been the use of optical recording devices. These types of devices encompass many different instruments of differing complexities, from simple cameras to 3D laser measuring tools. Regardless of form, their objective is to follow the location of specific points as a function of time and applied load. In many cases, landmarks are glued or otherwise attached to the specimen being tested so that movements are more easily recorded. In some studies, they are the only means through which strains are recorded. When correctly applied, this type of method is capable of obtaining a more comprehensive recording of strains and displacements than other methods limited to discrete position readings.

4.6 Measurement of Micro-motions Between Implant and Bone

With the objective of estimating micro-motions between a femur and a hip stem after implantation, Abdul-Kadir et al. [78] replicated the surgical insertion in four femur cadaver specimens. Through two holes in the bone, two points in the hip stem were marked with a linear variable differential transducer (LVDT), one in the proximal part and the other in the distal part. Using universal materials testing machine, micro-motions between the hip stem and the femur upon loading and unloading were visually measured. An FE model of bone and implant using CT scans was created that could afterwards be validated by the previously made experimental investigation.

Also intending to measure micro-motions after insertion of an implant, Chong et al. [79] also used LVDTs on a tibia following total knee replacement. In this case, the measurements were taken between the implant tray edge and the adjacent supporting bone at three different locations. Three load sequences were performed with intervening unloading periods, and the micro-motions obtained were averaged and used to validate a corresponding FE analysis.

5 Discussion

Bone finite element analyses pose a series of additional difficulties comparing with other structural or mechanical fields. As it is hopefully noticeable on the descriptions made throughout this chapter, there is a lot of variability of mechanical behaviours between anatomical parts. As described above, a single bone can be represented using multiple constitutive laws. Further to this variability, there is another issue that has been increasingly present in researchers' minds: inter-patient variability, e.g.: [4, 13, 80–87]. FE analyses are performed for well-defined geometries, constitutive laws and boundary conditions, and there is no way that a single FE analysis can account for anatomical differences between patients. This is, perhaps, where the frontier presently lies for this technique. Despite their complexity, there are already constitutive models suitably describing the mechanical behaviour of bony tissues. For most studies, researchers do not need to use the most complicated version of these models, and simpler models will suffice. Errors of interpretation of results may emerge from the lack of validation from a big enough sample in order to have high levels of confidence. During experimental validation, when using cadaver specimens, researchers know that the bigger the sample size, the higher the assurance they can get from the process. In a similar fashion, the FE method is also developing in order to account for patient variability and patient-specific analysis. With improving image technology, computer power and process automation, FE can continuously contribute to the development of diagnoses and surgical techniques.

Acknowledgements This work is supported by the project POCI-01-0145-FEDER-028424, funded by Programa Operacional Competitividade e Internacionalização (COMPETE 2020) on its component FEDER and by funding from FCT—Fundação para a Ciência e Tecnologia on its component OE.

References

1. Brekelmans W, Poort H, Slooff T (1972) A new method to analyse the mechanical behavior of skeletal parts. Acta Orthop Scand 43:301–317
2. Ascenzi A (1993) Biomechanics and galileo galilei. J Biomech 26(2):95–100
3. Huiskes R, Chao E (1983) A survey of finite element analysis in orthopedic biomechanics: the first decade. J Biomech 16(6):385–409
4. Taylor M, Prendergast P (2015) Four decades of finite element analysis of orthopaedic devices: where are we now and what are the opportunities. J Biomech 48:767–778
5. Schmidt H, Galbusera F, Rohlmann A, Shirazi-Adl A (2013) What have we learned from finite element model studies of lumbar intervertebral discs in the past four decades? J Biomech 46:2342–2355
6. Kumaresan S, Yoganandan N, Pintar F, Maiman D (1999) Finite element modeling of the cervical spine: role of the intervertebral disc under axial and eccentric loads. Med Eng Phys 21:689–700
7. Zander T, Rohlmann A, Calisse J, Bergmann G (2001) Estimation of muscle forces in the lumbar spine during upper-body inclination. Clin Biomech 16(1):S73–S80
8. Schmidt H et al (2007) Application of a calibration method provides more realistic results for a finite element model of a lumbar spinal segment. Clin Biomech 22:377–384
9. Papaioannou G et al (2008) Patient-specific knee joint finite element model validation with high-accuracy kinematics from biplane dynamic Roentgen. J Biomech 41:2633–2638
10. Faizan A et al (2009) Do design variations in the artificial disc influence cervical spine biomechanics? A finite element investigation. Eur Spine J 21(Suppl. 5):S653–S662
11. Wolfram U, Wilke H, Zysset P (2010) Valid u finite element models of vertebral trabecular bone can be obtained using tissue properties measured with nanoindentation under wet conditions. J Biomech 43:1731–1737
12. Zhang J, Wang F, Zhou RXQ (2011) A three-dimensional finite element model of the cervical spine: an investigation of whiplash injury. Med Biol Eng Comput 49:193–201
13. Niemeyer F, Wilke H, Schmidt H (2012) Geometry strongly influences the response of numerical models of the lumbar spine—a probabilistic finite element analysis. J Biomech 45:1414–1423
14. Kinzl M, Wolfram U, Pahr D (2013) Identification of a crushable foam material model and application to strength and damage prediction of human femur and vertebral body. J Mech Beha Biom Mat 26:136–147
15. Liu S et al (2014) Effect of bone material properties on effective region in screw-bone model: an experimental and finite element study. Biomed Eng Onl 13:83
16. Lughmani W, Marouf K, Ashcroft I (2015) Drilling in cortical bone: a finite element model and experimental investigations. J Mech Beha Biom Mat 42:32–42
17. Donald B (2011) Pracrical stress analysis with finite elements, 2nd edn. Glasnevin Publishing, Dublin
18. Burkhart T, Andrews D, Dunning C (2013) Finite element modeling mesh quality, energy balance and validation methods: a review with recommendations associated with the modeling of bone tissue. J Biomech 46:1477–1488
19. Gu K, Li L (2011) A human knee joint model considering fluid pressure and fiber orientation in cartilages and menisci. Med Eng Phys 33:497–503
20. Evans S et al (2012) Finite element analysis of a micromechanical model of bone and a new approach to validation. J Biomech 45:2702–2705
21. Hussain M et al (2012) Corpectomy versus discectomy for the treatment of multilevel cervical spine pathology: a finite element model analysis. Spine J. 12:401–408
22. Parr W et al (2013) Finite element micro-modelling of a human ankle bone reveals the importance of the trabecular network to mechanical performance: new methods for generation and comparison of 3D models. J Biomech 46:200–205
23. Taddei F et al (2006) Subject-specific finite element models of long bones: an in vitro evaluation of the overall accuracy. J Biomech 39:2457–2467

24. Austman R, Milner J, Holdsworth D, Dunning C (2008) The effect of the density-modulus relationship selected to apply material properties in a finite element model of long bone. J Biomech 41:3172–3176
25. Nazemi S et al (2015) Prediction of proximal tibial subchondral bone structural stiffness using subject-specific finite element modelling: effect of selected density-modulus relationship. Clin Biomech 30:703–712
26. Denozière G, Ku D (2006) Biomechanical comparison between fusion of two vertebrae and implantation of an artificial intervertebral disc. J Biomech 39:766–775
27. Ezquerro F et al (2011) Calibration of the finite element model of a lumbar functional spinal unit using an optimization technique based on differential evolution. Med Eng Phys 33:89–95
28. Beillas P, Lee S, Tashman S, Yang K (2007) Sensitivity of the tibio-femoral response to finite element modeling parameters. Comp Meth Biomech Biomed Eng 10(3):209–221
29. Bowden A et al (2008) Quality of motion considerations in numerical analysis of motion restoring implants of the spine. Clin Biomech 23:536–544
30. Little J, Adam C (2011) Effects of surgical joint destabilization on load sharing between ligamentous structures in the thoracic spine: a finite element investigation. Clin Biomech 26:895–903
31. Dong L et al (2013) Development and validation of a 10-year-old child ligamentous cervical spine finite element model. A Biomed Eng 41(12):2538–2552
32. Wang W, Zhang H, Sadeghipour K, Baran G (2013) Effect of posterolateral disc replacement on kinematics and stress distribution in the lumbar spine: a finite element study. Med Eng Phys 35:357–364
33. Bendjaballah M, Shirazi-Adl A, Zukor D (1995) Biomechanics of the human knee joint in compression: reconstruction, mesh generation and finite element analysis. Knee 2(2):69–79
34. Moglo K, Shirazi-Adl A (2003) On the coupling between anterior and posterior cruciate ligaments, and knee joint response under anterior femoral drawer in flexion: a finite element study. Clin Biomech 18:751–759
35. Donahue T, Hull M, Rashid M, Jacobs C (2003) How the stiffness of meniscal attachments and meniscal material properties affect tibio-femoral contact pressure computed using a validated finite element model of the human knee joint. J Biomech 36:19–34
36. Ramaniraka N, Terrier A, Theumann N, Siegrist O (2005) Effects of the posterior cruciate ligament reconstruction on the biomechanics of the knee joint: a finite element analysis. Clin Biomech 20:434–442
37. Peña E, Calvo B, Martínez M, Doblaré M (2006) A three-dimensional finite element analysis of the combined behavior of ligaments and menisci in the healthy human knee joint. J Boimech 39:1686–1701
38. Li L, Cheung J, Herzog W (2009) Three-dimensional fibril-reinforced finite element model of articular cartilage. Med Biol Eng Comput 47:607–615
39. Rohlmann A, Boustani H, Bergmann G, Zander T (2010) Effect of a pedicle-screw-based motion preservation system on lumbar spine biomechanics: A probabilistic finite element study with subsequent sensitivity analysis. J Biomech 43:2963–2969
40. Yue-fu D et al (2011) Accurate 3D reconstruction of subject-specific knee finite element model to simulate the articular cartilage defects. J Shang Jiaot Univ (Sci) 16(5):620–627
41. Carter D, Hayes W (1977) The compressive behaviour of bone as a two-phase porous structure. J Bone Joint Surg 954–962
42. Cowin SC (2001) Bone mechanics handbook, 2nd edn. CRC Press, Boca Raton
43. Mow V, Huiskes R (2005) Basic orthopaedic biomechanics and mechano-biology, 3rd edn. Lippincott Williams & Wilkins, Philadelphia
44. Dempster W, Liddicoat R (1952) Compact bone as a non-isotropic material. Am J Anat 91(3):331–362
45. Huiskes R, Janssen J, Slooff T (1981) A detailed comparison of experimental and theoretical stress-analyses of a human femur. Mech Proper Bone 45:211–234
46. Brown T, Ferguson A (1980) Mechanical property distributions in the cancellous bone of the human proximal femur. Acta Orthop Scand 51:429–437

47. Taylor M et al (1995) Cancellous bone stresses surrounding the femoral component of a hip prosthesis: an elastic-plastic finite element analysis. Med Eng Phys 17:544–550
48. Kopperdahl D, Keaveny T (1998) Yield strain behaviour of trabecular bone. J Biomech 31:601–608
49. Morgan E, Keaveny T (2001) Dependence of yield strain of human trabecular bone on anatomical site. J Biomech 34:569–577
50. Chang W et al (1999) Uniaxial yield strains for bovine trabecular bone are isotropic and asymmetric. J Orthop Res 17:582–585
51. Iyo T et al (2004) Anisotropic viscoelastic properties of cortical bone. J Biomech 37:1433–1437
52. Galante J, Rostoker W, Ray R (1970) Physical properties of trabecular bone. Calc Tiss Res 5:236–246
53. Lotz J, Gerhart T, Hayes W (1990) Mechanical properties of trabecular bone from the proximal femur: a quantitative CT study. J Comp Assist Tomogr 14:107–114
54. Hodgskinson R, Currey J (1992) Young's modulus, density and material properties in cancellous bone over a large density range. J Mat Sci Mat Med 3:377–381
55. Kalender W et al (1995) The European spine phantom—a tool for standardization and quality control in spinal bone mineral measurements by DXA and QCT. Eur J Radiol 20:83–92
56. Rho J, Hobatho M, Ashman R (1995) Relations of mechanical properties to density and CT numbers in human bone. Med Eng Phys 17:347–355
57. Wirtz D et al (2000) Critical evaluation of known bone material properties to realize anisotropic FE-simulation of the proximal femur. J Biomech 33:1325–1330
58. Adams J (2009) Quantitative computed tomography. Eur J Radiol 71:415–424
59. Carter D, Hayes W (1976) Bone compressive strength: the influence of density and strain rate. Science 194:1174–1175
60. Taylor W et al (2002) Determination of orthotropic bone elastic constants using FEA and modal analysis. J Biomech 35:767–773
61. Pancanti A, Bernakiewicz M, Viceconti M (2003) The primary stability of a cementless stem varies between subjects as much as between activities. J Biomech 36:777–785
62. Taddei F, Pancanti A, Viceconti M (2004) An improved method for the automatic mapping of computed tomography numbers onto finite element models. Med Eng Phys 26:61–69
63. Morgan E, Bayraktar H, Keaveny T (2003) Trabecular bone modulus-density relationships depend on anatomic site. J Biomech 36:897–904
64. Marangalou J et al (2013) A novel approach to estimate trabecular bone anisotropy using a database approach. J Biomech 46:2356–2362
65. Rietbergen B, Ito K (2015) A survey of micro-finite element analysis for clinical assessment of bone strength: the first decade. J Biomech 48:832–841
66. Scholz R et al (2013) Validation of density-elasticity relationships for finite element modelling of human pelvic bone by modal analysis. J Biomech 46:2667–2673
67. Liu S et al (2010) High-resolution peripheral quantitative computed tomography can assess microstructural and mechanical properties of human distal tibial bone. J Bone Min Res 25(4):746–756
68. Carnelli D et al (2010) A finite element model for direction-dependent mechanical response to nanoindentation of cortical bone allowing for anisotropic post-yield behaviour of the tissue. J Biomech Eng 132:081008.1–081008.10
69. Carnelli D et al (2011) Nanoindentation testing and finite element simulations of cortical bone allowing for anisotropic elastic and inelastic mechanical response. J Biomech 44:1852–1858
70. Koivumaki J et al (2012) Cortical bone finite element models in the estimation of experimentally measured failure loads in the proximal femur. Bone 51:737–740
71. Nelly N et al (2013) An investigation of the inelastic behaviour of trabecular bone during the press-fit implantation of a tibial component in total knee arthroplasty. Med Eng Phys 35:1599–1606
72. Viceconti M, Olsen S, Burton K (2005) Extracting clinically relevant data from finite element simulations. Clin Biomech 20:451–454

73. Neugebauer R et al (2011) Experimental modal analysis on fresh-frozen human hemipelvic bones employing a 3D laser vibrometer for the purpose of modal parameter identification. J Biomech 44:1610–1613
74. Cristofolini L, Viceconti M, Cappello A, Toni A (1996) Mechanical validation of whole bone composite femur models. J Biomech 29(4):525–535
75. Completo A, Fonseca F, Simões J (2007) Experimental validation of intact and implanted distal femur finite element models. J Biomech 40:2467–2476
76. Castro A, Completo A, Simões J, Flores P (2015) Biomechanical behaviour of cancellous bone on patellofemoral arthroplasty with Journey prosthesis: a finite element study. Comp Meth Biomech Biomed Eng 18(10):1090–1098
77. Meireles S, Completo A, Simões J, Flores P (2010) Strain shielding in distal femur after patellofemoral arthroplasty under different activity conditions. J Biomech 43:477–484
78. Abdul-Kadir M et al (2008) Finite element modelling of primary hip stem stability: the effect of interference fit. J Biomech 41:587–594
79. Chong D, Hansen U, Andrew A (2010) Analysis of bone-prosthesis interface micromotion for cementless tibial prosthesis fixation and the influence of loading conditions. J Biomech 43:1074–1080
80. Viceconti M, Davinelli M, Taddei F, Cappello A (2004) Automatic generation of accurate subject-specific bone finite element models to be used in clinical studies. J Biomech 37:1597–1605
81. Weiss J et al (2005) Three-dimensional finite element modeling of ligaments: technical aspects. Med Eng Phys 27:845–861
82. Laville A, Laporte S, Skalli W (2009) Parametric and subject-specific finite element modelling of the lower cervical spine. Influence of geometrical parameters on the motion patterns. J Biomech 42:1409–1415
83. Rothstock S et al (2010) Primary stability of uncemented femoral resurfacing implants for varying interface parameters and material formulations during walking and stair climbing. J Biomech 43:521–526
84. Taylor M, Bryan R, Galloway F (2013) Accounting for patient variability in finite element analysis of the intact and implanted hip and knee: a review. Int J Numer Methods Biomed Eng 29(2):273–292
85. Pankaj P (2013) Patient-specific modelling of bone and bone-implant systems: the challenges. J Numer Methods Biomed Eng 29(2):233–249
86. Arregui-Mena J, Margetts L, Mummery P (2014) Practical application of the stochastic finite element method. Comp Meth Eng. https://doi.org/10.1007/s11831-014-9139-3
87. Amirouche F et al (2014) Factors influencing initial cup stability in total hip arthroplasty. Clin Biomech 29:1177–1185

Computational Modelling of Tissue-Engineered Cartilage Constructs

Cátia Bandeiras and António M. G. Completo

Abstract Cartilage is a fundamental tissue to ensure proper motion between bones and damping of mechanical loads. This tissue often suffers damage and has limited healing capacity due to its avascularity. In order to replace surgery and replacement of joints by metal implants, tissue-engineered cartilage is seen as an attractive alternative. These tissues are obtained by seeding chondrocytes or mesenchymal stem cells in scaffolds and are given certain stimuli to improve the establishment of mechanical properties similar to the native cartilage. However, tissues with ideal mechanical properties were not obtained yet. Growth and remodelling (G&R) computational models of tissue-engineered cartilage are invaluable to interpret and predict the effects of experimental designs. The current model contribution in the field will be presented in this chapter, with a focus on the response to mechanical stimulation, and the development of fully coupled modelling approaches incorporating simultaneously solute transport and uptake, cell growth, production of extracellular matrix and remodelling of mechanical properties.

1 Introduction

Articular cartilage is a fundamental tissue that resides in the surface of bones, providing a smooth and lubricated surface for relative bone motion in the joints and for transmission of loads with low friction [1, 2]. Articular cartilage is generally between 2 and 4 mm thick. Unlike other tissues, it does not have surrounding blood vessels or nerves [2]. The cartilage is populated by chondrocytes, specialized cells

C. Bandeiras · A. M. G. Completo (✉)
Department of Mechanical Engineering, University of Aveiro, Aveiro, Portugal
e-mail: completo@ua.pt

C. Bandeiras
e-mail: catia.bandeiras@tecnico.ulisboa.pt

C. Bandeiras
Institute of Bioengineering and Biotechnology, Instituto Superior Técnico, University of Lisbon, Lisbon, Portugal

© Springer Nature Switzerland AG 2020
J. Belinha et al. (eds.), *The Computational Mechanics of Bone Tissue*,
Lecture Notes in Computational Vision and Biomechanics 35,
https://doi.org/10.1007/978-3-030-37541-6_8

for the production of extracellular matrix (ECM). This matrix is mostly composed of collagen fibres, proteoglycans, water and other less present components, such as noncollagenous proteins and glycoproteins. The ECM components are fundamental for water retention in the tissue, promoting a softer load transfer and motion [1, 2]. Water is the most abundant component of articular cartilage, generally accounting for 65–80% of the total tissue weight. The collagen of the articular cartilage is mostly type II, and it is related to the tensile resistance of cartilage. Collagen accounts for 10–20% of the total cartilage mass. The proteoglycans are composed of a protein core with glycosaminoglycans (GAGs) attached, being 5–10% of the cartilage mass. The GAGs have a global negative charge that helps to control the hydration of cartilage and to provide resistance to compression and shear. When under a mechanical load, the proteoglycans promote a redistribution of water in cartilage, leading to an increase in osmotic pressure with water flow. The osmotic pressure becomes larger than the applied load, which is fundamental to protect bones from loading [1, 2]. A fundamental characteristic of articular cartilage is the depth-dependent organization, with three zones with distinct functions and collagen architectures: the superficial zone, responsible for protection against shear stress and with collagen fibres parallel to the surface of the tissue; the middle zone, with oblique collagen fibres and providing the first resistance to compression and the deep zone, with the highest resistance to compressive forces with collagen fibres oriented perpendicularly to the cartilage surface. The cartilage is anchored to the subchondral bone by the calcified layer [1].

The most common pathology associated with articular cartilage is osteoarthritis (OA), a degenerative disease that causes loss of the smooth surface of cartilage with pain, inflammation and loss of motion amplitude. The highest risk factor for OA is increasing age, while other factors such as obesity, genetics and gender are also associated [3]. The worldwide prevalence of OA was estimated to be 3.8% in 2010 [4], and the direct and indirect costs of the disease are very high. In the USA only, the annual medical care expenditures with OA are of about $185 billion [5]. While traditionally seen as a disease of the cartilage only, more recently, OA has been identified as a multi-organ pathology, causing subsequent damage in bone marrow and bone, tendons, ligaments, muscles and neural tissues [6].

Since cartilage is an avascular tissue, the intrinsic regeneration capacity of articular cartilage is very limited, leading to increasing severity of damage. The current therapeutic solutions are the total joint replacement by a metal implant, which is more common in older patients with very advanced damage. Other solutions for younger patients are the microfracture and autologous chondrocyte implantation to promote the formation of new cartilaginous tissue in the injury site. These solutions have moderate short-term success rates, while long-term results are not satisfactory. The failure of these therapies is related with the formation of tissue with inferior mechanical properties to the native tissues, with possible fibrocartilage formation [7].

Tissue-engineered (TE) cartilage has been proposed as a prospective new treatment for osteoarthritis by the in vitro production of cartilaginous tissue with more similar structure, composition and properties to the native articular cartilage. TE cartilage is obtained by seeding chondrocytes or mesenchymal stem cells (MSCs) with chondrogenic cues, on a porous and biocompatible scaffold that is able to provide a favourable environment to maintain the differentiated phenotype of chondrocytes and to enable the production of extracellular matrix (ECM). Although promising, the translation of this approach to products has been hindered by some factors, such as insufficient mechanical properties, mainly due to the inability of the engineered tissues to have a type II collagen content similar to the native cartilage, difficulty in creating an anisotropic tissue structure with three layers with collagen fibres oriented as found in the native tissue, and heterogeneous mechanical properties with stiffer peripheries and softer cores [8–11].

In order to better predict the experimental conditions to subject the growing tissue to, either by mechanical, electrical and/or chemical stimuli, computational models of tissue-engineered cartilage are invaluable. Mathematical modelling in the context of TE cartilage has provided good insights on the nutrient distribution in the growing tissues [12–15], cell proliferation and death [12–14], synthesis of the main components of ECM, such as proteoglycans and collagen [16–18] and remodelling of biphasic mechanical properties [12, 14, 17, 19]. Most of these models attempt to solve one or two variables responsible for the full remodelling of TE cartilage. Recently, a new approach that couples all these factors in order to simulate spatiotemporal patterns of metabolic activity, biomass growth and remodelling properties simultaneously was developed with results for both unloaded and mechanical stimulated constructs [20–22].

This chapter aims to review the body of work in the computational modelling of tissue-engineered cartilage with a focus on metabolic, biomass growth and mechanical remodelling. It is organized into several sections that emphasize different relevant aspects of the biomechanical behaviour of the growing cartilaginous tissues. Thus, Sect. 2 emphasizes the transport, uptake and production of relevant metabolites or growth factors that impact the biosynthetic activity of chondrocytes or mesenchymal stem cells, and how thee mechanisms are affected by external stimuli. A particular focus will be given to the main metabolites involved in chondrocyte metabolism: glucose, oxygen and lactate. Then, Sect. 3 is related to the different models proposed to modulate the proliferation, death and migration of chondrocytes and in the case of MSCs, the proliferation of these and their differentiation into chondrocytes and other possible lineages. Section 4 is concerned with to the models of synthesis of the main components of the extracellular matrix (ECM), glycosaminoglycans (GAGs) and collagen taking into account the impact of different stimuli on the production rates, binding and degradation of the matrix, as well as the alignment of collagen fibres to establish the anisotropy of the cartilaginous tissue. Afterwards, Sect. 5 describes the models of the mechanical behaviour of cartilage and the remodelling of the mechanical properties of tissue-engineered cartilage based on the produced biomass and ECM. Section 6 presents the models that couple all the aforementioned concepts into simultaneous metabolic, biosynthetic and mechanical remodelling models.

2 Models for Solute Transport, Uptake and Release

In order to obtain tissue-engineered cartilage with a sufficient amount of extracellular matrix, cells need to consume high amounts of nutrients to support their anabolic activity. However, there are serious limitations to nutrient transport across the tissues, which become hindered by the increase of matrix accumulation and decrease of the tissue porosity. This limitation is particularly seen in the cores of the tissues, where supply of nutrients is limited and accumulation of toxic byproducts leads to increased heterogeneities in the growing tissues. As a consequence, tissues with cores with lower cell viability and ECM content are formed, leading to inferior mechanical properties [11, 12, 23, 24]. It has been postulated that, apart from the diffusive transport present in free swelling and unstrained constructs, advective nutrient transport may be helpful to reduce heterogeneities in nutrient supply, with a higher positive contribution for the transport of large solutes over small solutes [23, 25, 26].

The simplest modelling approach for solute transport is based on the diffusion-reaction equation, where solute diffuses through a porous tissue with a diffusivity that is a fraction of the diffusivity in the fluid phase, and with a reactive term correspondent to the consumption or release of solutes depending on the amount of solute and the cell density in the tissue. A typical way to demonstrate the decrease of diffusivity across a porous tissue is given by the Mackie–Mears relationship. The most common representation of the reactive term is based on the Michaelis–Menten kinetics, as shown below [12, 25, 27–30].

$$\frac{\partial c}{\partial t} - D_{\text{tissue}} \frac{\partial^2 c}{\partial x^2} = R \tag{1}$$

$$R = (+/-)\rho_{\text{cell}} \frac{V_{\text{max}} c}{K_{\text{m}} + c} \tag{2}$$

$$D_{\text{tissue}} = D_{\text{water}} \frac{n_{\text{f}}^2}{(2 - n_{\text{f}})^2} \tag{3}$$

When dynamic loading is involved, an advective term is included to represent the fluid flow mediated transport [27, 30–33].

$$\frac{\partial c}{\partial t} - D_{\text{tissue}} \frac{\partial^2 c}{\partial x^2} + v_{\text{f}} \frac{\partial c}{\partial x} = R \tag{4}$$

In the equations above, c represents the concentration of the nutrient, D_{tissue} represents the diffusion coefficient of the nutrient in the tissue, R the flux of the metabolite, ρ_{cell} the cell density, V_{max} the maximum uptake rate, and K_{m} is the half-maximum-rate concentration and n_{f} the fluid volume fraction. The solutes consumed by the cells that are typically simulated in previous works are glucose and oxygen, and several studies used this simplified assumption for consumption with good results [12, 22, 27, 29, 32]. However, particularly in cases where the culture medium has a high content in glucose, the deleterious effect of lactate production in cell proliferation cannot be ignored, and the release of lactate to the culture media is also modelled, considering both anaerobic and aerobic degradation of glucose depending on the experimental lactate to glucose ratios [12, 34]. Under dynamic loading conditions, deformation affects solute transport in several ways. In the first place, the diffusion coefficient depends on the porosity of the scaffold used for cell support. As stated before, a common law used in previous modelling contributions is the Mackie–Mears diffusion law. Another effect of dynamic loading in the cellular metabolism is related to the variation of cell density. Assuming, as a simplification, that the number of cells in a given volume is constant, under loading, the cell density is affected due to the change of volume of the constructs in a compressible scaffold. This volume change is described by the determinant of the deformation gradient tensor (J). This value describes the ratio between the volume of the deformed configuration and the undeformed configuration. Therefore, the deformed cell density is obtained as such [35, 36].

$$\rho_{\text{cell}} = \frac{\rho_{\text{cell},0}}{J} \tag{5}$$

3 Models for Cellular Dynamics

The cell population in tissue-engineered cartilage is highly dynamic and dependent on several metabolic and physical cues. The main mechanisms associated with cells populating the newly formed tissue are [37]:

- Proliferation—A fundamental factor to obtain ECM in proper amounts. However, the higher the cell population, the more likely the nutrient depletion and inhomogeneity in cell distributions across the tissue.
- Differentiation—When the tissue is seeded with mesenchymal stem cells (MSCs), an important factor to control besides their proliferation is the differentiation into chondrocytes. Since MSCs can also give rise to adipocytes and osteocytes, a precise control of the biochemical and biomechanical cues to favour differentiation into a given precise lineage is fundamental.
- Migration—spatial redistribution of cells in the scaffold can both occur due to random walks without a preferential direction or can happen directionally towards chemoattractants. In high-density scaffolds, cells can form colonies.
- Death—Apart from the regular lifespan of chondrocytes, lack of nutrients or aggressive external physical cues will speed up the process of death.

The typical model for cell dynamics in a tissue-engineered cartilage takes into account these factors as follows:

$$\frac{\partial \rho_{\text{cell}}}{\partial t} - D_{\text{cell}} \frac{\partial^2 \rho_{\text{cell}}}{\partial x^2} = \left(R_{\text{prol}} - R_{\text{death}} \right) \rho_{\text{cell}} \tag{6}$$

A diffusion coefficient, D_{cell}, is introduced due to the assumption that new chondrocytes have mobility due to random walks [38]. While the death rate per cell, R_{death}, is assumed constant, the proliferation rate, R_{prol}, is modulated both by metabolic and mechanical factors that decrease the actual proliferation from the maximum proliferation rate, μ_{max}. The simplest model for nutrient-limited cell proliferation (accounted for in R_{prol}) in tissue-engineered cartilage is given by the Monod kinetics. In this model, similar to the Michaelis–Menten kinetics for nutrient dynamics, growth is limited by the availability of a nutrient, for which a half-rate concentration controls the steepness until maximum growth.

$$R_{prol} = \frac{c_n}{K_n + c_n} \mu_{max} \tag{7}$$

Several models based on the Monod kinetics have good results in comparison with experimental data, either by using one solute only, such as glucose [39, 40] or oxygen [14, 41–44], or a combination of solutes, like models with glucose and collagen [45], or models inhibited by pH decrease simplified as accumulation of lactate [34].

Another model that is commonly used in the literature to describe the limitation of chondrocyte growth by substrate is the Contois kinetics. This representation differs from the Monod kinetics because the growth in the Contois kinetics is also inhibited by the cell density, implying saturation of growth due to spatial competition of cells for resources, as shown below:

$$R_{prol} = \frac{c_n}{\rho_{cell} K_n + c_n} \mu_{max} \tag{8}$$

The Contois kinetics has also provided good agreement to the growth of chondrocytes in different scaffolds, using glucose [13, 22, 38, 39], oxygen [46, 47], a combination of glucose and lactate [48] and a combination of glucose and lactate accounting for pH negative effects [49].

In terms of the metabolic modulation of cell growth, other less used equations are reported, such as the Heaviside step function [50, 51], the Moser or heterogeneous n-th order model [39] and logistic function [32].

In order to incorporate the impact of mechanical stimulation in cell growth, some models have built up from the aforementioned mechanical factors and introduced the impact of shear stress in cell growth. It was shown experimentally that articular chondrocytes show a dose and time-dependent response to shear stress [52, 53]. Two main modelling contributions have been proposed to incorporate this effect in mathematical models thus far. In the first place, a simple linear model of a linearly increase in the growth rate with increasing shear stress was proposed [14, 34]. An extension of this equation was proposed as a polynomial dependence with a non-integer factor by [47]. A more recent contribution resides in a piecewise function with shear stress, accounting for a maximum stimulatory range of shear stresses between 0.1 and 0.6 Pa as determined experimentally and assuming suppression of growth for stresses above 1 Pa [15].

All the models presented thus far consider growth on a homogeneous cell population with the same characteristics. Other models have focused on particular compartments of the cell population for modelling, with a cell in a given state having a different role in tissue homeostasis. A model reporting a proliferative, an extracellular matrix producing and a quiescent cell fraction was proposed with interchangeability between these compartments [54], having been recently expanded to include a transitional state between proliferative and ECM producing states and the possibility of quiescence and apoptosis [44]. Another compartmental modelling approach is related to the influence of the phase of the cell cycle during mitosis on the maturity of the cell and the possibility of undergoing protein synthesis [54]. While these contributions are valuable and in closer agreement with the inherent biology of the chondrocytes, the previously reported general chondrocyte growth models have shown good agreement with experiments and are, in most cases, a reasonable modelling approach.

4   Models for ECM Growth

The extracellular matrix of articular cartilage is a collagen fibre network, mainly composed by type II collagen, and of glycosaminoglycans (GAGs) that provide mechanical support to chondrocytes and resistance to the mechanical stimuli that cartilage is subjected to An and Martin [1, 16, 17]. In tissue-engineered cartilage, it is highly required to stimulate the new tissues to produce a network similar to the native cartilage, with a content of 5–10% of the total mass in GAGs and 10–20% of the total mass in collagen [1]. While there are reported studies able to produce GAGs in a concentration similar to the native cartilage, the collagen content is much lower than the native values, being this one of the most significant hurdles to surpass to obtain viable tissue for implantation [9, 55].

The main mechanisms behind ECM dynamics are

- Synthesis—Synthesis models are focused on the total cell population as a whole, assuming that all cells have the ability to produce ECM or on an ECM producing cell compartment, depending on the model for cell growth used. Synthesis rates may depend on the availability of a given substrate or on the mechanical stimuli that cells face.
- Binding—The newly synthesized ECM composing molecules are released into the media and then linked to the ECM.
- Degradation—Bound ECM molecules are degraded and diffused into the culture media at a given rate due to several forms of damage.

The first models appearing in the literature on this matter started to consider the ECM as a whole and did not provide distinction between collagen and proteoglycans. These models considered a linear synthesis rate modulated by the difference between current ECM concentrations and the steady-state concentration of ECM, due to the experimental observation that synthesis rates decay with the accumulation of ECM [16]. This model formulation was later adapted to include the impact of cell density in the growth rates and a separation by type of ECM component, the differentiation between bound and unbound ECM, and the rates of degradation [17, 18, 22, 56–59]. The typical formulation for these three ECM mechanisms is depicted in the equations below:

$$\frac{\partial \text{ECM}_{ub}}{\partial t} - D_{\text{ECM},ub}\frac{\partial^2 \text{ECM}_{ub}}{\partial x^2} = k_S\left(\text{ECM}_{b,ss} - \text{ECM}_b\right) - k_b\text{ECM}_{ub} \qquad (9)$$

$$\frac{\partial \text{ECM}_b}{\partial t} = k_b\,\text{ECM}_{ub} - k_d\,\text{ECM}_b \qquad (10)$$

$$\frac{\partial \text{ECM}_d}{\partial t} - D_{\text{ECM},d}\frac{\partial^2 \text{ECM}_d}{\partial x^2} = k_d\,\text{ECM}_d \qquad (11)$$

In the equations above, the unbound (ECM_{ub}), bound (ECM_b) and the degradation products (ECM_d) are controlled by the respective diffusion coefficients and by the synthesis (k_s), binding (k_b) and degradation rates (k_d). It is assumed that, in tissue-engineered cartilage, ECM growth will saturate at a given concentration, hence the dependence of the synthesis rate on the steady-state concentration $\text{ECM}_{b,ss}$. Apart from the commonly used linear dependence of the synthesis rate on the concentration of the specific ECM entity to model, other dependences were reported, such as a logistic dependence on the unbound GAG concentration [44, 60, 61] or the impact of levels of relevant solutes, like glucose [20, 62] and oxygen [20, 44, 60, 61], on the synthesis rates of GAG. For collagen, a dependence on the cell proliferation time derivative, instead of the typical linear dependence on cell density, was also reported [11].

The impact of mechanical stimuli on the synthesis rates of extracellular matrix has been less explored so far, but some contributions are provided in the literature. Fluid velocity levels were considered in a previous level to directly affect the GAG synthesis rate, based on experimental observations that fluid velocity has a stimulatory effect on GAG synthesis, while for collagen, an attempt in the same work was performed considering augmented synthesis when the maximum principal strain is above a given threshold value [63]. The dependence on fluid velocity for GAG synthesis and on maximum principal strain for collagen synthesis was implemented in previous research works [20, 21] although with a different formulation. Fluid velocity and shear stress were also considered as stimulatory for cartilage growth above a given threshold for both proteoglycans and collagen [64]. More recently, a modular function assuming that there is an optimal cell volume for synthesis of GAG by chondrocytes was implemented as a function of tissue deformation [65].

Another factor that is relevant for the establishment of extracellular matrix and of the anisotropic properties is the remodelling and reorientation of the collagen fibres. It is known that collagen fibres align in preferred directions between the maximum principal strain directions [10, 59, 66]. Therefore, the application of mechanical stimuli can be used to drive the desired orientation of tissue-engineered cartilage. The first work that applied the remodelling theory of collagen fibres in cartilage assumed that collagen fibres rotated with an angular velocity controlled by the angle between the collagen fibre directions in the undeformed configuration and the preferred fibril directions, taking into account the magnitude of the three possible spatial principal strains to establish these directions [66]. Another approach to computational modelling of fibre reorientation was provided through an anisotropic tensor describing the degree of structural anisotropy and an ellipsoid representation for the fibre material parameters. Here, the reorientation is described through the angle between the current anisotropic tensor and the Cauchy–Green strain tensor and mediated through a time constant [18]. Furthermore, a probability density approach for the distribution of collagen fibres that can change over time with respect to the mechanical stimuli was also proposed [67]. These concepts are inherently coupled with the constitutive relationship used to describe the mechanical behaviour of growing cartilage.

5 Models for Description of the Mechanical Behaviour and Remodelling of Mechanical Properties

Cartilaginous tissue obtained through support of a porous scaffold is a material with a very high water content, like the native articular cartilage. For this reason, a simple monophasic constitutive material is not sufficient to explain the viscoelastic behaviour of cartilage, caused by fluid flow-dependent and independent mechanisms inherent to the properties of the solid material [68]. For this reason, mixture models based on the biphasic theory were proposed. These models describe the total stress in the tissue with a solid stress σ^S and a hydraulic pore pressure component p [20–22, 68–70].

$$\sigma = \sigma^S - p \tag{12}$$

Fluid flow is governed by the Darcy's law, which states that fluid velocity, v^f (m s^{-1}), relative to the solid matrix, v^s, is proportional to the gradient of the pore pressure Vp (Pa) and controlled by the permeability, k (m^2), of the porous scaffold, accounting as well for the porosity, n, of the material [56].

$$n\left(v^f - v^S\right) = -k\,V\mathrm{p} \tag{13}$$

Due to the very high fluid content of these native cartilage, both the solid and fluid phases are generally described as incompressible or nearly incompressible for simplicity (that is, with a Poisson's ratio close to 0.5). However, in the tissue-engineered cartilage, some polymers are described as compressive solids in equilibrium with an incompressible fluid [8, 10, 22, 71]. These models can partly describe the fluid flow-related viscoelasticity related to the low permeability of the material.

As an extension to the biphasic model for cartilage behaviour, triphasic models were developed to take into account the swelling behaviour due to gradients in osmotic pressure. This model accounts for a fluid with ionic particles, inducing or limiting chemical expansion of the negatively charged proteoglycan chains due to electrostatic repulsion. The osmotic pressure gradient ($\Delta\pi$) and the chemical potential of the fluid (μf) that drive the ionic phase stress contribution are caused by differences in ion concentrations of the cartilage and the surrounding fluid [68].

$$\sigma = \sigma^s - (\Delta\pi + \mu f)I \tag{14}$$

For the solid phase of these models, several constitutive relationships were proposed to describe the mechanical behaviour. The simples theory to describe the behaviour of porous and viscoelastic materials is the poroelastic theory, where the solid phase is linear elastic and the fluid is viscous. In this model, the stress–strain relationship of the solid phase is provided by the Hooke's Law:

$$\sigma^s = H_a\varepsilon \tag{15}$$

In the equation above, H_a is the aggregate modulus, which is a measure of the stiffness of the material in equilibrium when fluid flow through the material ceases. This quantity is related to the Young's modulus (E) and the Poisson's ratio (υ) through the following relationship:

$$H_a = \frac{E(1-\upsilon)}{(1+\upsilon)(1-2\upsilon)} \tag{16}$$

For mechanical modelling of scaffolds impregnated with chondrocytes, the poroelastic theory has been widely used [22, 23, 71, 72]. Another reported theory for modelling of the solid phase is the porohyperelastic theory. Similarly, to the poroelastic theory, the fluid is viscous but the solid phase has an hyperelastic constitutive relationship, such as the stress–strain behaviour is modelled by a nonlinear relationship dependent on the strain energy (W) and on the deformation gradient tensor (F) and its determinant, J.

$$\sigma^S = \frac{1}{J}\frac{\partial W}{\partial F}F^T \tag{17}$$

Models with a solid phase described by the neo–Hookean hyperelastic relationship have been widely used to modulate the mechanical behaviour of articular cartilage [25, 70, 73, 74]. Modelling of the hydrogel solid phase has used either the neo–Hookean law or the Odgen law [75, 76].

The mentioned hyperelastic models until now are isotropic with the same mechanical properties in all dimensions. However, articular cartilage is an anisotropic material, with stress–strain behaviour dependent on the orientation of the collagen fibres [68]. Therefore, it is more appropriate in long-term studies of cartilage growth to model the growing tissue with an anisotropic model with augmented tensile response in the loading directions equal to the fibre directions. A proposed model for this is the Holzapfel–Gasser–Odgen model [69], which divided the ECM into a non-fibrilar component, explained by the neo–Hookean model, and a fibrilar component with strain energy dependent on material parameters and the degree of anisotropy of the tissue. A similar relationship was reported, with the difference of taking into account a continuous exponential angular fibre distribution [18, 67].

Another theory reported to describe the mechanical behaviour of hydrogels is the poroviscoelastic model, with a non-viscous fluid and a viscoelastic solid phase [77, 78]. This model is not used, to our knowledge, to model hydrogels with growing cartilage.

A highly relevant parameter for the description of the biphasic behaviour of cartilage is the hydraulic permeability. This parameter is related to changes in the porosity and void ratio of the material. Several exponential relationships between the permeability and porosity or void ratio were presented, with two of the most common ones being the Holmes and Mow law [33, 79, 80] and the Carman–Kozeny law [14, 22] for isotropic permeability remodelling. However, with the growth of collagen fibres, the permeability also becomes anisotropic, with different values according to the orientation parallel or perpendicular to the fibres. Studies for modelling of articular cartilage explants have already included this dependency [67, 69].

While the newly formed tissue is growing and ECM is deposited, the mechanical properties of the tissue are changing. The target average values of mechanical properties of tissue-engineered cartilage are between 450 and 800 kPa of compressive Young's modulus and 10^{-16} to 10^{-15} m^4. N^{-1} s^{-1} in hydraulic permeability [81]. A tissue that combines these two ranges of mechanical parameters has not yet been established. In long-term tissue-engineered cartilage modelling, establishing remodelling algorithms to simulate and account for the change of the mechanical properties is fundamental to determine with accuracy the intrinsic mechanical response of the tissues to external stimuli.

Few studies have reported relationships for the modelling of the solid matrix properties under linear elastic assumptions. The Young's modulus remodelling was previously described by a linear model for the aggregate modulus with the concentration of GAGs and collagen derived from experimental data on bovine cartilage in different ages [19]. An extension of this relationship was reported as well with a fourth-order polynomial dependence on the collagen concentration [22]. If the Poisson's ratio is assumed constant, the Young's modulus can be derived directly from such relationships. However, a possible remodelling relationship for the Poisson's ratio related to the porosity of the material was adapted from [82] since a compressible material, with the growth of ECM, tends to approach incompressibility.

In anisotropic models, the remodelling of the non-fibrilar part is controlled by the concentration of GAGs and the remodelling of fibrilar part controlled by the concentration of collagen. One reported relationship relates the rate of remodelling with the ratio of the current concentrations to the expected steady-state concentrations [18]. Regarding the remodelling of permeability, the lower availability of experimental permeability measurements compared to the modulus measurements hinders the fitting to mechanistic models. However, as a proper simplification, previously reported models relate the decrease of permeability with the increase of the volumetric fraction of cells and ECM throughout the construct, leading to a decrease in porosity [11].

6 Coupled Metabolic and Mechanical Remodelling Models

Currently, most of the modelling contributions for tissue-engineered cartilage are focused in up to three of the modelling dimensions presented. While all these contributions are very valuable, for a complete description of the behaviour of the tissue and the time and spatial evolution of mechanical properties, all four dimensions need to be included. The creation of a validated model with explanatory and predictive power with the dimensions of solute transport and consumption, cell dynamics, extracellular matrix growth and remodelling of mechanical properties will allow to explain in a more quantitative way the histologically observed differences in the distribution of the modelled quantities across the tissues, as well as being invaluable to recommend changes to the processes of tissue culture in order to obtain better results. Finally, the complete coupled model can be used to predict the impact of envisioned changes to the culture protocol, such as the type and geometry of the scaffold material, dynamic loading, culture exchange, seeding densities, among others. The general full modelling scheme flow is represented in Fig. 1.

For the simulation of free swelling constructs with different scaffold geometries, the model was applied to simulate short-term effects in the Young's modulus and hydraulic permeability of constructs with cylindrical and cubic geometries, either solid or with a central channel [83]. Despite the short culture period of 72 h, it was possible to determine that the channelled constructs had a large increase of nutrient availability related to the solid counterparts, with an up to 136-fold increase in minimum glucose concentrations and up to 220-fold increase in minimum oxygen concentrations. Under the model assumptions, ECM matrix synthesis increased up to 50% in the constructs with channel, favouring already a small positive impact on the mechanical properties after 72 h as a result of improved homogeneity across the tissues (Fig. 2).

This work was expanded to a long-term culture case by simulating an experiment with both a solid and a channelled 2% w/w agarose construct with chondrocytes during 56 days. The model was calibrated with solid construct data on GAG and collagen concentrations, as well as on the compressive Young's modulus and validated by reproducing well the experimental data for the channelled construct [22]. This modelling effort allowed to gain quantitative insights on the spatial heterogeneity of the constructs, showing that the degree of spatial heterogeneity of the Young's modulus the constructs with a central diffusion channel is 23% of the control value, while for permeability, the heterogeneity is 27% of the control one, showing a significant improvement for the channelled condition (Fig. 3). The degree of spatial heterogeneity and the insufficient permeability remodelling in the simulated solid constructs affects significantly the mechanical response to compressive strain, with nominal

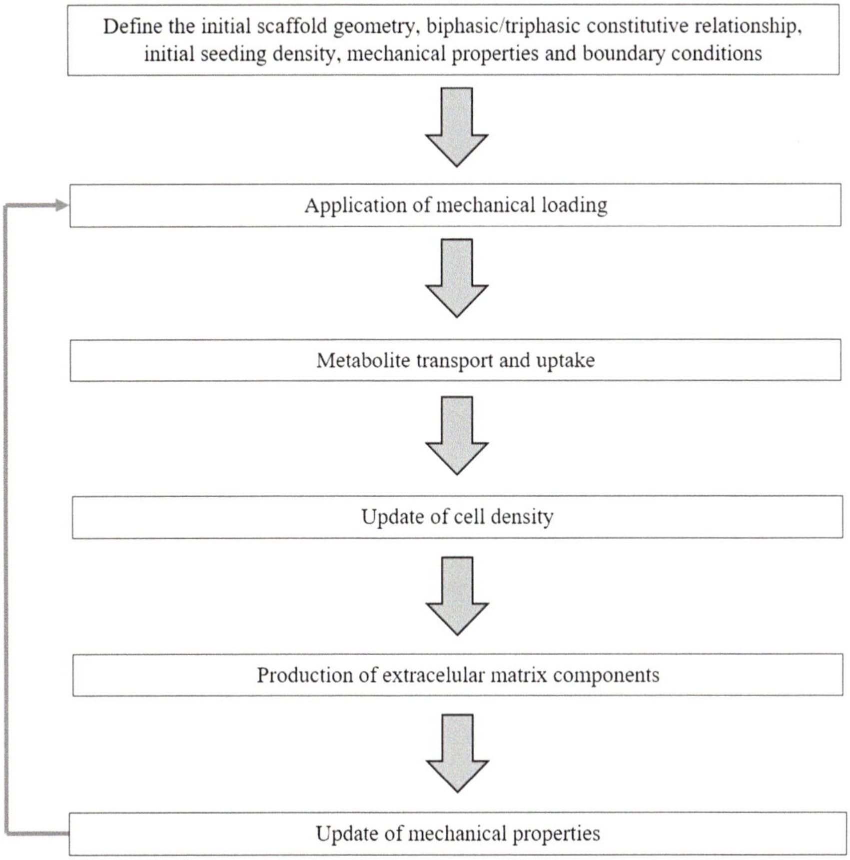
Define the initial scaffold geometry, biphasic/triphasic constitutive relationship, initial seeding density, mechanical properties and boundary conditions
Application of mechanical loading
Metabolite transport and uptake
Update of cell density
Production of extracelular matrix components
Update of mechanical properties

Fig. 1 Modelling scheme workflow for fully coupled tissue-engineered cartilage growth and remodelling. Scheme employed by Bandeiras et al. [20–22]

Young's modulus (kPa)
46.60
45.28
43.96
42.64
41.32
40.00
Permeability
($10^{12}m^4.N^{-1}.s^{-1}$)
4.90
4.85
4.80
4.75
4.70
4.65

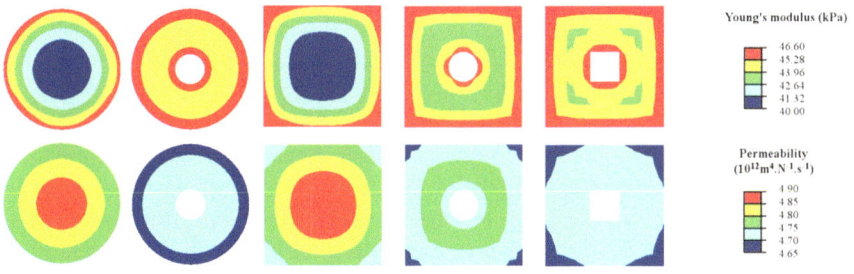

Fig. 2 Impact of several construct geometrical configurations in free swelling culture conditions on the radial distribution of biphasic mechanical properties after 72 h in culture. Reprinted from [83]

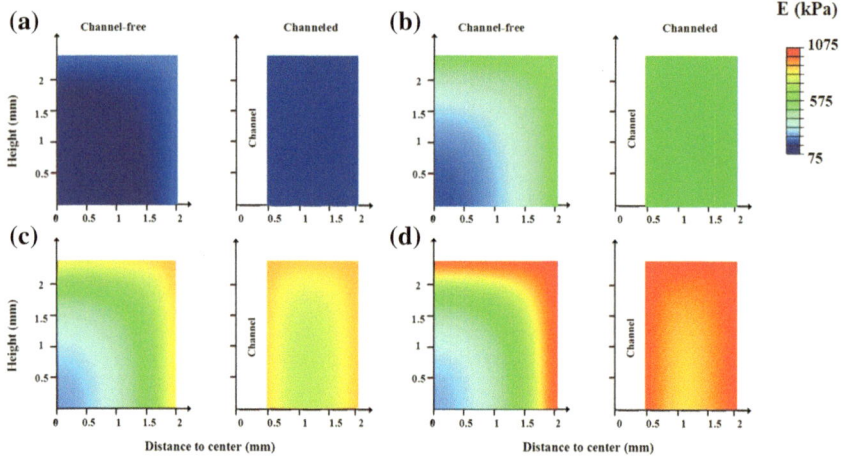

Fig. 3 Spatial distributions of Young's modulus at **a** 14 days, **b** 28 days, **c** 42 days, **d** 56 days. Reprinted Bandeiras and Ramos [22]

stresses for the simulated heterogeneous TE cartilage 57% lower than for native articular cartilage and pore pressures 53% lower than the native case. Therefore, permeability is the main parameter to be improved to get a more similar mechanical response, calling for new scaffold material designs and stimulation protocols.

The developed complete model was also applied to simulate dynamic loading conditions. A model parameterized with the literature parameters was used to simulate the distribution of cell density and ECM in cubic constructs subjected to either compression, shear or bending at 5% of height, 1 Hz for 6 h continuously [20]. While the simulation time is very short for relevant differences in the mechanical properties to be seen, bending was, under these conditions, the more favourable regime for cell proliferation and the spatial distributions are relevant with the establishment of cartilage with different structural organizations due to different maximum principal strain directions (Fig. 4). Current work is related to the model validation for the estimation of cell proliferation, ECM growth and mechanical properties remodelling under several different regimes of cyclic unconfined compressive loading.

All the reported works until now have been focused on articular cartilage. However, the complete modelling scheme was also successfully applied to estimate the remodelling of temporomandibular joint disc, an area where tissue engineering is still in a very early phase. The application of static hydrostatic pressure for 72 h on PEDGA-condylar constructs promoted a very slight improvement of the mechanical properties. This preliminary study provided a future basis to estimate the impact of long-term static or dynamic hydrostatic pressure on the growth of new temporomandibular joint discs [21].

The presented model applications for simulation of growth and remodelling under dynamic loading represent short-term loading with limited differences in the mechanical properties between conditions. Future work involves the simulation of long-term

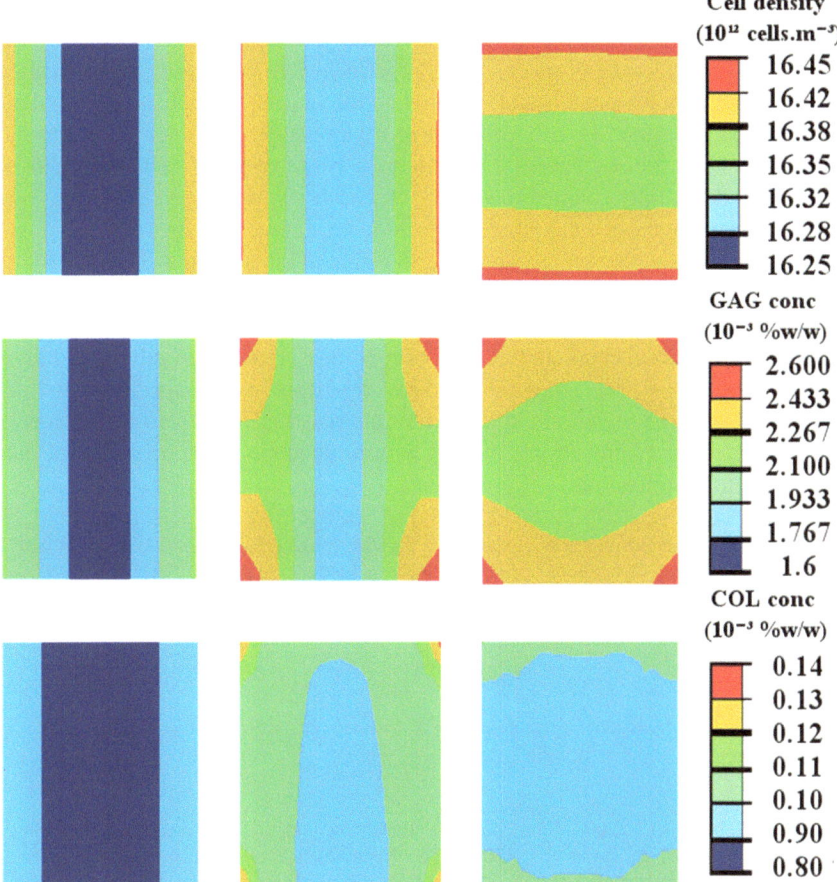

Fig. 4 Values of the cell density, GAG and COL outputs for the compression (left), shear (centre) and bending (right) stimulus. Values obtained for the $x = 0$ plane. Reprinted from Bandeiras et al. [20]

intermitted compressive loading. On long-term regimes, the collagen content is relevant in terms of fibre organization, and therefore, the expansion of the constitutive relationships to include the anisotropic behaviour of tissue-engineered cartilage and its time dynamics is a future goal.

7   Concluding Remarks

This chapter provided an overview of the current state of the computational models used for simulation of growth of tissue-engineered cartilage, namely the models based in mixture theory. The several individual contributions for the underlying phenomena, such as solute transport and uptake, cell growth, production of extracellular matrix and remodelling of mechanical properties, were presented, with a focus on, when applicable, the models that incorporate the impact of mechanical stimulation on these phenomena. Several constitutive relationships to model the mechanical behaviour of tissue-engineered cartilage have been proposed and integrated into these models. A fully coupled modelling approach was developed to accommodate all these phenomena in a simultaneous fashion for a more realistic representation of the biomechanical phenomena and used to estimate the growth and remodelling of mechanical properties under free swelling and mechanical loading of tissues. Future challenges on this area include refinement of the developed equations through validation with experimental studies with proper measurements of all the underlying variables when possible, study the degree of robustness and/or specificity with different cell-matrix systems, simulation of long-term mechanical loading cultures and the incorporation of anisotropic models with reorientation of collagen fibres in the fully coupled formulation.

Acknowledgements This work is supported by the project POCI-01-0145-FEDER-028424, funded by Programa Operacional Competitividade e Internacionalização (COMPETE 2020) on its component FEDER and by funding from FCT—Fundação para a Ciência e Tecnologia on its component OE.

References

1. An YH, Martin KL (2003) Handbook of histology methods for bone and cartilage. Springer
2. Fox AJS, Bedi A, Rodeo SA (2009) The basic science of articular cartilage: structure, composition, and function. Sports Health: A Multidisciplinary Approach 1(6):461–468
3. Loeser RF, Collins JA, Diekman BO (2016) Ageing and the pathogenesis of osteoarthritis. Nat Rev Rheumatol
4. Cross M, Smith E, Hoy D, Nolte S, Ackerman I, Fransen M, Bridgett L, Williams S, Guillemin F, Hill CL et al (2014) The global burden of hip and knee osteoarthritis: estimates from the global burden of disease 2010 study. Annals of the rheumatic diseases pp.annrheumdis–2013
5. Hiligsmann M, Reginster J (2013) The economic weight of osteoarthritis in Europe. Medicographia 35:197–202
6. Martin JA, Buckwalter JA (2002) Aging, articular cartilage chondrocyte senescence and osteoarthritis. Biogerontology 3(5):257–264
7. Kock L, van Donkelaar CC, Ito K (2012) Tissue engineering of functional articular cartilage: the current status. Cell Tissue Res 347(3):613–627
8. Mesallati T, Buckley CT, Nagel T, Kelly DJ (2013) Scaffold architecture determines chondrocyte response to externally applied dynamic compression. Biomech Model Mechanobiol 12(5):889–899
9. Bian L, Angione S, Ng K, Lima E, Williams D, Mao D, Ateshian G, Hung C (2009) Influence of decreasing nutrient path length on the development of engineered cartilage. Osteoarthritis Cartilage 17(5):677–685
10. Khoshgoftar M, Wilson W, Ito K, van Donkelaar CC (2013) The effect of tissue-engineered cartilage biomechanical and biochemical properties on its post-implantation mechanical behavior. Biomech Model Mechanobiol 12(1):43–54

11. Bandeiras C, Completo A (2015) Comparison of mechanical parameters between tissue-engineered and native cartilage: a numerical study. Comput Methods Biomech Biomed Eng 18(S1):1876–1877
12. Sengers BG, Heywood HK, Lee DA, Oomens CW, Bader DL (2005) Nutrient utilization by bovine articular chondrocytes: a combined experimental and theoretical approach. J Biomech Eng 127(5):758–766
13. Chung C, Chen C, Chen C, Tseng C (2007) Enhancement of cell growth in tissue-engineering constructs under direct perfusion: Modeling and simulation. Biotechnol Bioeng 97(6):1603–1616
14. Sacco R, Causin P, Zunino P, Raimondi MT (2011) A multiphysics/multiscale 2d numerical simulation of scaffold-based cartilage regeneration under interstitial perfusion in a bioreactor. Biomech Model Mechanobiol 10(4):577–589
15. Nava MM, Raimondi MT, Pietrabissa R (2013) A multiphysics 3d model of tissue growth under interstitial perfusion in a tissue-engineering bioreactor. Biomech Model Mechanobiol 12(6):1169–1179
16. Wilson CG, Bonassar LJ, Kohles SS (2002) Modeling the dynamic composition of engineered cartilage. Arch Biochem Biophys 408(2):246–254
17. Haider MA, Olander JE, Arnold RF, Marous DR, McLamb AJ, Thompson KC, Woodruff WR, Haugh JM (2011) A phenomenological mixture model for biosynthesis and linking of cartilage extracellular matrix in scaffolds seeded with chondrocytes. Biomech Model Mechanobiol 10(6):915–924
18. Nagel T, Kelly DJ (2011) Mechanically induced structural changes during dynamic compression of engineered cartilaginous constructs can potentially explain increases in bulk mechanical properties. J R Soc Interf 9(69):777–789
19. Bandeiras C, Completo A (2017) A mathematical model of tissue-engineered cartilage development under cyclic compressive loading. Biomech Model Mechanobiol 16(2):651–666
20. Bandeiras C, Completo A, Ramos A (2014) Compression, shear and bending on tissue-engineered cartilage: a numerical study. Comput Methods Biomech Biomed Eng 17(S1):2–3
21. Bandeiras C, Completo A, Ramos A (2014) Simulation of remodeling of tissue engineered condylar cartilage under static hydrostatic pressure. Biodental Engineering III p 83
22. Bandeiras C, Completo A, Ramos A (2015) Influence of the scaffold geometry on the spatial and temporal evolution of the mechanical properties of tissue-engineered cartilage: insights from a mathematical model. Biomech Model Mechanobiol 14(5):1057–1070
23. Mauck RL, Hung CT, Ateshian GA (2003) Modeling of neutral solute transport in a dynamically loaded porous permeable gel: implications for articular cartilage biosynthesis and tissue engineering. J Biomech Eng 125(5):602–614
24. Aristotelous AC, Haider MA (2014) Use of hybrid discrete cellular models for identification of macroscopic nutrient loss in reaction–diffusion models of tissues. Int J Numer Method Biomed Eng 30(8):767–780
25. Zhang L, Szeri A (2005) Transport of neutral solute in articular cartilage: effects of loading and particle size. In: Proceedings of the royal society of London A: mathematical, physical and engineering sciences, vol 461. The Royal Society, pp 2021–2042
26. Chahine NO, Albro MB, Lima EG, Wei VI, Dubois CR, Hung CT, Ateshian GA (2009) Effect of dynamic loading on the transport of solutes into agarose hydrogels. Biophys J 97(4):968–975
27. Zhou S, Cui Z, Urban JP (2004) Factors influencing the oxygen concentration gradient from the synovial surface of articular cartilage to the cartilage–bone interface: a modeling study. Arthritis Rheum 50(12):3915–3924
28. Zhou S, Cui Z, Urban JP (2008) Nutrient gradients in engineered cartilage: metabolic kinetics measurement and mass transfer modeling. Biotechnol Bioeng 101(2):408–421
29. Devarapalli M, Lawrence BJ, Madihally SV (2009) Modeling nutrient consumptions in large flow-through bioreactors for tissue engineering. Biotechnol Bioeng 103(5):1003–1015
30. Unnikrishnan G, Unnikrishnan V, Reddy J (2012) Finite element model for nutrient distribution analysis of a hollow fiber membrane bioreactor. Int J Numer Method Biomed Eng 28(2):229–238

31. Evans RC, Quinn TM (2006) Dynamic compression augments interstitial transport of a glucose-like solute in articular cartilage. Biophys J 91(4):1541–1547

32. Shakeel M (2013) 2-d coupled computational model of biological cell proliferation and nutrient delivery in a perfusion bioreactor. Math Biosci 242(1):86–94

33. Urciuolo F, Imparato G, Netti P (2008) Effect of dynamic loading on solute transport in soft gels implication for drug delivery. AIChE J 54(3):824–834

34. Hossain MS, Bergstrom D, Chen X (2015) Modelling and simulation of the chondrocyte cell growth, glucose consumption and lactate production within a porous tissue scaffold inside a perfusion bioreactor. Biotechnol Rep 5:55–62

35. Kam KK (2011) Poroelastic finite element analysis of a heterogeneous articular cartilage explant under dynamic compression in abaqus

36. Malandrino A, Noailly J, Lacroix D (2011) The effect of sustained compression on oxygen metabolic transport in the intervertebral disc decreases with degenerative changes. PLoS Comput Biol 7(8):e1002112

37. Sengers BG, Taylor M, Please CP, Oreffo RO (2007) Computational modelling of cell spreading and tissue regeneration in porous scaffolds. Biomaterials 28(10):1926–1940

38. Chung C, Yang C, Chen C (2006) Analysis of cell growth and diffusion in a scaffold for cartilage tissue engineering. Biotechnol Bioeng 94(6):1138–1146

39. Galban CJ, Locke BR (1999) Analysis of cell growth kinetics and substrate diffusion in a polymer scaffold. Biotechnol Bioeng 65(2):121–132

40. Cheng G, Markenscoff P, Zygourakis K (2009) A 3d hybrid model for tissue growth: the interplay between cell population and mass transport dynamics. Biophys J 97(2):401–414

41. Croll TI, Gentz S, Mueller K, Davidson M, O'Connor AJ, Stevens GW, Cooper-White JJ (2005) Modelling oxygen diffusion and cell growth in a porous, vascularising scaffold for soft tissue engineering applications. Chem Eng Sci 60(17):4924–4934

42. Malda J, Brink PVD, Meeuwse P, Grojec M, Martens D, Tramper J, Riesle J, Blitterswijk CV (2004) Effect of oxygen tension on adult articular chondrocytes in microcarrier bioreactor culture. Tissue Eng 10(7–8):987–994

43. Flaibani M, Magrofuoco E, Elvassore N (2009) Computational modeling of cell growth heterogeneity in a perfused 3D scaffold. Ind Eng Chem Res 49(2):859–869

44. Lelli C, Sacco R, Causin P, Raimondi MT (2015) A poroelastic mixture model of mechanobiological processes in tissue engineering. part i: Mathematical formulation. arXiv preprint arXiv: 1512.02182

45. Chung C, Ho SY (2010) Analysis of collagen and glucose modulated cell growth within tissue engineered scaffolds. Ann Biomed Eng 38(4):1655–1663

46. Coletti F, Macchietto S, Elvassore N (2006) Mathematical modeling of three-dimensional cell cultures in perfusion bioreactors. Ind Eng Chem Res 45(24):8158–8169

47. Liu D, Chua CK, Leong KF (2013) A mathematical model for fluid shear-sensitive 3d tissue construct development. Biomech Model Mechanobiol 12(1):19–31

48. Yan X, Bergstrom D, Chen X (2012) Modeling of cell cultures in perfusion bioreactors. IEEE Trans Biomed Eng 59(9):2568–2575

49. Chung C, Chen C, Lin T, Tseng C (2008) A compact computational model for cell construct development in perfusion culture. Biotechnol Bioeng 99(6):1535–1541

50. Lewis MC, MacArthur BD, Malda J, Pettet G, Please CP (2005) Heterogeneous proliferation within engineered cartilaginous tissue: the role of oxygen tension. Biotechnol Bioeng 91(5):607–615

51. Landman KA, Cai AQ (2007) Cell proliferation and oxygen diffusion in a vascularising scaffold. Bull Math Biol 69(7):2405–2428

52. Lane Smith R, Trindade M, Ikenoue T, Mohtai M, Das P, Carter D, Goodman S, Schurman D (2000) Effects of shear stress on articular chondrocyte metabolism. Biorheology 37(1,2):95–107

53. Raimondi MT, Moretti M, Cioffi M, Giordano C, Boschetti F, Laganà K, Pietrabissa R (2006) The effect of hydrodynamic shear on 3d engineered chondrocyte systems subject to direct perfusion. Biorheology 43(3,4):215–222

54. Sengers B, Oomens C, Nguyen T, Bader D (2006) Computational modeling to predict the temporal regulation of chondrocyte metabolism in response to various dynamic compression regimens. Biomech Model Mechanobiol 5(2–3):111

55. Byers BA, Mauck RL, Chang IE, Tuan RS (2008) Transient exposure to TGF-β3 under serum-free conditions enhances the biomechanics and biochemical maturation of tissue-engineered cartilage. Tissue engineering Part A 14(11):1821

56. Sengers B, Van Donkelaar C, Oomens C, Baaijens F (2004) The local matrix distribution and the functional development of tissue engineered cartilage, a finite element study. Ann Biomed Eng 32(12):1718–1727

57. van Donkelaar C, Chao G, Bader D, Oomens C (2011) A reaction–diffusion model to predict the influence of neo-matrix on the subsequent development of tissue-engineered cartilage. Comput Methods Biomech Biomed Eng 14(05):425–432

58. DiMicco MA, Sah RL (2003) Dependence of cartilage matrix composition on biosynthesis, diffusion, and reaction. Transport Porous Med 50(1–2):57–73

59. Baaijens F, Bouten C, Driessen N (2010) Modeling collagen remodeling. J Biomech 43(1):166–175

60. Nikolaev N, Obradovic B, Versteeg HK, Lemon G, Williams DJ (2010) A validated model of GAG deposition, cell distribution, and growth of tissue engineered cartilage cultured in a rotating bioreactor. Biotechnol Bioeng 105(4):842–853

61. Obradovic B, Meldon JH, Freed LE, Vunjak-Novakovic G (2000) Glycosaminoglycan deposition in engineered cartilage: experiments and mathematical model. AIChE J 46(9):1860–1871

62. Myers K, Ateshian GA (2014) Interstitial growth and remodeling of biological tissues: tissue composition as state variables. J Mech Behav Biomed Mater 29:544–556

63. Yamauchi KA (2012) Prediction of articular cartilage remodeling during dynamic compression with a finite element model

64. Ficklin TP, Davol A, Klisch SM (2009) Simulating the growth of articular cartilage explants in a permeation bioreactor to aid in experimental protocol design. J Biomech Eng 131(4):041008

65. Gao X, Zhu Q, Gu W (2015) Analyzing the effects of mechanical and osmotic loading on glycosaminoglycan synthesis rate in cartilaginous tissues. J Biomech 48(4):573–577

66. Wilson W, Driessen N, Van Donkelaar C, Ito K (2006) Prediction of collagen orientation in articular cartilage by a collagen remodeling algorithm. Osteoarthritis Cartilage 14(11):1196–1202

67. Federico S, Herzog W (2008) On the anisotropy and inhomogeneity of permeability in articular cartilage. Biomech Model Mechanobiol 7(5):367–378

68. Julkunen P, Wilson W, Isaksson H, Jurvelin JS, Herzog W, Korhonen RK (2013) A review of the combination of experimental measurements and fibril-reinforced modeling for investigation of articular cartilage and chondrocyte response to loading. Comput Math Methods Med 2013

69. Pierce DM, Ricken T, Holzapfel GA (2013) A hyperelastic biphasic fibre-reinforced model of articular cartilage considering distributed collagen fibre orientations: continuum basis, computational aspects and applications. Comput Methods Biomech Biomed Eng 16(12):1344–1361

70. Bandeiras C, Completo A (2013) Comparison between constitutive models for the solid phase of biphasic agarose/chondrocytes constructs for knee cartilage engineering. Comput Methods Biomech Biomed Eng 16(S1):262–263

71. Tasci A, Ferguson SJ, Büchler P (2011) Numerical assessment on the effective mechanical stimuli for matrix-associated metabolism in chondrocyte-seeded constructs. J Tissue Eng Regen Med 5(3):210–219

72. Sengers BG, Oomens CW, Baaijens FP (2004) An integrated finite-element approach to mechanics, transport and biosynthesis in tissue engineering. J Biomech Eng 126(1):82–91

73. Wu J, Herzog W (2000) Finite element simulation of location-and time-dependent mechanical behavior of chondrocytes in unconfined compression tests. Ann Biomed Eng 28(3):318–330

74. Vahdati A, Wagner DR (2012) Finite element study of a tissue-engineered cartilage transplant in human tibiofemoral joint. Comput Methods Biomech Biomed Eng 15(11):1211–1221

75. Khoshgoftar M, van Donkelaar CC, Ito K (2011) Mechanical stimulation to stimulate formation of a physiological collagen architecture in tissue-engineered cartilage: a numerical study. Comput Methods Biomech Biomed Eng 14(02):135–144

76. Chen J, Irianto J, Inamdar S, Pravincumar P, Lee D, Bader DL, Knight M (2012) Cell mechanics, structure, and function are regulated by the stiffness of the three-dimensional microenvironment. Biophys J 103(6):1188–1197
77. Kalyanam S, Yapp RD, Insana MF (2009) Poro-viscoelastic behavior of gelatin hydrogels under compression-implications for bioelasticity imaging. J Biomech Eng 131(8):081005
78. Roberts JJ, Earnshaw A, Ferguson VL, Bryant SJ (2011) Comparative study of the viscoelastic mechanical behavior of agarose and poly (ethylene glycol) hydrogels. J Biomed Mater Res B Appl Biomater 99(1):158–169
79. Gu W, Yao H, Huang C, Cheung H (2003) New insight into deformation-dependent hydraulic permeability of gels and cartilage, and dynamic behavior of agarose gels in confined compression. J Biomech 36(4):593–598
80. Holmes M, Mow V (1990) The nonlinear characteristics of soft gels and hydrated connective tissues in ultrafiltration. J Biomech 23(11):1145–1156
81. Mansour JM (2003) Biomechanics of cartilage. Kinesiology: the mechanics and pathomechanics of human movement, pp 66–79
82. Wilson W, Huyghe J, Van Donkelaar C (2007) Depth-dependent compressive equilibrium properties of articular cartilage explained by its composition. Biomech Model Mechanobiol 6(1–2):43–53
83. Bandeiras C, Completo A, Ramos A (2014) Scaffold geometry influences the mechanical properties of tissue engineered cartilage. In: Proceedings of the 11th World Congress on Computational Mechanics (WCCM XI), Barcelona, 20–25 July 2014

On the Computational Biomechanics of the Intervertebral Disc

A. P. G. Castro, P. Flores, J. C. P. Claro, António M. G. Completo and J. L. Alves

Abstract The intervertebral disc (IVD) is a central piece for spine biomechanics. When the IVD fails, there is a high chance that one is suffering from degenerative disc disease (DDD), which is one of the largest health problems faced worldwide. However, DDD and back pain are also strictly related to the other structures in the spine, such as the vertebral bodies (VBs) or the connecting ligaments. An important amount of experimental and numerical works have studied the spine, focusing on the IVD, the VB or the whole spinal segment, but questions on how degeneration occurs and what causes it are still to be fully answered. This chapter deals with finite element (FE) simulations of the non-degenerated human IVD time-dependent behaviour, using a generic IVD + VB FE model. The outcomes are inside the scope of different sources of experimental and numerical literature data, proving that this model is useful to distinguish between healthy and unhealthy loading levels (shown here as above 600–800 N in activity periods for human spine). In other words, the numerical simulations with this FE model demonstrated potential to mimic the IVD. The biomechanical behaviour of the spine is still dependent on multiple factors, but this increased knowledge on overload levels definitely helps to reduce the risk of DDD and other spine-related diseases to occur.

A. P. G. Castro (✉)
IDMEC, Instituto Superior Técnico, University of Lisbon, 1049-001 Lisbon, Portugal
e-mail: andre.castro@tecnico.ulisboa.pt

P. Flores · J. C. P. Claro · J. L. Alves
Department of Mechanical Engineering, University of Minho, 4800-058 Guimarães, Portugal
e-mail: pflores@dem.uminho.pt

J. C. P. Claro
e-mail: jcpclaro@dem.uminho.pt

J. L. Alves
e-mail: jlalves@dem.uminho.pt

A. M. G. Completo
Department of Mechanical Engineering, University of Aveiro, 3810-193 Aveiro, Portugal
e-mail: completo@ua.pt

© Springer Nature Switzerland AG 2020
J. Belinha et al. (eds.), *The Computational Mechanics of Bone Tissue*,
Lecture Notes in Computational Vision and Biomechanics 35,
https://doi.org/10.1007/978-3-030-37541-6_9

223

1 Introduction

1.1 Spine Anatomy

The human spine is a complex system, anchored on an advanced neuromuscular control and consisting of four major zones, with respect to its vertebral bodies (VBs): the neck with the cervical VBs (C1–C7), the thoracic VBs (T1–T12), the lumbar VBs (L1–L5) and the sacral VB (S1). Figure 1 shows a representation of the complete human spine, which is also denominated as vertebral column or backbone.

This is a stable structure, even if highly mobile. The major functions of the spine are weight bearing, allowing motion between upper torso and pelvis and protection of the spinal cord and nerve roots [1, 2]. The functional unit of the spine is the motion segment (MS), composed by one central intervertebral disc (IVD), the two adjacent VBs (with the facets) and the peripheral structures (muscles, ligaments and organ-covering membranes) [3, 4].

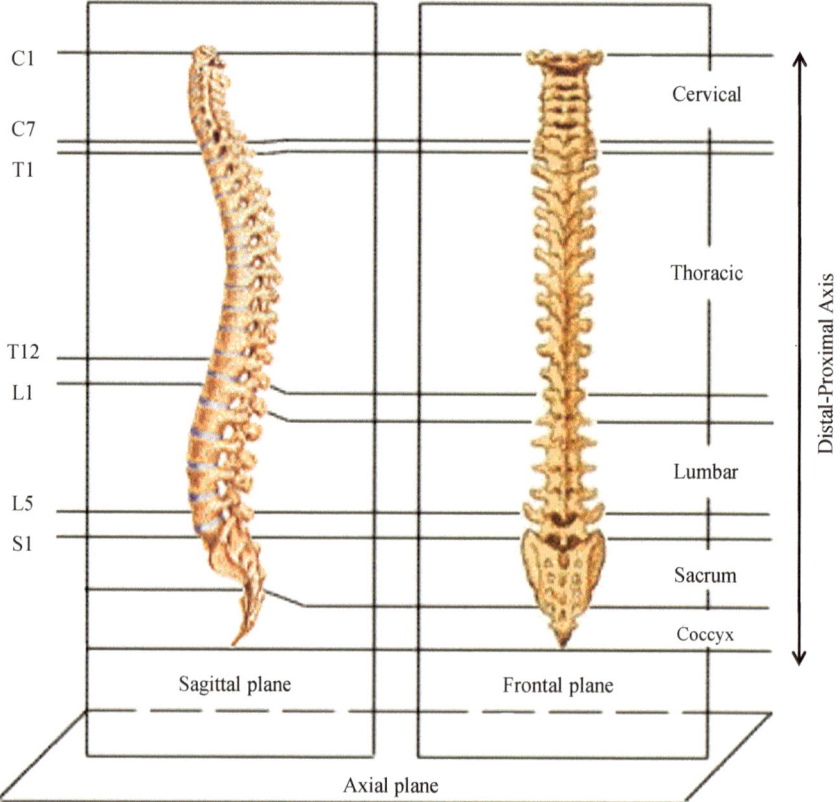

Fig. 1 Complete human spine, in both sagittal and frontal views. Adapted from Noailly [1]

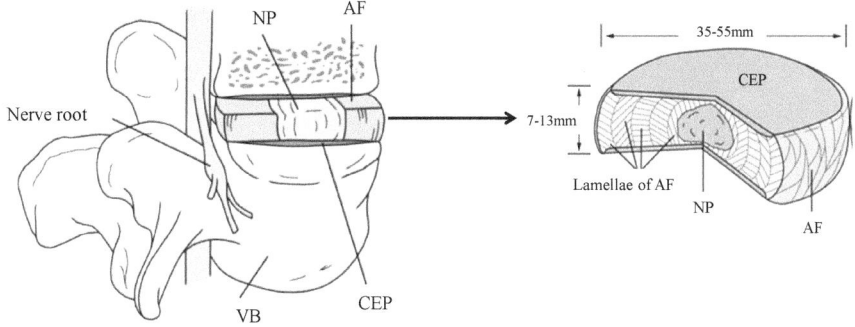

Fig. 2 Anatomy of an MS, with emphasis on the IVD dimensions. Adapted from Raj [5]

The IVD is an avascular highly inhomogeneous porous structure. Its central structures, the nucleus pulposus (NP) and the annulus fibrosus (AF), are paired structures, vertically limited by the cartilage endplates (CEPs, covering the NP and one-third of the AF) and also the vertebral endplates (VEPs, covering the CEP and two-thirds of the AF). The NP, a jelly structure with embedded fibres, occupies the core of the IVD. Surrounding it, emerge an amount of concentrically arranged fibres supported on a porous matrix, which is the AF [5, 6]. The CEP is a layer of hyaline cartilage that is responsible for most of the nutrients exchange with the VB. Each IVD has approximately 7–13 mm in height and 35–55 mm in diameter (axial plane). If one considers the 23 IVDs stacked, this construct would comprise approximately one-fourth to one-third of the total height of the spine [3, 5, 7]. Figure 2 shows an example of an MS, with the IVD anatomy highlighted.

The VB is substantially stiffer than the IVD. It is a highly porous and vascularized portion of bone tissue, containing both trabecular (TB) and cortical bone (CB) [8]. The VEP is mostly composed by CB. The discrimination between the VEP and the CEP is not always clear, as some authors only refer to the "endplate" and do not separate these two structures [9]. However, the CEP only covers the internal one-third of the extension of the AF, which means that this IVD component is also in direct contact with the VB. From the biomechanical point of view, the interaction between VB and IVD is essential to keep the healthy functioning of the spine, as both components support each other, i.e. the flexibility of the IVD compensates the strength of the VB, and vice versa [10–12].

The regular daily loads acting on the human spine are averagely 200 N during rest (lying prone) and between 600 and 800 N during activity, if one considers the load as independent from the type of solicitation, i.e. all the described activities are equalled to a compressive load. Moderate activities such as level walking, sitting or carrying light objects are within the group of typical daily loading profiles [13]. Harsher activities, such as lifting and carrying heavy objects, may be represented through 1500 or even 2000 N loads [12, 14]. Nevertheless, angular movements are of topmost importance for the spine. Typical moments on the spine for pure flexion and extension movements are averagely 4–10 Nm, mostly associated with position changes [15, 16].

1.2 Intervertebral Disc Degeneration

The degeneration of the IVD is firmly associated with the diseases of the spine, particularly low back pain. For many years, a considerable amount of studies have developed efforts to trace the causes and possible solutions for such issue, given that spine problems are a major cause of disability in western societies. Recent reports show that the frequency of these diseases tends to augment every year [17, 18].

The causes for IVD degeneration are not yet fully understood, and the complaints from the patients are also miscellaneous. An assortment of pathways for degeneration may be numbered: it may start with a fracture (or other damage) on the CEP, as a result of abnormal loading or calcification [19]. Proteoglycan loss, first on the CEP and then on the NP, may precipitate degeneration as well [20]. Moreover, it must be highlighted that CEP failure is one of the most important triggers for degenerative disc disease (DDD), as nutrition can be interrupted and so the viability of the IVD cells is compromised. It is thus reinforced that healthy VBs (through the VEPs) are important to keep the nutrients flow, via outer AF, but this pathway is not enough to maintain the integrity of the whole IVD [21].

Ageing is one of the most reported factors for IVD degeneration, i.e. problems such as IVD herniation are associated with senescence of IVD cells. Blood irrigation of the IVD after CEP calcification and cracking is also a collateral effect of ageing, as the IVD is not vascularized. This is another example where the VB is determinant for the healthy functioning of the MS, or how does the VB's condition directly affects the IVD [5, 22, 23]. Nowadays, it is also widely accepted that mechanical stress and inflammatory response are directly connected. At the cellular level, abnormal events (as chronic loading or severe acute efforts) trigger metabolic reactions (which begin with lower nutrients supply) that lead to loss of extracellular matrix integrity, and consequently, losses on IVD functions [19, 23].

Furthermore, the pathway for IVD degeneration may depend on the time of the day: it is recognized that early hours of the day are more suitable to spine injuries, so IVDs become more vulnerable to degeneration if subjected recurrently to efforts during that time. Accumulation of repetitive efforts or, in other words, fatigue has also an important role on promoting the degeneration due to the low metabolic rate presented by the IVD [1, 24, 25].

1.3 Objectives

The main goal of this chapter is to identify the mechanical loads and mechanical properties of the healthy IVD, accordingly to the state of the art of soft tissue and IVD constitutive modelling. This objective will be completed with the analysis of several groups of results, namely short creep (computational tests with duration inferior to 3 h, associated with short-term activities) and long creep tests (computational tests with the duration of 48 h, related to the human daily activities).

In order to do so, a custom finite element (FE) solver, V-Biomech®, was adopted [26, 27]. This solver includes an innovative biphasic poroelastic formulation particularly developed for soft tissues, such as the IVD, coupled with strain-dependent osmotic swelling behaviour and fibre reinforcement [27–30]. The adoption of this FE solver offers major advantages over commercial software packages, as the rigidity of a proprietary commercial code hampers the freedom of the researcher, when the complexity of the model increases. Given that the researcher has control over the source code, the verifiability of the software is increased.

Numerical simulations on the IVD biomechanical behaviour include comparison with experimental data from various sources [31–33], in order to establish an IVD characteristics framework.

2 Materials and Methods

2.1 Finite Element Model

The FE simulations were performed with the partial MS FE model shown in Fig. 3, which included L3 and L4 VB (without facets) and the L3–L4 IVD. The most relevant material constitutive modelling and properties of the osmo-poro-hyper-viscoelastic

Fig. 3 Sagittal cut of the partial human L3–L4 MS FE model, which contains 1892 27-node quadratic hexahedral elements and 16425 nodes. Adapted from Castro et al. [34]

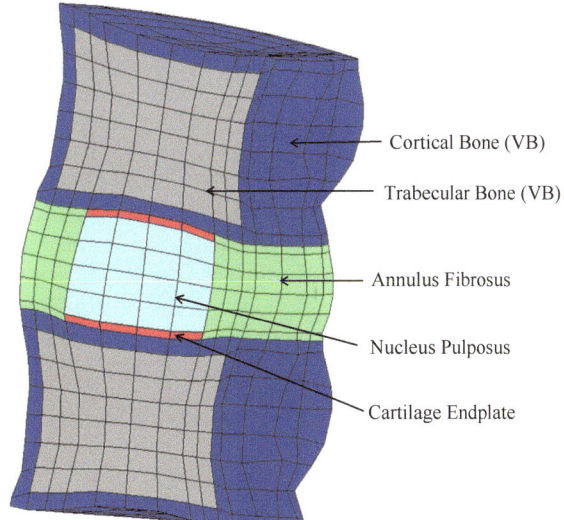

Cortical Bone (VB)

Trabecular Bone (VB)

Annulus Fibrosus

Nucleus Pulposus

Cartilage Endplate

Table 1 Constitutive models adopted

Isotropy	Mooney-Rivlin	$\overline{W}_{MR}(\overline{\mathbf{C}}) = C_{10}(\bar{I}_1 - 3) + C_{01}(\bar{I}_2 - 3)$
Permeability	van der Voet	$K^*(J) = K_0^* \left(\frac{1-n_{f,0}}{1-n_f}\right)^M = K_0^* J^M$
Anisotropy	Holzapfel	$\overline{W}_{Hoizapfel}(\overline{\mathbf{C}}, \mathbf{a}_1, \mathbf{a}_2) =$ $\frac{1}{2k_2}\left\{k_1\left[e^{k_2(\bar{I}_4-1)^2} - 1\right] + k_1\left[e^{k_2(\bar{I}_6-1)^2} - 1\right]\right\}$
Viscoelasticity	Maxwell	$G(t) = 1 + \sum\limits_{n-1}^{N} a_n \exp\left(-\frac{t}{\tau_n}\right)$
Swelling	Wilson	$\Delta\pi = \pi_{int} - \pi_{ext} = \phi_{int}RT(\sqrt{c_F^2 + 4c_{ext}^2}) - 2\phi_{ext}RTc_{ext}$
		$c_F = c_{F,0}\frac{n_{f,0}}{n_{f,0}-1+J}$

Adapted from Castro et al. [34]. Please check this reference for further information on each model

and fibre-reinforced model are summarized in Table 1 and Table 2, respectively [27, 34].

Several studies have shown the importance of osmotic swelling behaviour to the IVD biomechanics, namely for the height recovery during rest periods and the maintenance of healthy IDP levels; in agreement, the model was enhanced with Wilson's swelling model [29, 30, 35–37].

Mechanical properties of the AF fibres are assumed to evolve linearly through the axial plane, both in radial and circumferential directions [28, 38]. Fibre angle also varies from ±23.2° at ventral position to ±46.6° at dorsal position [28, 39].

For a more detailed description of the custom FE solver, the biphasic IVD constitutive modelling and MS FE model, the authors would like to refer to Castro et al. [27, 34].

2.2 Numerical Simulations

Simulations were divided into short- and long-term creep tests. The works of Heuer et al. [31] and O'Connell et al. [32] were taken into consideration as reference for short-term tests, while the bioreactor data reported by Castro et al. [34] and Paul et al. [40] was considered for the long-term tests. The non-degenerated MS FE model ("FE Nat") was therefore subjected to different load magnitudes and load rates, in order to distinguish between healthy and unhealthy load levels. It must be highlighted that free fluid flow is allowed between the MS components and also on the MS external boundaries at all times so that no artificial barriers are created to the natural IVD behaviour.

In the first test, corresponding to the work of Heuer et al. [31], a uniaxial vertical load of 500 N was applied on the top VB during 5 min (slow loading to allow proper

Table 2 Material properties of the MS components

		NP	AF	CEP	TB	CB
Isotropy	C_{10} [MPa]	0.15	0.18	1.00	41.67	3846.15
	C_{01} [MPa]	0.03	0.045	0.00	0.00	0.00
Permeability	K_0^* [mm^4 N^{-1} s^{-1}]	7.5e-4	7.5e-4	7.5e-3	1.0e-1	1.0e-1
	M	8.50	8.50	8.50	18.0	22.0
Anisotropy	$\bar{k}$	–	300.0	–		
	$k_4 = k_6$ [MPa]	–	12.0	–		
Viscoelasticity	a_1	1.7	–	–		
	τ_1 [s]	11.765	–	–		
	a_2	1.2	–	–		
	τ_2 [s]	1.100	–	–		
	a_3	2.0	–	–		
	τ_3 [s]	0.132	–	–		
Swelling	R [N mm mmol^{-1} K^{-1}]	8.31450	8.31450	–		
	T [K]	298.0	298.0	–		
	ϕ_{int}	0.83	0.83	–		
	ϕ_{ext}	0.92	0.92	–		
	C_{ext} [mmol.mm^{-3}]	0.00015	0.00015	–		
	$C_{F,0}$ [mmol.mm^{-3}]	0.00030	0.00018	–		
	$n_{f,0}$	0.80	0.70	–		

Isotropic (MS ground substances), permeability, anisotropic (AF fibres), viscoelastic and swelling properties were considered. Adapted from Castro et al. [27]

stabilization of the model) and then held for 15 min. The bottom VB was kept fully constrained. Lateral and sagittal movements were allowed.

For the second test, three stages were considered: (i) a short free swelling preconditioning period (1 h), (ii) a loading period of 2000 N at 1 N/s (in agreement with the experimental test of O'Connell et al. [32] and, finally, (iii) a creep stage (1 h). The global configuration of the boundary conditions was maintained from the previous test.

The third test had two variants: physiological loading and overloading. This involves more complex data and longer-term experiments. Experimental bioreactor data from the loaded disc culture system (LDCS), developed by the Department of Orthopaedic Surgery of VUmc (Amsterdam, The Netherlands), is here considered (Fig. 4). In short, the LDCS maintains the IVDs alive for approximately three weeks after the sacrifice of the animals, under physiological loading conditions. This mechanical stimulation keeps the IVD within its biomechanical and physiological properties, allowing for degeneration or overloading evaluation [40, 42, 43].

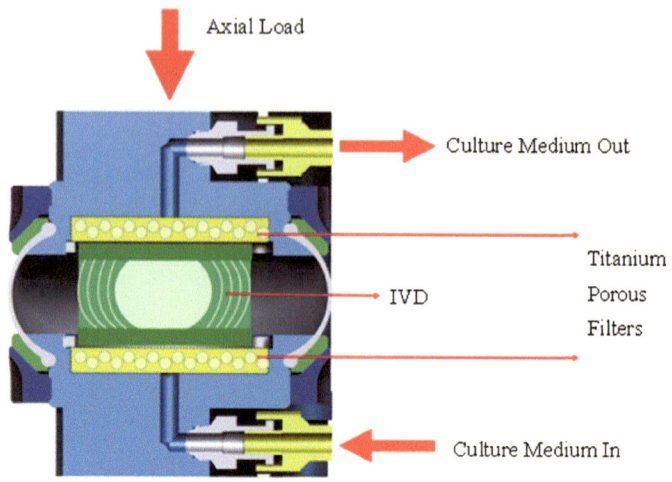

Fig. 4 Schematic representation of the LCDS system from VUmc. Adapted from Paul et al. [41]

In order to compare the human MS FE model with LDCS data, a normalization factor of four was applied, i.e. the axial cross section of the human IVD is averagely four times larger than the goat IVD, so it was assumed that the loads to be applied on the MS FE model should be four times higher (in magnitude) than the original LDCS loads. In what concerns to height, no normalization was needed, i.e. the IVD FE model has an average height of 12 mm, while the goat IVDs registered an average height of 9 mm. The validity of this approach is supported by the previous reports that human and goat IVDs produce similar internal stresses, regardless of the geometric differences [44–47].

Five goat IVDs were considered for this analysis, having that four were kept under a physiological loading profile (from Goat 1 to Goat 4 Native, abbreviated as "G1–G4 Nat"), and the fifth was overloaded ("G5 Ove"). The physiological loading profile resembles activities such as lying down and walking in goats, equivalent to relaxed standing and unsupported sitting in humans. It consists of a sinusoidal load (1 Hz) of 150 N average and 100 N amplitude for 16 h (activity period), followed by other sinusoidal loads (1 Hz) of 50 N average and 10 N amplitude for 8 h (resting period). It must be highlighted that the transition between the activity and resting periods is performed with 1 h of triangular loading (0.25 Hz) of 200 N average and 100 N amplitude. These were compared with "FE Nat" model.

For "G5 Ove", the resting period and the transitions between the two major periods were kept, but the activity loading profile consists of a sinusoidal load (1 Hz) of 300 N average and 100 N amplitude for 16 h. This loading profile simulates jumping on a haystack in goats, equivalent to lifting objects in humans. This experiment was compared with the equivalent numerical model, "FE Ove".

3 Results

Figure 5 shows the comparison between experimental results of Heuer et al. [31] and the MS FE model outcomes during a 15-min creep test at 500 N of compression. Figures 6 and 7 show the numerical outcomes of the three-stage short-term creep test based on the work of O'Connell et al. [32]. Figure 8 shows the comparison between LDCS experimental results and the MS FE model outcomes, in both native and overloaded configurations, during two daily cycles.

The assessed parameters, along with the diverse numerical tests, are displacement, hydrostatic pressure and volume variations. Disc height variation (DHV) refers to the height difference calculated between the two VBs at each time. Intradiscal pressure (IDP) is the internal pressure in the IVD tissue, namely in the NP. Osmotic pressure (OsmP) refers to the osmotic swelling pressure, which regulates the IVD osmotic balance and healthy pressure levels. Finally, relative volume variations are calculated with respect to the volume changes occurring during each simulation in the different IVD components.

Fig. 5 Results of a 15-min creep test at 500 N of compression: **a** DHV; **b** IDP. The experimental work of Heuer et al. [31] is compared with the present MS FE model

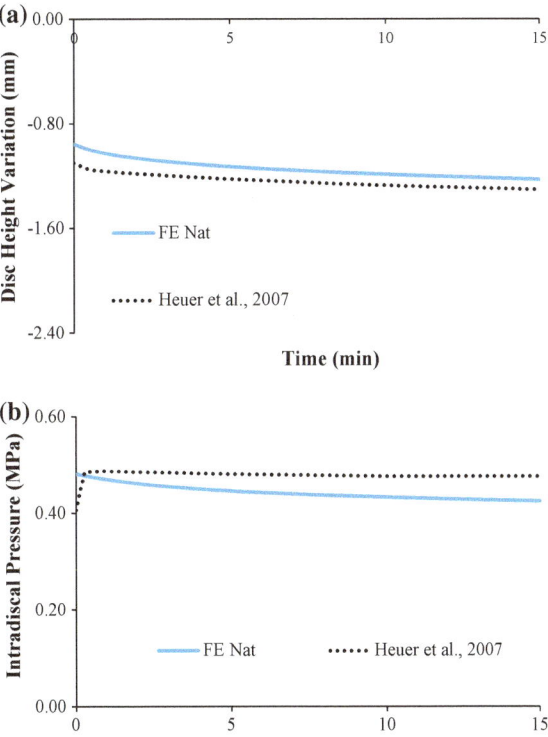

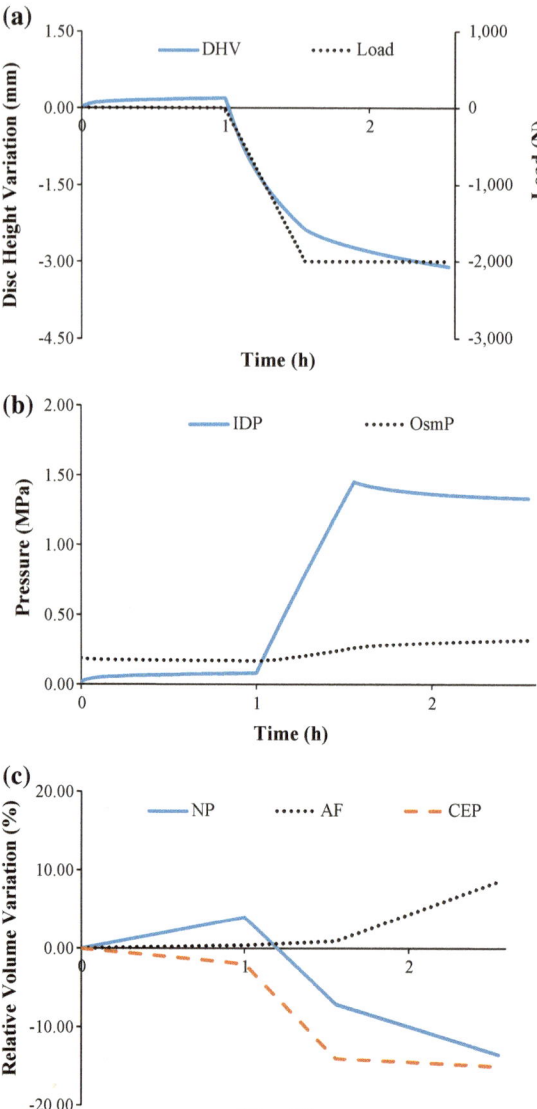

Fig. 6 Three stages loading test with the MS FE model, based on the work of O'Connell et al. [32], involving: (i) a preconditioning period, (ii) a loading period of 2000 N at 1 N/s and, finally, (iii) a creep stage. The following parameters were assessed, considering the full length of the test: **a** DHV versus load; **b** IDP and osmotic pressure of the NP (OsmP); **c** volume variation of NP, AF and CEP

(a)   **(b)**

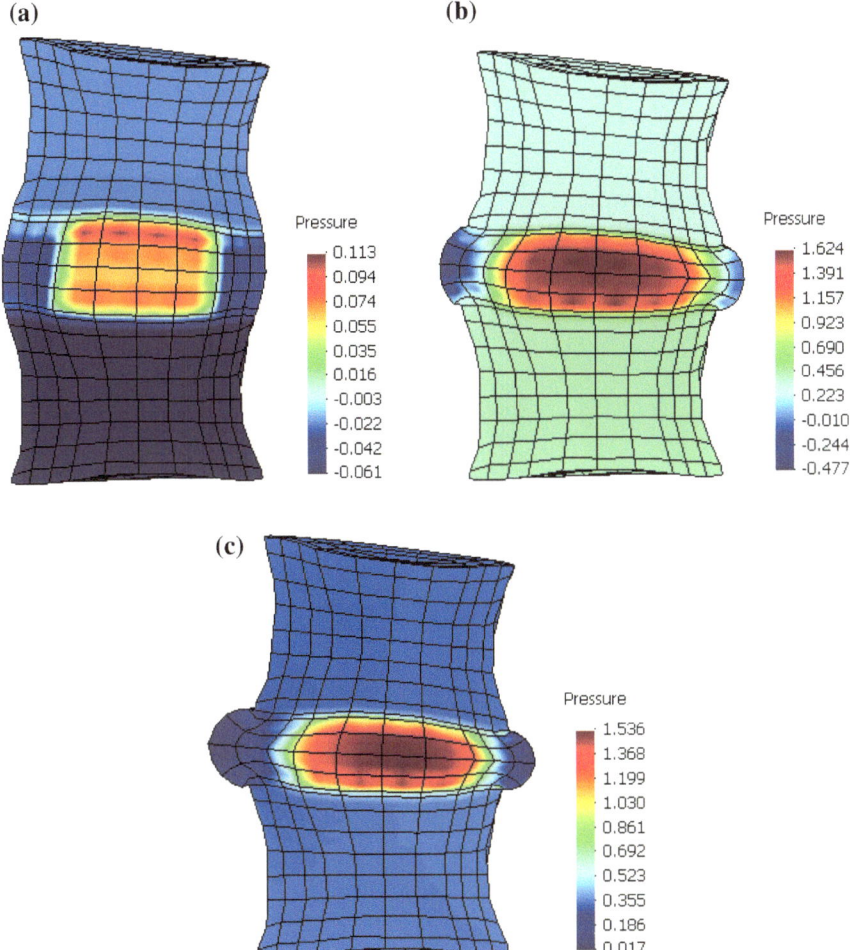

(c)

Fig. 7 Sagittal cuts of the MS FE model, showing the average hydrostatic pressure distribution inside the model at the end of each one of the three stages: **a** after the preconditioning; **b** after the ramp loading; **c** at the end of the test, after the creep phase. The scale is presented on the images (in MPa)

4 Discussion

4.1 Short Creep Tests

Heuer et al. [31] experimentally measured a range of −1.08 to −1.57 mm of DHV (average of −1.32 mm), at the end of the 15 min with a sustained 500 N load. At the same time, the range of IDP was between 0.36 and 0.52 MPa (average of 0.44 MPa). The correspondent numerical outcomes of the MS FE model are a DHV of −1.21 mm

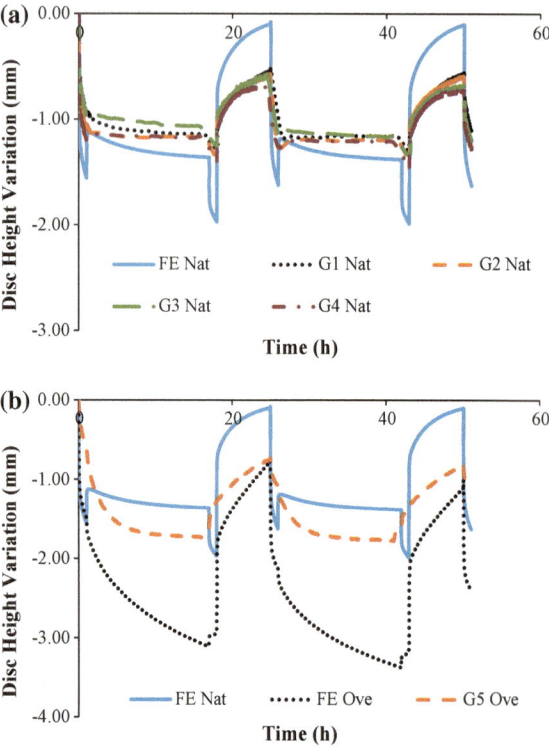

Fig. 8 DHV outcomes of the LDCS goat IVDs and correspondent MS FE model for two daily cycles. **a** The four native IVDs compared with the native FE model; **b** the overloaded IVD compared with both native and overloaded FE model

and an IDP of 0.41 MPa (Fig. 5). The differences between the values obtained with the model and those from Heuer et al. [31] are probably related to specimen-to-specimen variability and material properties [48, 49]. Nevertheless, the numerical model showed an overall good agreement with the experimental benchmark, having that the 500 N load allowed the IVD to be kept under physiological conditions [50].

The second short creep test is associated with the experimental test of O'Connell et al. [32], but considering: (i) a preconditioning period of 1 h, (ii) a ramp loading period (compressive uniaxial load) of 2000 N at 1 N/s and (iii) a creep of 1 h. Overall, this test also showed the expected IVD behaviour, with a strong deformation after 1 h of sustained 2000 N (mostly visible in Figs. 6a and 7c). After the ramp loading period, the DHV is 2.55 mm and the IDP is 1.45 MPa (Fig. 6a, b). The comparison of these values with the results of the previous tests indicates that the loading rate influences the stress state of the IVD, but only in what concerns to the pressure measurements. It must be highlighted that the IDP is lower than the OsmP during the first hour of the test (Fig. 6b), due to the free swelling conditions.

The final IDP is 1.32 MPa, which is in agreement with the literature data [33, 51]. The evolution of IDP and OsmP reveals a straight response to the applied load, from the preconditioning to the creep period. The DHV outcomes represent an average reduction of 20% of the initial IVD height, even if the final DHV calculation was 3.29 mm (or 26% of the initial IVD height). These measurements are in good agreement with the 23% of average axial compression experimentally determined by O'Connell et al. [32], but they probably represent an (almost) irreversible overloading effect [40, 43, 52].

Figure 6c shows that both NP and CEP were losing volume during the loading phases of this test. The final relative volume variation values were −13.5% and −15.1%, respectively. In contrast, the volume of the AF is increasing throughout the test (8.6% of relative volume variation). This volume increase is due to the fluid flow and pressure gradients within the IVD. The physiological direction of the flow is from the CEP into the inner structures of the IVD, so these volume variation outcomes are most likely in accordance with the literature [53, 54].

Figure 7 confirms that the 2000 N compression is excessive for the MS. Figure 7b shows a clear bulging effect after the ramp loading, even if it does not seem exaggerated. However, after 1 h of sustained 2000 N loading (Fig. 7c), the bulging is visibly unhealthy, as the IVD becomes highly deformed. Nevertheless, the NP still contains most of the pressure, in comparison with the other MS components. The combined analysis of the outcomes shown in Figs. 6 and 7 suggests that the IVD may be able to bear up the ramp loading of 2000 N at 1 N/s, on an (almost) instantaneous or short-term point of view. In that period, the numerical outcomes are inside the range of the results from the work of O'Connell et al. [32], even if the overloading effect is already noticeable. The sustainment of those 2000 N for 1 h seems to be noticeably out of the physiological IVD loading range.

4.2 Long Creep Tests

In what concerns to the third and last test, the comparison of the first two daily cycles of the native goat IVDs with the equivalent period of the "FE Nat" model (Fig. 8a) shows that the numerical model is able to reproduce the physiological behaviour of the goat IVDs, particularly during the activity period. The maximum calculated DHV during this period was −1.36 mm, while the average experimental measurement was −1.20 mm. Regarding the resting periods, an important difference is noticed, as the MS FE model is able to regain all the fluid lost during the activity period (DHV close to 0). The goat IVDs do not complete the recovery cycle in the same way, as their average DHV value on the resting period is −0.58 mm.

Nonetheless, these four IVDs maintain the same DHV recovery level from the first to the second daily cycle, which is a sign of no degeneration [23, 34, 40, 55]. In other words, the DHV results indicate no degeneration but incomplete recovery. This fact is probably related to the intrinsic behavioural differences between the goat and human IVDs, namely the specific biomechanical stimuli. The MS FE model helped to understand that the ideal situation is to fully recover the fluid loss during activity on the following resting periods. However, the action of AF fibres may also be limiting the range of DHV, and this limit situation (stretch constrainment after alteration of the loading profile) is not predicted by the MS FE model. This action shall not be considerably different in human and goat spines.

The overloading case is quite different. The DHV calculated from "FE Ove" model is unexpectedly distant from the experimental case ("G5 Ove", as seen in Fig. 8b), namely during the activity periods. This finding probably indicate that the MS FE model is excessively sensitive to the applied loads, having that the goat IVDs presented a limited range of height variation, as previously mentioned. If these experimental tests were performed in vivo, one might argue about the influence of the ligaments, but these structures were removed before the IVDs were tested in the LDCS. Therefore, this noticeable behavioural difference is almost certainly related to the intrinsic biochemical and biomechanical properties of the goat IVDs, counting as well with the contribution of the stretch limit of AF fibres. One may also argue that the human IVDs could have a larger range of DHV, due to their larger cross sections and initial height, but such indication would require further research. However, it must be highlighted that the DHV after recovery is similar for both experimental reference and MS FE model.

5   Conclusions

The literature review endorsed the IVD as an inhomogeneous porous tissue. The IVDs provide six degrees of freedom to each spinal MS, serving as central axial cushions for the diverse loading efforts, working together with the VBs to keep spinal healthiness. In vivo studies are problematic and potentially harmful, mostly due to the proximity of the spinal canals. Therefore, most of the data on IVD biochemistry and biomechanics comes from in vitro studies. Despite the advances in the experimental techniques, which are essential to the description of the biomechanical features of the tissues, the numerical methods are an essential tool to analyse and dissect the IVD behaviour, as they allow unlimited repetitions and complete control over the testing conditions.

In order to achieve valid numerical studies, the IVD was accurately modelled as osmo-hyper-poro-visco-elastic and reinforced with anisotropic fibres. The short and long creep tests have shown that the behaviour of the native MS FE model is aligned with experimental and numerical literature data. It was proved that the MS FE model here developed is able of reproducing experimental studies and also concluded that loads higher than 600–800 N are potentially harmful to the IVD, despite several studies indicating otherwise. The creep response of the MS FE model denoted a clear overload effect when loads of that magnitude were applied, both on displacement and pressure outcomes, i.e. the recovery process of the IVD is impaired for loads higher than 800 N, meaning that higher loads will probably accelerate IVD degeneration.

The analysis of LDCS data showed that the overloading mode caused IVD degeneration, i.e. the overloaded IVD has shown degradation signs, through unrecovered height levels, from the first overloading cycle. Nevertheless, the comparison with the MS FE model was fruitful, also for the physiological situations. Some discrepancies were noticed between experiments and simulations, probably due to the undisclosed physiological damage-prevention mechanism.

In resume, the numerical simulations with the MS FE model showed its validity and potential to contribute to the understanding of the IVD biomechanics.

References

1. Noailly J (2009) Model developments for in silico studies of the lumbar spine biomechanics. PhD thesis, Universitat Politècnica de Catalunya, Spain. http://hdl.handle.net/2117/93381
2. Niosi CA, Oxland TR (2004) Degenerative mechanics of the lumbar spine. Spine J. 4:202–208. https://doi.org/10.1016/j.spinee.2004.07.013
3. Ebraheim NA, Hassan A, Lee M, Xu R (2004) Functional anatomy of the lumbar spine. Semin Pain Med 2:131–137. https://doi.org/10.1016/j.spmd.2004.08.004
4. Grumme T, Bittl M (1998) Imaging and therapy of degenerative spine diseases—a neurosurgeon's view. Eur J Radiol 27:235–240
5. Raj P (2008) Intervertebral disc: anatomy physiology pathophysiology treatment. Pain Pract 8:18–44. https://doi.org/10.1111/j.1533-2500.2007.00171.x
6. Shankar H, Scarlett JA, Abram SE (2009) Anatomy and pathophysiology of intervertebral disc disease. Tech Reg Anesth Pain Manag 13:67–75. https://doi.org/10.1053/j.trap.2009.05.001
7. Whatley BR, Wen X (2012) Intervertebral disc (IVD): structure, degeneration, repair and regeneration. Mater Sci Eng C 32:61–77. https://doi.org/10.1016/j.msec.2011.10.011
8. Fields AJ, Lee GL, Keaveny TM (2010) Mechanisms of initial endplate failure in the human vertebral body. J Biomech 43:3126–3131. https://doi.org/10.1016/j.jbiomech.2010.08.002
9. Swider P, Accadbled F, Laffosse JM, Sales de Gauzy J (2012) Influence of fluid-flow direction on effective permeability of the vertebral end plate: an analytical model. Comput Methods Biomech Biomed Eng 15:151–156. https://doi.org/10.1080/10255842.2010.518960
10. Adams MA, Dolan P, McNally DS (2009) The internal mechanical functioning of intervertebral discs and articular cartilage, and its relevance to matrix biology. Matrix Biol 28:384–389. https://doi.org/10.1016/j.matbio.2009.06.004
11. Hussain M, Natarajan RN, An HS, Andersson GBJ (2012) Progressive disc degeneration at C5–C6 segment affects the mechanics between disc heights and posterior facets above and below the degenerated segment: a flexion-extension investigation using a poroelastic C3-T1 finite element model. Med Eng Phys 34:552–558. https://doi.org/10.1016/j.medengphy.2011.08.014
12. Shirazi-Adl A, Schmidt H, Kingma I (2016) Spine loading and deformation—from loading to recovery. J Biomech 49:813–816. https://doi.org/10.1016/j.jbiomech.2016.02.024
13. Sato K, Kikuchi S, Yonezawa T (1999) In vivo intradiscal pressure measurement in healthy individuals and in patients with ongoing back problems. Spine (Phila. Pa. 1976). 24:2468–2474
14. Pollintine P, van Tunen MSLM, Luo J, Brown MD, Dolan P, Adams MA (2010) Time-dependent compressive deformation of the ageing spine: relevance to spinal stenosis. Spine (Phila. Pa. 1976). 35:386–394. https://doi.org/10.1097/BRS.0b013e3181b0ef26
15. Guan Y, Yoganandan N, Moore J, Pintar FA, Zhang J, Maiman DJ, Laud P (2007) Moment-rotation responses of the human lumbosacral spinal column. J Biomech 40:1975–1980. https://doi.org/10.1016/j.jbiomech.2006.09.027
16. Rohlmann A, Petersen R, Schwachmeyer V, Graichen F, Bergmann G (2012) Spinal loads during position changes. Clin Biomech 27:754–758. https://doi.org/10.1016/j.clinbiomech.2012.04.006
17. Stannard JT, Edamura K, Stoker AM, O'Connell GD, Kuroki K, Hung CT, Choma TJ, Cook JL (2016) Development of a whole organ culture model for intervertebral disc disease. J Orthop Transl 5:1–8. https://doi.org/10.1016/j.jot.2015.08.002

18. Taher F, Essig D, Lebl DR, Hughes AP, Sama AA, Cammisa FP, Girardi FP (2012) Lumbar degenerative disc disease: current and future concepts of diagnosis and management. Adv Orthop 2012:970752. https://doi.org/10.1155/2012/970752

19. Colombini A, Lombardi G, Corsi MM, Banfi G (2008) Pathophysiology of the human intervertebral disc. Int J Biochem Cell Biol 40:837–842. https://doi.org/10.1016/j.biocel.2007.12.011

20. Massey CJ, Van Donkelaar CC, Vresilovic E, Zavaliangos A, Marcolongo M (2012) Effects of aging and degeneration on the human intervertebral disc during the diurnal cycle: a finite element study. J Orthop Res 30:122–128. https://doi.org/10.1002/jor.21475

21. Adams MA, Roughley PJ (2006) What is intervertebral disc degeneration, and what causes it? Spine (Phila. Pa. 1976). 31:2151–2161. https://doi.org/10.1097/01.brs.0000231761.73859.2c

22. Hadjipavlou AG, Tzermiadianos MN, Bogduk N, Zindrick MR (2008) The pathophysiology of disc degeneration: a critical review. J Bone Joint Surg Br 90:1261–1270. https://doi.org/10.1302/0301-620X.90B10.20910

23. Vergroesen P-PA, Kingma I, Emanuel KS, Hoogendoorn RJW, Welting TJ, van Royen BJ, van Dieën JH, Smit TH (2015) Mechanics and biology in intervertebral disc degeneration: a vicious circle. Osteoarthr Cartil 23:1057–1070. https://doi.org/10.1016/j.joca.2015.03.028

24. Martin MD, Boxell CM, Malone DG (2002) Pathophysiology of lumbar disc degeneration: a review of the literature. Neurosurg Focus 13:E1. https://doi.org/10.3171/foc.2002.13.2.2

25. Qasim M, Natarajan RN, An HS, Andersson GBJ (2012) Initiation and progression of mechanical damage in the intervertebral disc under cyclic loading using continuum damage mechanics methodology: a finite element study. J Biomech 45:1934–1940. https://doi.org/10.1016/j.jbiomech.2012.05.022

26. Castro APG, Laity P, Shariatzadeh M, Wittkowske C, Holland C, Lacroix D (2016) Combined numerical and experimental biomechanical characterization of soft collagen hydrogel substrate. J Mater Sci Mater Med 27:1–9. https://doi.org/10.1007/s10856-016-5688-3

27. Castro APG, Wilson W, Huyghe JM, Ito K, Alves JL (2014) Intervertebral disc creep behavior assessment through an open source finite element solver. J Biomech 47:297–301. https://doi.org/10.1016/j.jbiomech.2013.10.014

28. Cavalcanti C, Correia H, Castro APG, Alves JL (2013) Constitutive modelling of the annulus fibrosus: numerical implementation and numerical analysis. IEEE 3rd Port Meet Bioeng 7:3–6. https://doi.org/10.1109/ENBENG.2013.6518408

29. Wilson W, van Donkelaar CC, Huyghe JM (2005) A comparison between mechano-electrochemical and biphasic swelling theories for soft hydrated tissues. J Biomech Eng 127:158–165. https://doi.org/10.1115/1.1835361

30. Wilson W, Van Donkelaar CC, Van Rietbergen B, Huiskes R (2005) A fibril-reinforced poroviscoelastic swelling model for articular cartilage. J Biomech 38:1195–1204. https://doi.org/10.1016/j.jbiomech.2004.07.003

31. Heuer F, Schmitt H, Schmidt H, Claes L, Wilke HJ (2007) Creep associated changes in intervertebral disc bulging obtained with a laser scanning device. Clin Biomech 22:737–744. https://doi.org/10.1016/j.clinbiomech.2007.04.010

32. O'Connell GD, Jacobs NT, Sen S, Vresilovic EJ, Elliott DM (2011) Axial creep loading and unloaded recovery of the human intervertebral disc and the effect of degeneration. J Mech Behav Biomed Mater 4:933–942. https://doi.org/10.1016/j.jmbbm.2011.02.002

33. Wilke H-JJ, Neef P, Caimi M, Hoogland T, Claes LELE (1999) New in vivo measurements of pressures in the intervertebral disc in daily life. Spine (Phila. Pa. 1976). 24:755–762

34. Castro APG, Paul CPL, Detiger SEL, Smit TH, van Royen BJ, Pimenta Claro JC, Mullender MG, Alves JL (2014) Long-term creep behavior of the intervertebral disk: comparison between bioreactor data and numerical results. Front Bioeng Biotechnol 2:56. https://doi.org/10.3389/fbioe.2014.00056

35. Huyghe JM, Houben GB, Drost MR, van Donkelaar CC (2002) An ionised/non-ionised dual porosity model of intervertebral disc tissue. Biomech Model Mechanobiol 2:3–19. https://doi.org/10.1007/s10237-002-0023-y

36. Riches PE, Dhillon N, Lotz J, Woods AW, McNally DS (2002) The internal mechanics of the intervertebral disc under cyclic loading. J Biomech 35:1263–1271

37. Schroeder Y, Huyghe JM, Van Donkelaar CC, Ito K (2010) A biochemical/biophysical 3D FE intervertebral disc model. Biomech Model Mechanobiol 9:641–650. https://doi.org/10.1007/s10237-010-0203-0

38. Eberlein R, Holzapfel GA, Schulze-Bauer CAJ (2001) An anisotropic model for annulus tissue and enhanced finite element analyses of intact lumbar disc bodies. Comput Methods Biomech Biomed Eng 4:209–229. https://doi.org/10.1080/10255840108908005

39. Holzapfel GA, Schulze-Bauer CAJ, Feigl G, Regitnig P (2005) Single lamellar mechanics of the human lumbar anulus fibrosus. Biomech Model Mechanobiol 3:125–140. https://doi.org/10.1007/s10237-004-0053-8

40. Paul CPL, Schoorl T, Zuiderbaan HA, Zandieh Doulabi B, van der Veen AJ, van de Ven PM, Smit TH, van Royen BJ, Helder MN, Mullender MG (2013) Dynamic and static overloading induce early degenerative processes in caprine lumbar intervertebral discs. PLoS One 8. https://doi.org/10.1371/journal.pone.0062411

41. Paul CPL, Zuiderbaan HA, Zandieh Doulabi B, van der Veen AJ, van de Ven PM, Smit TH, Helder MN, van Royen BJ, Mullender MG (2012) Simulated-physiological loading conditions preserve biological and mechanical properties of caprine lumbar intervertebral discs in EX vivo culture. PLoS ONE 7:29–34. https://doi.org/10.1371/journal.pone.0033147

42. Detiger S, de Bakker J, Emanuel K, Schmitz M, Vergroesen P, van der Veen A, Mazel C, Smit T (2015). Translational challenges for the development of a novel nucleus pulposus substitute: experimental results from biomechanical and in vivo studies. J Biomater Appl 0:1–12. https://doi.org/10.1177/0885328215611946

43. Vergroesen PPA, Van Der Veen AJ, Van Royen BJ, Kingma I, Smit TH (2014) Intradiscal pressure depends on recent loading and correlates with disc height and compressive stiffness. Eur Spine J 23:2359–2368. https://doi.org/10.1007/s00586-014-3450-4

44. Alini M, Eisenstein SM, Ito K, Little C, Kettler AA, Masuda K, Melrose J, Ralphs J, Stokes I, Wilke HJ (2008) Are animal models useful for studying human disc disorders/degeneration? Eur Spine J 17:2–19. https://doi.org/10.1007/s00586-007-0414-y

45. Ayotte DC, Ito K, Tepic S (2001) Direction-dependent resistance to flow in the endplate of the intervertebral disc: an ex vivo study. J Orthop Res 19:1073–1077. https://doi.org/10.1016/S0736-0266(01)00038-9

46. Hoogendoorn RJW, Helder MN, Kroeze RJ, Bank RA, Smit TH, Wuisman PIJM (2008) Reproducible long-term disc degeneration in a large animal model. Spine (Phila. Pa. 1976). 33:949–954. https://doi.org/10.1097/BRS.0b013e31816c90f0

47. Schmidt H, Reitmaier S (2012) Is the ovine intervertebral disc a small human one? A finite element model study. J Mech Behav Biomed Mater 17:229–241. https://doi.org/10.1016/j.jmbbm.2012.09.010

48. Johannessen W, Elliott DM (2005) Effects of degeneration on the biphasic material properties of human nucleus pulposus in confined compression. Spine (Phila. Pa. 1976). 30:E724–E729. https://doi.org/10.1097/01.brs.0000192236.92867.15

49. Périé D, Korda D, Iatridis JC (2005) Confined compression experiments on bovine nucleus pulposus and annulus fibrosus: sensitivity of the experiment in the determination of compressive modulus and hydraulic permeability. J Biomech 38:2164–2171. https://doi.org/10.1016/j.jbiomech.2004.10.002

50. Araujo ARG, Peixinho N, Pinho A, Claro JCP (2015) The intradiscal failure pressure on porcine lumbar intervertebral discs: an experimental approach. Mech Sci 6:255–263. https://doi.org/10.5194/ms-6-255-2015

51. Bashkuev M, Vergroesen PPA, Dreischarf M, Schilling C, van der Veen AJ, Schmidt H, Kingma I (2016) Intradiscal pressure measurements: a challenge or a routine? J Biomech 49:864–868. https://doi.org/10.1016/j.jbiomech.2015.11.011

52. Araujo ARG, Peixinho N, Pinho ACM, Claro JCP (2014) A novel methodology to assess the relaxation rate of the intervertebral disc by increments on intradiscal pressure. Appl Mech Mater 664:379–383. https://doi.org/10.4028/www.scientific.net/AMM.664.379

53. Ferguson SJ, Ito K, Nolte LP (2004) Fluid flow and convective transport of solutes within the intervertebral disc. J Biomech 37:213–221. https://doi.org/10.1016/S0021-9290(03)00250-1
54. Schmidt H, Schilling C, Reyna ALP, Shirazi-Adl A, Dreischarf M (2016) Fluid-flow dependent response of intervertebral discs under cyclic loading: on the role of specimen preparation and preconditioning. J Biomech 49:846–856. https://doi.org/10.1016/j.jbiomech.2015.10.029
55. Lai A, Moon A, Purmessur D, Skovrlj B, Winkelstein BA, Cho SK, Hecht AC, Iatridis JC (2015) Assessment of functional and behavioral changes sensitive to painful disc degeneration. J Orthop Res 33:755–764. https://doi.org/10.1002/jor.22833

Author Index

© Springer Nature Switzerland AG 2020
J. Belinha et al. (eds.), *The Computational Mechanics of Bone Tissue*,
Lecture Notes in Computational Vision and Biomechanics 35,
https://doi.org/10.1007/978-3-030-37541-6